ORDINARY DIFFERENTIAL EQUATIONS

RICHARD K. MILLER
Department of Mathematics
Iowa State University
Ames, Iowa

ANTHONY N. MICHEL
Department of Electrical Engineering
University of Notre Dame
Notre Dame, Indiana

DOVER PUBLICATIONS, INC.
Mineola, New York

Copyright

Copyright © 1982 by Richard K. Miller and Anthony N. Michel
All rights reserved.

Bibliographical Note

This Dover edition, first published in 2007, is an unabridged republication of the work published by Academic Press, Inc., in 1982. A new errata list has been prepared for the Dover edition.

Library of Congress Cataloging-in-Publication Data

Miller, Richard K.
  Ordinary differential equations / Richard K. Miller, Anthony N. Michel.
    p. cm.
  Includes bibliographical references and index.
  ISBN-13: 978-0-486-46248-6
  ISBN-10: 0-486-46248-X
  1. Differential equations. I. Michel, Anthony N. II. Title.

QA372.M655 2007
515'.352—dc22

2007022373

Manufactured in the United States of America
Dover Publications, Inc., 31 East 2nd Street, Mineola, N.Y. 11501

To
C. A. DESOER and J. A. NOHEL

CONTENTS

PREFACE	ix
ACKNOWLEDGMENTS	xi
ERRATA	xiii

1 INTRODUCTION — 1

1.1	Initial Value Problems	1
1.2	Examples of Initial Value Problems	7
	Problems	35

2 FUNDAMENTAL THEORY — 39

2.1	Preliminaries	40
2.2	Existence of Solutions	45
2.3	Continuation of Solutions	49
2.4	Uniqueness of Solutions	53
2.5	Continuity of Solutions with Respect to Parameters	58
2.6	Systems of Equations	63
2.7	Differentiability with Respect to Parameters	68
2.8	Comparison Theory	70
2.9	Complex Valued Systems*	74
	Problems	75

3 LINEAR SYSTEMS — 80

3.1	Preliminaries	80
3.2	Linear Homogeneous and Nonhomogeneous Systems	88
3.3	Linear Systems with Constant Coefficients	100
3.4	Linear Systems with Periodic Coefficients	112
3.5	Linear nth Order Ordinary Differential Equations	117
3.6	Oscillation Theory	125
	Problems	130

4 BOUNDARY VALUE PROBLEMS* — 137

4.1	Introduction	137
4.2	Separated Boundary Conditions	143
4.3	Asymptotic Behavior of Eigenvalues	147
4.4	Inhomogeneous Problems	152
4.5	General Boundary Value Problems	159
	Problems	164

5 STABILITY — 167

5.1	Notation	168
5.2	The Concept of an Equilibrium Point	169
5.3	Definitions of Stability and Boundedness	172
5.4	Some Basic Properties of Autonomous and Periodic Systems	178
5.5	Linear Systems	179
5.6	Second Order Linear Systems	186
5.7	Lyapunov Functions	194
5.8	Lyapunov Stability and Instability Results: Motivation	202
5.9	Principal Lyapunov Stability and Instability Theorems	205
5.10	Linear Systems Revisited	218

Contents vii

5.11	Invariance Theory	221
5.12	Domain of Attraction	230
5.13	Converse Theorems	234
5.14	Comparison Theorems	239
5.15	Applications: Absolute Stability of Regulator Systems	243
	Problems	250

6 PERTURBATIONS OF LINEAR SYSTEMS 258

6.1	Preliminaries	258
6.2	Stability of an Equilibrium Point	260
6.3	The Stable Manifold	265
6.4	Stability of Periodic Solutions	273
6.5	Asymptotic Equivalence	280
	Problems	285

7 PERIODIC SOLUTIONS OF TWO-DIMENSIONAL SYSTEMS 290

7.1	Preliminaries	290
7.2	Poincaré–Bendixson Theory	292
7.3	The Levinson–Smith Theorem	298
	Problems	302

8 PERIODIC SOLUTIONS OF SYSTEMS 305

8.1	Preliminaries	306
8.2	Nonhomogeneous Linear Systems	306
8.3	Perturbations of Nonlinear Periodic Systems	312
8.4	Perturbations of Nonlinear Autonomous Systems	317
8.5	Perturbations of Critical Linear Systems	319
8.6	Stability of Systems with Linear Part Critical	324
8.7	Averaging	330

8.8	Hopf Bifurcation	333
8.9	A Nonexistence Result*	335
	Problems	338

BIBLIOGRAPHY 342

INDEX 346

PREFACE

This book is an outgrowth of courses taught for a number of years at Iowa State University in the mathematics and the electrical engineering departments. It is intended as a text for a first graduate course in differential equations for students in mathematics, engineering, and the sciences. Although differential equations is an old, traditional, and well-established subject, the diverse backgrounds and interests of the students in a typical modern-day course cause problems in the selection and method of presentation of material. In order to compensate for this diversity, we have kept prerequisites to a minimum and have attempted to cover the material in such a way as to be appealing to a wide audience.

The prerequisites assumed include an undergraduate ordinary differential equations course that covers, among other topics, separation of variables, first and second order linear systems of ordinary differential equations, and elementary Laplace transformation techniques. We also assume a prerequisite course in advanced calculus and an introductory course in matrix theory and vector spaces. All of these topics are standard undergraduate fare for students in mathematics, engineering, and most sciences. Occasionally, in sections of the text or in problems marked by an asterisk (*), some elementary theory of real or complex variables is needed. Such material is clearly marked (*) and has been arranged so that it can easily be omitted without loss of continuity. We think that this choice of prerequisites and this arrangement of material allow maximal flexibility in the use of this book.

The purpose of Chapter 1 is to introduce the subject and to briefly discuss some important examples of differential equations that arise in science and engineering. Section 1.1 is needed as background for Chapter 2, while Section 1.2 can be omitted on the first reading. Chapters 2 and 3 contain the fundamental theory of linear and nonlinear differential equations. In particular, the results in Sections 2.1–2.7 and 3.1–3.5 will be required as background for any of the remaining chapters. Linear boundary value problems are studied in Chapter 4. We concentrate mainly on the second order, separated case. In Chapter 5 we deal with Lyapunov stability theory, while in Chapter 6 we consider perturbations of linear systems. Chapter 5 is required as background for Sections 6.2–6.4. In Chapter 7 we deal with the Poincaré–Bendixson theory and with two-dimensional van der Pol type equations. It is useful, but not absolutely essential, to study Chapter 7 before proceeding to the study of periodic solutions of general order systems in Chapter 8. Chapter 5, however, contains required background material for Section 8.6.

There is more than enough material provided in this text for use as a one-semester or a two-quarter course. In a full-year course, the instructor may need to supplement the text with some additional material of his or her choosing. Depending on the interests and on the backgrounds of a given group of students, the material in this book could be edited or supplemented in a variety of ways. For example, if the students all have taken a course in complex variables, one might add material on isolated singularities of complex-valued linear systems. If the students have sufficient background in real variables and functional analysis, then the material on boundary value problems in Chapter 4 could be expanded considerably. Similarly, Chapter 8 on periodic solutions could be supplemented, given a background in functional analysis and topology. Other topics that could be considered include control theory, delay-differential equations, and differential equations in a Banach space.

Chapters are numbered consecutively with arabic numerals. Within a given chapter and section, theorems and equations are numbered consecutively. Thus, for example, while reading Chapter 5, the terms "Section 2," "Eq. (3.1)," and "Theorem 3.1" refer to Section 2 of Chapter 5, the first equation in Section 3 of Chapter 5, and the first theorem in Section 3 of Chapter 5, respectively. Similarly, while reading Chapter 5 the terms "Section 3.2," "Eq. (2.3.1)," "Theorem 3.3.1," and "Fig. 3.2" refer to Section 2 of Chapter 3, the first equation in Section 3 of Chapter 2, the first theorem in Section 3 of Chapter 3, and the second figure in Chapter 3, respectively.

ACKNOWLEDGMENTS

We gratefully acknowledge the contributions of the students at Iowa State University and at Virginia Polytechnic Institute and State University, who used the classroom notes that served as precursor to this text. We especially wish to acknowledge the help of Mr. D. A. Hoeflin and Mr. G. S. Krenz. Special thanks go to Professor Harlan Steck of Virginia Polytechnic Institute who taught from our classroom notes and then made extensive and valuable suggestions. We would like to thank Professors James W. Nilsson, George Sell, George Seifert, Paul Waltman, and Robert Wheeler for their help and advice during the preparation of the manuscript. Likewise, thanks are due to Professor J. O. Kopplin, Chairman of the Electrical Engineering Department at Iowa State University for his continued support, encouragement, and assistance to both authors. We appreciate the efforts and patience of Miss Shellie Siders and Miss Gail Steffensen in the typing and manifold correcting of the manuscript. In conclusion, we are grateful to our wives, Pat and Leone, for their patience and understanding.

ERRATA

Page	Line	For:	Read:
Page 7	Line 15	For: $\ldots, z_n) \in C^n$	Read: $\ldots, z_n)^T \in C^n$
Page 14	Line −3	For: $i = v/R$.	Read: $i = v_R/R$.
Page 42	Line −11	For: For any rational	Read: For any number
Page 42	Line −5	For: R	Read: R^n
Page 44	Line −6	For: $\bigcap_{m=1}^{\infty} \bigcup_{k \geq m}$	Read: $\bigcup_{m=1}^{\infty} \bigcap_{k \geq m}$
Page 48	Line 7	For: $\int_a^t$	Read: $\int_\tau^t$
Page 55	Line 11	For: ξ	Read: ξ_0
Page 56	Line 19	For: accomplished be	Read: accomplished by
Page 56	Line −1	Add: Define $\Phi_0(t) = \varphi_0(t) - \xi$.	
Page 57	Line 1	For: Now define...	Read: Also define...
Page 59	Line −11	For: $b - b_m < A/(2M)$.	Read: $b - b_m < A/(4M)$.
Page 59	Line −10	For: $b' = b + A/(2M)$.	Read: $b' = b + A/(4M)$.
Page 59	Line −7	For: $\ldots \leq MA/M = A$	Read: $\ldots \leq MA/2M = A/2$
Page 59	Line −6	For: $|t - b_m| \leq A/M$.	Read: $|t - b_m| \leq A/2M$.
Page 59	Line −5	For: on $\tau \leq t \leq b_m + A/M$. Moreover $b_m + A/M > b'$ when m is large.	Read: on $\tau \leq t \leq b_m + A/2M$. Moreover $b_m + A/2M > b'$ when m is large.
Page 66	Line 7	For: (M6) $\;\; |A| \leq \sum_{i=1}^{m} \left(\max_{1 \leq j \leq m} |a_{ij}| \right) \leq \sum_{i=1}^{m} \sum_{j=1}^{n} |a_{ij}|$	Read: (M6) $\;\; |A| \leq \sum_{i=1}^{m} \sum_{j=1}^{n} |a_{ij}|$ for the norm $|x|_2$.
Page 68	Line −6	For: definine	Read: define
Page 69	Line −1	For: $\lim_{k \to 0} z(t, \ldots$	Read: $\lim_{k \to \infty} z(t, \ldots$
Page 73	Line −12	For: ...functions exist.	Read: ...functions exist and $t \geq \tau$.
Page 73	Line −4	For: $\ldots (\eta - B/A) + B/A$.	Read: $\ldots (\eta + B/A) - B/A$.
Page 74	Line 1	For: 8.3	Read: 8.4
Page 76	Line −16	For: $\phi \in C[\tau, \infty) \ldots$	Read: $\phi \in C[\tau, \infty)$ to $R^n \ldots$
Page 76	Line −4	For: to x up to and...	Read: to (t, x) up to and...
Page 79	Line −5	For: $a = \tau, \alpha = L$,	Read: $a = \tau, \alpha = -L$,
Page 79	Line −1	For: Theorem 4.1	Read: Theorem 4.6
Page 95	Line −4	For: if ψ is *any*	Read: if Ψ is *any*
Page 96	Line 12	For: $[\psi_1(\tau)]^{-1}$.	Read: $[\Psi_1(\tau)]^{-1}$.

xiii

Errata

Page 114	Line −15	For: determined over $(0, T)$.	
		Read: determined on $[t_0, t_0 + T]$.	
Page 125	Line 7	For: $0, 1, \ldots, n - 1$	
		Read: $0, 1, \ldots, n - 1, a_n = 1$	
Page 125	Line −11	For: $= \sum_{k=0}^{n-1} \left[\sum_{j=0}^{k} \ldots \right.$	Read: $= \sum_{k=1}^{n} \left[\sum_{j=0}^{k-1} \ldots \right.$
Page 127	Line 16	For: $-k(t_1)\phi_1'(t_1)\phi_2(t_2) = \ldots$	
		Read: $-k(t_1)\phi_1'(t_1)\phi_2(t_1) = \ldots$	
Page 128	Line 2	For: $\ldots - k_1(t_1) \ldots$	Read: $\ldots - k(t_1) \ldots$
Page 129	Line −2	For: decreasing	Read: increasing
Page 129	Line −1	For: increases	Read: decreases
Page 140	Line 10	For: $\ldots [f(s)/w(\phi_1, \phi_2)(s)] \, ds$.	
		Read: $\ldots [f(s)/W(\phi_1, \phi_2)(s)k(s)] \, ds$.	
Page 145	Line −9	For: $t \in [a, b]$,	Read: $t \in (a, b]$,
Page 146	Line 4	For: $\psi(t, \lambda)$	Read: (ρ, ψ)
Page 146	Line 11	For: ψ will have	Read: y will have
Page 146	Line 12	For: arbitary	Read: arbitrary
Page 146	Line 18	For: $\theta'(t, \lambda) \leq G + \lambda R$	Read: $\theta'(t, \lambda) \leq (G + \lambda R)$
Page 146	Line 19	For: $\ldots - G - 1/K\} \ldots$	
		Read: $\ldots - G \sin^2 \varepsilon - 1/K\} \ldots$	
Page 147	Line 4	For: Theorem 3.6.1.	Read: Theorem 3.6.2.
Page 147	Line −2	For: $q = (Q_1''/Q_1) + \ldots$	Read: $q = (Q_1''/Q_1) - \ldots$
Page 147	line −1	For: Q_2.	Read: $K^2 Q_2$.
Page 151	Line −10	For: $[cs/m + O(m^{-2})]$,	Read: $[cs/m\pi + O(m^{-2})]$,
Page 151	Line −9	For: $[cs/m + O(m^{-2})]$,	Read: $[cs/m\pi + O(m^{-2})]$,
Page 151	Line −5	For: $\ldots dv - cs$.	Read: $\ldots dv - cs/\pi$.
Page 153	Line −8	Delete: , i.e., $L_2(y_2) \neq 0$	
Page 153	Line −3	For: (4.5) if and only if	
		Read: (4.5) at λ_m if and only if	
Page 154	Line 14	For: $\Delta(\lambda) \neq 0$,	Read: $\Delta(\lambda) = 0$,
Page 161	Line 6	For: $S = \{(t, \tau, \lambda) \ldots$	Read: $S = \{(t, s, \lambda) \ldots$
Page 163	Line 1	For: $G(t, \tau, \lambda)$	Read: $G(t, s, \lambda)$
Page 163	Line −12	For: $Ly + \lambda \rho y = f$	Read: $Ly + \lambda \rho y = -f$

Errata

Page 163	Line −11	For: $\langle f, \mathcal{G}_{\bar{\lambda}} h \rangle = \langle Ly + \lambda\rho y, z \rangle = \langle y, Lz + \bar{\lambda}\rho z \rangle = \langle \mathcal{G}_\lambda f, h \rangle$	
		Read: $-\langle f, \mathcal{G}_{\bar{\lambda}} h \rangle = \langle Ly + \lambda\rho y, z \rangle = \langle y, Lz + \bar{\lambda}\rho z \rangle = -\langle \mathcal{G}_\lambda f, h \rangle$	
Page 163	Line −2	For: $L\mathcal{G}f = f$ and $gLy = y$	
		Read: $L\mathcal{G}f = -f$ and $-\mathcal{G}Ly = y$	
Page 164	Line 2	For: $y + \lambda\mathcal{G}(y\rho) = F$,	Read: $-y + \lambda\mathcal{G}(y\rho) = F$,
Page 164	Line 5	For: $y(t) = F(t) - \lambda \ldots$	Read: $y(t) = -F(t) + \lambda \ldots$
Page 177	Line −5	For: p. 84	Read: p. 191
Page 180	Line −9	For: (iii) $\lim_{t\to\infty} |\Phi(t, t_0)| = 0$.	
		Read: (iii) $\lim_{t\to\infty} |\Phi(t, t_0)| = 0$ for all $t_0 \geq 0$.	
Page 182	Line −11	For: t^2	Read: $t^2/2$
Page 186	Line 15	For: SECOND ORDER	Read: TWO DIMENSIONAL
Page 207	Line 17	For: $v_{(9.3)}(t, x_1, x_2) = \ldots$	Read: $v'_{(9.3)}(t, x_1, x_2) = \ldots$
Page 207	Line −10	For: $\ldots = -e^t x_2^2$.	Read: $\ldots = -2e^t x_2^2$.
Page 209	Line 8	For: $0 < \psi_1(\delta_2) \leq \ldots$	Read: $0 < \psi_1(\delta_1) \leq \ldots$
Page 209	Line −14	For: $(t, x) \Pi R^+ \times R^n$,	Read: $(t, x) \in R^+ \times R^n$,
Page 209	Line −5	For: Suppose that no $T(\alpha, \varepsilon)$ exists. Then for some x_0, $\eta > 0$.	
		Read: Suppose that $\eta > 0$ for some x_0 and t_0.	
Page 215	Line −12	For: $t_1 \geq 0$	Read: $t_1 \geq t_0$
Page 215	Line −4	For: $|\phi_1(t)| \leq h$. If $|\phi_1(t)| \leq h$	
		Read: $|\phi_1(t)| < h$. If $|\phi_1(t)| < h$	
Page 221	Line 1	For: Assume det $A \neq 0$.	
		Read: Assume no eigenvalue of A has real part zero.	
Page 226	Line −16	For: Since H_k is compact and invariant, then	
		Read: Since H_k is compact, v is bounded there.	
Page 226	Line −15	For: by the remarks in...	Read: By the remarks in...
Page 238	Line 5	For: $dy/ds = f_x(s, \ldots$	Read: $dy/ds = f_x(s + t, \ldots$
Page 251	Line 9	For: is uniformly stable	
		Read: is uniformly asymptotically stable	
Page 253	Line 1	For: $G(t, y) = G(t, -y)$	Read: $G(t, y) = -G(t, -y)$
Page 256	Line 18	For: $a_{jj} - \ldots$	Read: $a_{jj} + \ldots$
Page 260	Line 14	For: (E)	Read: (G)
Page 267	Line 8	For: (t, x)	Read: t

Errata

Page 267	Line 17	For: ... $Q(x-v)$ ...	Read: ... $Q(x+v)$ ...
Page 267	Line 21	For: F satisfy hypothesis (3.1)	Read: F be C^1 and satisfy (3.1)
Page 269	Line -10	For: $\leq (K\varepsilon/\sigma)\|\psi - \psi_j\| \to 0, \quad j \to \infty.$	
		Read: $\to 0, \quad j \to \infty.$	
Page 270	Line 4	For: (τ, ε)	Read: (τ, ξ)
Page 270	Line -2	For: Equation	Read: With a projection,
Page 270	-1	For: ... $= -\int_\tau^\infty$ ...	Read: ... $= -P_{n-k}\int_\tau^\infty$ ...
Page 271	Line 3	For: $L \leq \frac{1}{4}$.	Read: $L < 1$.
Page 272	Line 3	For: $\lambda = \sqrt{ab} > 0$ and $\lambda = -\sqrt{ab} < 0$.	
		Read: $\lambda = \sqrt{ad} > 0$ and $\lambda = -\sqrt{ad} < 0$.	
Page 274	Line 21	For: characteristic	Read: Floquet
Page 275	Line 1	For: characteristic	Read: Floquet
Page 275	Line -9	For: characteristic	Read: Floquet
Page 275	Line -5	For: $\Phi(t) \triangleq \text{diag}(k^{-1}, 1, \ldots, 1)\Phi_1(t)$,	
		Read: $\Phi(t) \triangleq \Phi_1(t)\text{diag}(k^{-1}, 1, \ldots, 1)$,	
Page 286	Line -6	For: characteristic	Read: Floquet
Page 286	Line -2	For: characteristic	Read: Floquet
Page 295	Line 22	For: $\phi(0) = \xi$.	Read: $\phi(0) = \xi^*$.
Page 295	Line 23	For: (ξ)	Read: (ξ^*)
Page 297	Line 2	For: (ξ)	Read: (ξ_0)
Page 302	Line -9	For: ... $-12x)x' + x = 0$	Read: ... $-12x)x\,x' + x = 0$
Page 303	Line -7	For: $\iint_{F(D)} d\xi = \ldots$	Read: $\iint_{F(D)} dx = \ldots$
Page 304	Line 5	For: (See Problem 9.)	Read: (See Problem 10.)
Page 313	Line -8	For: If the real parts of the characteristic...	
		Read: If the characteristic...	
Page 321	Line 5	For: ... $+ 2\pi b_{k-2} + O(\varepsilon)$,	Read: ... $+ 2\pi b_{k-1} + O(\varepsilon)$,
Page 322	Line -6	For: $\gamma_0 A \sin 0$	Read: $A \sin 0$
Page 324	Line 3	For: ... $b_{k+1}) \in R^n$.	Read: ... $b_{k+1})T \in R^n$.
Page 324	Line 6	For: $\det(\partial G_K/\partial c)(c_0, 0) \neq 0$.	
		Read: $\det(\partial G_k/\partial c)(c_0, 0) \neq 0$.	

INTRODUCTION

1

In the present chapter we introduce the initial value problem for differential equations and we give several examples of initial value problems.

1.1 INITIAL VALUE PROBLEMS

The purpose of this section, which consists of five parts, is to introduce and classify initial value problems for ordinary differential equations. In Section A we consider first order ordinary differential equations, in Section B we present systems of first order ordinary differential equations, in Section C we give a classification of systems of first order differential equations, in Section D we consider nth order ordinary differential equations, and in Section E we present complex valued ordinary differential equations.

A. First Order Ordinary Differential Equations

Let R denote the set of real numbers and let $D \subset R^2$ be a domain (i.e., an open connected nonempty subset of R^2). Let f be a real valued function which is defined and continuous on D. Let $x' = dx/dt$ denote the derivative of x with respect to t. We call

$$x' = f(t, x) \tag{E'}$$

an **ordinary differential equation of the first order**. By a **solution** of the differential equation (E') on an open interval $J = \{t \in R : a < t < b\}$, we mean a real valued, continuously differentiable function ϕ defined on J such that the points $(t, \phi(t)) \in D$ for all $t \in J$ and such that

$$\phi'(t) = f(t, \phi(t))$$

for all $t \in J$.

Definition 1.1. Given $(\tau, \xi) \in D$, the **initial value problem** for (E') is

$$x' = f(t, x), \qquad x(\tau) = \xi. \tag{I'}$$

A function ϕ is a **solution** of (I') if ϕ is a solution of the differential equation (E') on some interval J containing τ and $\phi(\tau) = \xi$.

A typical solution of an initial value problem is depicted in Fig. 1.1.

We can represent the initial value problem (I') equivalently by an integral equation of the form

$$\phi(t) = \xi + \int_\tau^t f(s, \phi(s))\, ds. \tag{V}$$

To prove this equivalence, let ϕ be a solution of the initial value problem (I'). Then $\phi(\tau) = \xi$ and

$$\phi'(t) = f(t, \phi(t))$$

for all $t \in J$. Integrating from τ to t, we have

$$\int_\tau^t \phi'(s)\, ds = \int_\tau^t f(s, \phi(s))\, ds$$

or

$$\phi(t) - \xi = \int_\tau^t f(s, \phi(s))\, ds.$$

Therefore, ϕ is a solution of the integral equation (V).

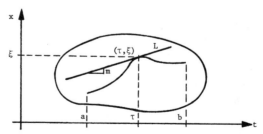

FIGURE 1.1 *Solution of an initial value problem; t interval $J = (a, b)$, m (slope of line L) $= f(\tau, \phi(\tau))$.*

1.1 Initial Value Problems

Conversely, let ϕ be a solution of the integral equation (V). Then $\phi(\tau) = \xi$, and differentiating both sides of (V) with respect to t, we have

$$\phi'(t) = f(t, \phi(t)).$$

Therefore, ϕ is also a solution of the initial value problem (I').

B. Systems of First Order Ordinary Differential Equations

We can extend the preceding to initial value problems involving a system of first order ordinary differential equations. Here we let $D \subset R^{n+1}$ be a domain, i.e., an open, nonempty, and connected subset of R^{n+1}. We shall often find it convenient to refer to R^{n+1} as the $(t, x_1, \ldots, x_n)$ space. Let $f_1, \ldots, f_n$ be n real valued functions which are defined and continuous on D, i.e., $f_i: D \to R$ and $f_i \in C(D)$, $i = 1, \ldots, n$. We call

$$x_i' = f_i(t, x_1, \ldots, x_n), \qquad i = 1, \ldots, n, \tag{E_i}$$

a **system of n ordinary differential equations of the first order**. By a **solution** of the system of ordinary differential equations (E_i) we shall mean n real continuously differentiable functions $\phi_1, \ldots, \phi_n$ defined on an interval $J = (a, b)$ such that $(t, \phi_1(t), \ldots, \phi_n(t)) \in D$ for all $t \in J$ and such that

$$\phi_i'(t) = f_i(t, \phi_1(t), \ldots, \phi_n(t)), \qquad i = 1, \ldots, n,$$

for all $t \in J$.

Definition 1.2. Let $(\tau, \xi_1, \ldots, \xi_n) \in D$. Then the **initial value problem** associated with (E_i) is

$$\begin{aligned} x_i' &= f_i(t, x_1, \ldots, x_n), & i = 1, \ldots, n, \\ x_i(\tau) &= \xi_i, & i = 1, \ldots, n. \end{aligned} \tag{I_i}$$

A set of functions $(\phi_1, \ldots, \phi_n)$ is a **solution** of (I_i) if $(\phi_1, \ldots, \phi_n)$ is a solution of the system of equations (E_i) on some interval J containing τ and if $(\phi_1(\tau), \ldots, \phi_n(\tau)) = (\xi_1, \ldots, \xi_n)$.

In dealing with systems of equations, it is convenient to use vector notation. To this end, we let

$$x = \begin{bmatrix} x_1 \\ \vdots \\ x_n \end{bmatrix}, \qquad \xi = \begin{bmatrix} \xi_1 \\ \vdots \\ \xi_n \end{bmatrix}, \qquad \phi = \begin{bmatrix} \phi_1 \\ \vdots \\ \phi_n \end{bmatrix},$$

$$f(t, x) = \begin{bmatrix} f_1(t, x_1, \ldots, x_n) \\ \vdots \\ f_n(t, x_1, \ldots, x_n) \end{bmatrix} = \begin{bmatrix} f_1(t, x) \\ \vdots \\ f_n(t, x) \end{bmatrix},$$

and we define $x' = dx/dt$ componentwise, i.e.,

$$x' = \begin{bmatrix} x'_1 \\ \vdots \\ x'_n \end{bmatrix}.$$

We can now express the initial value problem (I_i) by

$$x' = f(t,x), \qquad x(\tau) = \xi. \tag{I}$$

As in the scalar case, it is possible to rephrase the preceding initial value problem (I) in terms of an equivalent integral equation.

Now suppose that (I) has a unique solution ϕ defined for t on an interval J containing τ. By the **motion through** (τ, ξ) we mean the set

$$\{(t, \phi(t)) : t \in J\}.$$

This is, of course, the graph of the function ϕ. By the **trajectory** or **orbit through** (τ, ξ) we mean the set

$$C(\xi) = \{\phi(t) : t \in J\}.$$

The **positive semitrajectory** (or **positive semiorbit**) is defined as

$$C^+(\xi) = \{\phi(t) : t \in J \text{ and } t \geq \tau\}.$$

Also, the **negative trajectory** (or **negative semiorbit**) is defined as

$$C^-(\xi) = \{\phi(t) : t \in J \text{ and } t \leq \tau\}.$$

C. Classification of Systems of First Order Differential Equations

There are several special classes of differential equations, resp., initial value problems, which we shall consider. These are enumerated in the following discussion.

1. If in (I), $f(t,x) = f(x)$ for all $(t,x) \in D$, i.e., $f(t,x)$ does not depend on t, then we have

$$x' = f(x). \tag{A}$$

We call (A) an **autonomous system** of first order ordinary differential equations.

2. If in (I), $(t + T, x) \in D$ when $(t, x) \in D$ and if f satisfies $f(t, x) = f(t + T, x)$ for all $(t, x) \in D$, then

$$x' = f(t, x) = f(t + T, x). \tag{P}$$

1.1 Initial Value Problems

Such a system is called a **periodic system** of first order differential equations of period T. The smallest number $T > 0$ for which (P) is true is the **least period** of this system of equations.

3. If in (I), $f(t, x) = A(t)x$, where $A(t) = [a_{ij}(t)]$ is a real $n \times n$ matrix with elements $a_{ij}(t)$ which are defined and at least piecewise continuous on a t interval J, then we have

$$x' = A(t)x \qquad \text{(LH)}$$

and we speak of a **linear homogeneous system** of ordinary differential equations.

4. If for (LH) $A(t)$ is defined for all real t and if there is a $T > 0$ such that $A(t) = A(t + T)$ for all t, then we have

$$x' = A(t)x = A(t + T)x. \qquad \text{(LP)}$$

This system is called a **linear periodic system** of ordinary differential equations.

5. If in (I), $f(t, x) = A(t)x + g(t)$, where $g(t)^T = [g_1(t), \ldots, g_n(t)]$, and where $g_i : J \to R$, then we have

$$x' = A(t)x + g(t). \qquad \text{(LN)}$$

In this case we speak of a **linear nonhomogeneous system** of ordinary differential equations.

6. If in (I), $f(t, x) = Ax$, where $A = [a_{ij}]$ is a real $n \times n$ matrix with constant coefficients, then we have

$$x' = Ax. \qquad \text{(L)}$$

This type of system is called a **linear, autonomous, homogeneous system** of ordinary differential equations.

D. *n*th Order Ordinary Differential Equations

It is also possible to characterize initial value problems by means of nth order ordinary differential equations. To this end, we let h be a real function which is defined and continuous on a domain D of the real $(t, y_1, \ldots, y_n)$ space and we let $y^{(k)} = d^k y/dt^k$. Then

$$y^{(n)} = h(t, y, y^{(1)}, \ldots, y^{(n-1)}) \qquad (E_n)$$

is an nth **order ordinary differential equation**. A **solution** of (E_n) is a real function ϕ which is defined on a t interval $J = (a, b) \subset R$ which has n continuous derivatives on J and satisfies $(t, \phi(t), \ldots, \phi^{(n-1)}(t)) \in D$ for all $t \in J$ and

$$\phi^{(n)}(t) = h(t, \phi(t), \ldots, \phi^{(n-1)}(t))$$

for all $t \in J$.

Definition 1.3. Given $(\tau, \xi_1, \ldots, \xi_n) \in D$, the **initial value problem** for (E_n) is

$$y^{(n)} = h(t, y, y^{(1)}, \ldots, y^{(n-1)}), \qquad y(\tau) = \xi_1, \ldots, y^{(n-1)}(\tau) = \xi_n. \qquad (I_n)$$

A function ϕ is a **solution** of (I_n) if ϕ is a solution of Eq. (E_n) on some interval containing τ and if $\phi(\tau) = \xi_1, \ldots, \phi^{(n-1)}(\tau) = \xi_n$.

As in the case of systems of first order equations, we single out several special cases.

First we consider equations of the form

$$a_n(t)y^{(n)} + a_{n-1}(t)y^{(n-1)} + \cdots + a_1(t)y^{(1)} + a_0(t)y = g(t),$$

where $a_n(t), \ldots, a_0(t)$ are real continuous functions defined on the interval J and where $a_n(t) \neq 0$ for all $t \in J$. Without loss of generality, we shall consider in this book the case when $a_n(t) \equiv 1$, i.e.,

$$y^{(n)} + a_{n-1}(t)y^{(n-1)} + \cdots + a_1(t)y^{(1)} + a_0(t)y = g(t). \qquad (1.1)$$

We refer to Eq. (1.1) as a **linear nonhomogeneous ordinary differential equation of order** n.

If in Eq. (1.1) we let $g(t) \equiv 0$, then

$$y^{(n)} + a_{n-1}(t)y^{(n-1)} + \cdots + a_1(t)y^{(1)} + a_0(t)y = 0. \qquad (1.2)$$

We call Eq. (1.2) a **linear homogeneous ordinary differential equation of order** n. If in Eq. (1.2) we have $a_i(t) \equiv a_i$, $i = 0, 1, \ldots, n-1$, so that (1.2) reduces to

$$y^{(n)} + a_{n-1}y^{(n-1)} + \cdots + a_1 y^{(1)} + a_0 y = 0, \qquad (1.3)$$

then we speak of a **linear, autonomous, homogeneous ordinary differential equation of order** n.

We can, of course, also define **periodic** and **linear periodic ordinary differential equations of order** n in the obvious way.

We now show that the theory of nth order ordinary differential equations reduces to the theory of a system of n first order ordinary differential equations. To this end, we let $y = x_1$, $y^{(1)} = x_2, \ldots, y^{(n-1)} = x_n$ in Eq. (I_n). Then we have the system of first order ordinary differential equations

$$\begin{aligned} x_1' &= x_2, \\ x_2' &= x_3, \\ &\vdots \\ x_n' &= h(t, x_1, \ldots, x_n). \end{aligned} \qquad (1.4)$$

1.2 Examples of Initial Value Problems

This system of equations is clearly defined for all $(t, x_1, \ldots, x_n) \in D$. Now assume that the vector $\phi = (\phi_1, \ldots, \phi_n)^T$ is a solution of Eq. (1.4) on an interval J. Since $\phi_2 = \phi_1'$, $\phi_3 = \phi_2'$, $\ldots$, $\phi_n = \phi_1^{(n-1)}$, and since

$$h(t, \phi_1(t), \ldots, \phi_n(t)) = h(t, \phi_1(t), \ldots, \phi_1^{(n-1)}(t)) = \phi_1^{(n)}(t),$$

it follows that the first component ϕ_1 of the vector ϕ is a solution of Eq. (E_n) on the interval J. Conversely, assume that ϕ_1 is a solution of Eq. (E_n) on the interval J. Then the vector $\phi = (\phi, \phi^{(1)}, \ldots, \phi^{(n-1)})^T$ is clearly a solution of the system of equations (1.4). Note that if $\phi_1(\tau) = \xi_1, \ldots, \phi_1^{(n-1)}(\tau) = \xi_n$, then the vector ϕ satisfies $\phi(\tau) = \xi$, where $\xi = (\xi_1, \ldots, \xi_n)^T$. The converse is also true.

E. Complex Valued Ordinary Differential Equations

Thus far, we have concerned ourselves with initial value problems characterized by **real ordinary differential equations**. There are also initial value problems involving **complex ordinary differential equations**. For example, let t be real and let $z = (z_1, \ldots, z_n) \in C^n$, i.e., z is a complex vector with components of the form $z_k = u_k + iv_k$, $k = 1, \ldots, n$, where u_k and v_k are real and where $i = \sqrt{-1}$. Let D be a domain in the (t, z) space $R \times C^n$ and let $f_1, \ldots, f_n$ be continuous complex valued functions on D (i.e., $f_i: D \to C$). Let $f = (f_1, \ldots, f_n)^T$ and let $z' = dz/dt$. Then

$$z' = f(t, z) \tag{C}$$

is a **system of n complex ordinary differential equations of the first order**. The definition of solution and of the initial value problem are essentially the same as in the real cases already given. It is, of course, possible to consider various special cases of (C) which are analogous to the autonomous, periodic, linear, systems, and nth order cases already discussed for real valued equations. It will also be of interest to replace t in (C) by a complex variable and to consider the behavior of solutions of such systems.

1.2 EXAMPLES OF INITIAL VALUE PROBLEMS

In this section, which consists of seven parts, we give several examples of initial value problems. Although we concentrate here on simple examples from mechanics and electric circuits, it is emphasized that initial value problems of the type considered here arise in virtually all branches of the physical sciences, in engineering, in biological sciences, in economics, and in other disciplines.

In Section A we consider mechanical translation systems and in Section B we consider mechanical rotational systems. Both of these types of systems are based on Newton's second law. In Section C we give examples of electric circuits obtained from Kirchhoff's voltage and current laws. The purpose of Section D is to present several well-known ordinary differential equations, including some examples of Volterra population growth equations. We shall have occasion to refer to some of these examples later. In Section E we consider the Hamiltonian formulation of conservative dynamical systems, while in Section F we consider the Lagrangian formulation of dynamical systems. In Section G we present examples of electromechanical systems.

A. Mechanical Translation Systems

Mechanical translation systems obey Newton's second law of motion which states that the sum of the applied forces (to a point mass) must equal the sum of the reactive forces. In linear systems, which we consider presently, it is sufficient to consider only inertial elements (i.e., point masses), elastance or stiffness elements (i.e., springs), and damping or viscous friction terms (e.g., dashpots).

When a force f is applied to a point mass, an acceleration of the mass results. In this case the reactive force f_M is equal to the product of the mass and acceleration and is in the opposite direction to the applied force. In terms of displacement x, as shown in Fig. 1.2, we have velocity $= v = x' = dx/dt$, acceleration $= a = x'' = d^2x/dt^2$, and

$$f_M = Ma = Mv' = Mx'',$$

where M denotes the mass.

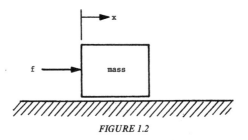

FIGURE 1.2

1.2 Examples of Initial Value Problems

The stiffness terms in mechanical translation systems provide restoring forces, as modeled, for example, by springs. When compressed, the spring tries to expand to its normal length, while when expanded, it tries to contract. The reactive force f_K on each end of the spring is the same and is equal to the product of the stiffness K and the deformation of the spring, i.e.,

$$f_K = K(x_1 - x_2),$$

where x_1 is the position of end 1 of the spring and x_2 the position of end 2 of the spring, measured from the original equilibrium position. The direction of this force depends on the relative magnitudes and directions of positions x_1 and x_2 (Fig. 1.3).

FIGURE 1.3

The damping terms or viscous friction terms characterize elements that dissipate energy while masses and springs are elements which store energy. The damping force is proportional to the difference in velocity of two bodies. The assumption is made that the viscous friction is linear. We represent the damping action by a dashpot as shown in Fig. 1.4. The reaction damping force f_B is equal to the product of damping B and the relative velocity of the two ends of the dashpot, i.e.,

$$f_B = B(v_1 - v_2) = B(x'_1 - x'_2).$$

The direction of this force depends on the relative magnitudes and directions of the velocities x'_1 and x'_2.

FIGURE 1.4

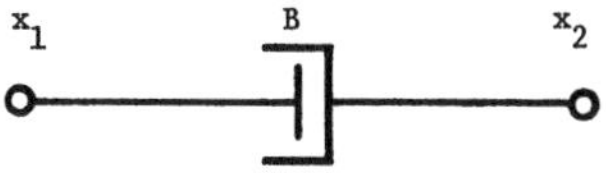

The preceding relations must be expressed in a consistent set of units. For example, in the MKS system, we have the following set of units: time in seconds; distance in meters; velocity in meters per second; acceleration in meters per (second)^2; mass in kilograms; force in newtons; stiffness coefficient K in newtons per meter; and damping coefficient B in newtons per (meter/second).

In a mechanical translation system, the (kinetic) energy stored in a mass is given by

$$T = \tfrac{1}{2}M(x')^2,$$

the (potential) energy stored by a spring is given by

$$W = \tfrac{1}{2}K(x_1 - x_2)^2,$$

while the energy dissipation due to viscous damping (as represented by a dashpot) is given by

$$2D = B(x_1' - x_2')^2.$$

In arriving at the differential equations which describe the behavior of a mechanical translation system, we may find it convenient to use the following procedure:

1. Assume that the system originally is in equilibrium. (In this way, the often troublesome effect of gravity is eliminated.)
2. Assume that the system is given some arbitrary displacement if no disturbing force is present.
3. Draw a "free-body diagram" of the forces acting on each mass of the system. A separate diagram is required for each mass.
4. Apply Newton's second law of motion to each diagram, using the convention that any force acting in the direction of the assumed displacement is positive.

Let us now consider a specific example.

Example 2.1. The mechanical system of Fig. 1.5 consists of two point masses M_1 and M_2 which are acted upon by viscous damping forces (due to B and due to the friction terms B_1 and B_2) and spring forces (due to the terms K_1, K_2, and K), and external forces $f_1(t)$ and $f_2(t)$. The

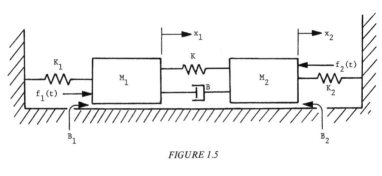

FIGURE 1.5

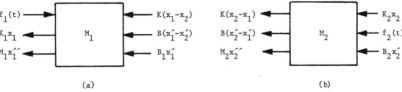

FIGURE 1.6 Free body diagrams for (a) M_1 and (b) M_2.

1.2 Examples of Initial Value Problems

initial displacements of masses M_1 and M_2 are given by $x_1(0) = x_{10}$ and $x_2(0) = x_{20}$, respectively, and their initial velocities are given by $x'_1(0) = x'_{10}$ and $x'_2(0) = x'_{20}$. The arrows in this figure establish positive directions for displacements x_1 and x_2.

The free-body diagrams for masses M_1 and M_2 are depicted in Fig. 1.6. From these figures, there now result the following equations which describe the system of Fig. 1.5.

$$\begin{aligned} M_1 x''_1 + (B + B_1)x'_1 + (K + K_1)x_1 - Bx'_2 - Kx_2 &= f_1(t), \\ M_2 x''_2 + (B + B_2)x'_2 + (K + K_2)x_2 - Bx'_1 - Kx_1 &= -f_2(t), \end{aligned} \quad (2.1)$$

with initial data $x_1(0) = x_{10}$, $x_2(0) = x_{20}$, $x'_1(0) = x'_{10}$, and $x'_2(0) = x'_{20}$.

Letting $y_1 = x_1, y_2 = x'_1, y_3 = x_2$, and $y_4 = x'_2$, we can express Eq. (2.1) equivalently by a system of four first order ordinary differential equations given by

$$\begin{bmatrix} y'_1 \\ y'_2 \\ y'_3 \\ y'_4 \end{bmatrix} = \begin{bmatrix} 0 & 1 & 0 & 0 \\ -[(K_1 + K)/M_1] & -[(B_1 + B)/M_1] & (K/M_1) & (B/M_1) \\ 0 & 0 & 0 & 1 \\ (K/M_2) & (B/M_2) & -[(K + K_2)/M_2] & -[(B + B_2)/M_2] \end{bmatrix}$$
$$\times \begin{bmatrix} y_1 \\ y_2 \\ y_3 \\ y_4 \end{bmatrix} + \begin{bmatrix} 0 \\ (1/M_1)f_1(t) \\ 0 \\ -(1/M_2)f_2(t) \end{bmatrix} \quad (2.2)$$

with initial data given by $(y_1(0)\ y_2(0)\ y_3(0)\ y_4(0))^T = (x_{10}\ x'_{10}\ x_{20}\ x'_{20})^T$.

B. Mechanical Rotational Systems

The equations which describe mechanical rotational systems are similar to those already given for translation systems. In this case forces are replaced by torques, linear displacements are replaced by angular displacements, linear velocities are replaced by angular velocities, and linear accelerations are replaced by angular accelerations. The force equations are replaced by corresponding torque equations and the three types of system elements are, again, inertial elements, springs, and dashpots.

The torque applied to a body having a moment of inertia J produces an angular acceleration $\alpha = \omega' = \theta''$. The reaction torque T_J is opposite to the direction of the applied torque and is equal to the product of moment of inertia and acceleration. In terms of angular displacement θ, angular velocity ω, or angular acceleration α, the torque equation is given by

$$T_J = J\alpha = J\omega' = J\theta''.$$

When a torque is applied to a spring, the spring is twisted by an angle θ and the applied torque is transmitted through the spring and appears at the other end. The reaction spring torque T_K that is produced is equal to the product of the stiffness or elastance K of the spring and the angle of twist. By denoting the positions of the two ends of the spring, measured from the neutral position, as θ_1 and θ_2, the reactive torque is given by

$$T_K = K(\theta_1 - \theta_2).$$

Once more, the direction of this torque depends on the relative magnitudes and directions of the angular displacements θ_1 and θ_2.

The damping torque T_B in a mechanical rotational system is proportional to the product of the viscous friction coefficient B and the relative angular velocity of the ends of the dashpot. The reaction torque of a damper is

$$T_B = B(\omega_1 - \omega_2) = B(\theta'_1 - \theta'_2).$$

Again, the direction of this torque depends on the relative magnitudes and directions of the angular velocities ω_1 and ω_2.

The expressions for T_J, T_K, and T_B are clearly counterparts to the expressions for f_M, f_K, and f_B, respectively.

The foregoing relations must again be expressed in a consistent set of units. In the MKS system, these units are as follows: time in seconds; angular displacement in radians; angular velocity in radians per second; angular acceleration in radians per second^2; moment of inertia in kilogram-meters^2; torque in newton-meters; stiffness coefficient K in newton-meters per radian; and damping coefficient B in newton-meters per (radians/second).

In a mechanical rotational system, the (kinetic) energy stored in a mass is given by

$$T = \tfrac{1}{2}J(\theta')^2,$$

the (potential) energy stored in a spring is given by

$$W = \tfrac{1}{2}K(\theta_1 - \theta_2)^2,$$

and the energy dissipation due to viscous damping (in a dashpot) is given by

$$2D = B(\theta'_1 - \theta'_2)^2.$$

Let us consider a specific example.

Example 2.2. The rotational system depicted in Fig. 1.7 consists of two masses with moments of inertia J_1 and J_2, two springs with stiffness constants K_1 and K_2, three dissipation elements with dissipation coefficients B_1, B_2, and B, and two externally applied torques T_1 and T_2.

1.2 Examples of Initial Value Problems

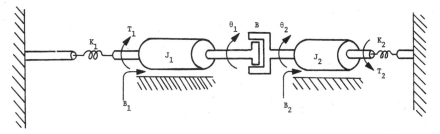

FIGURE 1.7

The initial angular displacements of the two masses are given by $\theta_1(0) = \theta_{10}$ and $\theta_2(0) = \theta_{20}$, respectively, and their initial angular velocities are given by $\theta_1'(0) = \theta_{10}'$ and $\theta_2'(0) = \theta_{20}'$.

The free-body diagrams for this system are given in Fig. 1.8. These figures yield the following equations which describe the system of Fig. 1.7.

FIGURE 1.8

$$J_1\theta_1'' + B_1\theta_1' + B(\theta_1' - \theta_2') + K_1\theta_1 = T_1,$$
$$J_2\theta_2'' + B_2\theta_2' + B(\theta_2' - \theta_1') + K_2\theta_2 = -T_2. \quad (2.3)$$

Letting $x_1 = \theta_1$, $x_2 = \theta_1'$, $x_3 = \theta_2$, and $x_4 = \theta_2'$, we can express these equations by the four equivalent first order ordinary differential equations

$$\begin{bmatrix} x_1' \\ x_2' \\ x_3' \\ x_4' \end{bmatrix} = \begin{bmatrix} 0 & 1 & 0 & 0 \\ -K_1/J_1 & -(B_1+B)/J_1 & 0 & B/J_1 \\ 0 & 0 & 0 & 1 \\ 0 & B/J_2 & -K_2/J_2 & -(B_2+B)/J_2 \end{bmatrix} \begin{bmatrix} x_1 \\ x_2 \\ x_3 \\ x_4 \end{bmatrix}$$
$$+ \begin{bmatrix} 0 \\ T_1(t)/J_1 \\ 0 \\ -T_2(t)/J_2 \end{bmatrix} \quad (2.4)$$

with initial data given by $[x_1(0)\ x_2(0)\ x_3(0)\ x_4(0)]^T = [\theta_{10}\ \theta_{10}'\ \theta_{20}\ \theta_{20}']^T$.

C. Electric Circuits

In describing electric circuits, we utilize Kirchhoff's voltage law (KVL) and Kirchhoff's current law (KCL) which state:

(a) The algebraic sum of potential differences around any closed loop in a circuit equals zero, i.e., in traversing any closed loop in a circuit, the sum of the voltage rises equals the sum of the voltage drops.

(b) The algebraic sum of currents at a junction or node in a circuit equals zero, i.e., the sum of the currents entering the junction equals the sum of the currents leaving the junction.

In the present discussion we concern ourselves with linear circuits consisting of voltage sources, current sources, capacitors, inductors, resistors, transformers, and the like. We shall discuss only those elements which we shall require.

Voltage (current) sources are modeled by voltage (current) generators. Direct current (dc) voltage sources are often modeled by batteries (see Fig. 1.9).

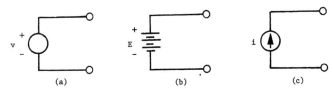

FIGURE 1.9 (a) *Voltage source*, (b) *dc voltage source*, (c) *current source*.

The voltage drop across a resistor is given by Ohm's law which states that the voltage drop across a resistor is equal to the product of the current i through the resistor and the resistance R (see Fig. 1.10), i.e.,

$$v_R = Ri \quad \text{or} \quad i = v/R.$$

The voltage drop across an inductor is equal to the product of the inductance L and the time rate of change of current, di/dt (see Fig. 1.10),

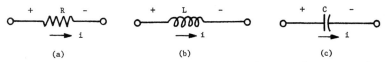

FIGURE 1.10 *Voltage drop across* (a) *a resistor R,* (b) *an inductor L, and* (c) *a capacitor C.*

1.2 Examples of Initial Value Problems

i.e.,

$$v_L = L\frac{di}{dt} \quad \text{or} \quad i = \frac{1}{L}\int_0^t v_L(\tau)\,d\tau + i_L(0).$$

The initial current $i_L(0)$ in the inductor carries its own algebraic sign, i.e., if $i_L(0)$ is in the same direction as i, then it is positive; otherwise it is negative.

The positively directed voltage drop across a capacitor is defined in magnitude as the ratio of the magnitude of the positive electric charge q on its positive plate to the value of its capacitance C. Its direction is from the positive plate to the negative plate. The charge on a capacitor plate equals the time integral from the initial instant to the arbitrary time instant t of the current $i(t)$ entering the plate, plus the initial value of the charge q_0 (see Fig. 1.10). Thus,

$$v_C = \frac{q}{C} = \frac{1}{C}\int_0^t i(\tau)\,d\tau + \frac{q_0}{C} = \frac{1}{C}\int_0^t i(\tau)\,d\tau + v_C(0) \quad \text{or} \quad i = C\frac{dv}{dt}.$$

The initial voltage $v_C(0)$ on the capacitor carries its own algebraic sign, i.e., if $v_C(0)$ has the same polarity as v_C, then it is positive; otherwise it is negative.

In using the foregoing relations, we need to use a consistent set of units. In the MKS system these are: charge, coulombs; current, amperes; voltage, volts; inductance, henrys; capacitance, farads; and resistance, ohms.

The energy dissipated in a resistor R is given by $i^2R = v_R^2/R$, where v_R is the applied voltage and i is the resulting current. The energy stored in an inductor L is given by $\frac{1}{2}Li^2$, where i is the current through the inductor. Also, the energy stored in a capacitor is given by $[q^2/(2C)]$, where q is the charge on the capacitor C.

There are several methods of analyzing electric circuits. We shall consider two of these, the *Maxwell mesh current method* (also called the *loop current method*) and the *nodal analysis method*.

The loop current method is based on Kirchhoff's voltage law and it consists of assuming that currents, termed "loop currents," flow in each loop of a multiloop network. In this method, the algebraic sum of the voltage drops around each loop, obtained by the use of the loop currents, is set equal to zero. The following procedure may prove useful:

1. Assume loop currents in a clockwise direction. Be certain that a current flows through every element and that the number of currents assumed is sufficient.

2. Around each loop, write an equation obtained from Kirchhoff's voltage law.

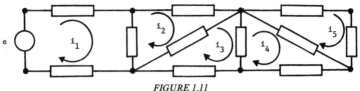

FIGURE 1.11

One method to ensure that a sufficient number of currents have been assumed in a network is as indicated in Fig. 1.11. (This method is applicable to "planar networks," i.e., networks that can be drawn with no wires crossing.) The currents are selected in such a fashion that through every element there is a current, and no element crosses a loop. This is the case in Fig. 1.11, but not in Fig. 1.12, where i_2 is crossed by an element.

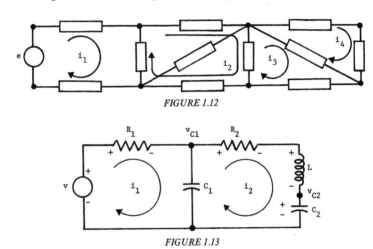

FIGURE 1.12

FIGURE 1.13

Example 2.3. As an example of loop analysis, consider the circuit depicted in Fig. 1.13. The loop currents i_1 and i_2 will suffice. Note that the current flow through capacitor C_1 is determined by both i_1 and i_2. In view of Kirchhoff's voltage law we have the following integrodifferential equations (see Fig. 1.14):

$$v = i_1 R_1 + \frac{1}{C_1} \int_0^t (i_1 - i_2) \, dt + v_{C1}(0) \tag{2.5}$$

$$0 = i_2 R_2 + L \frac{di_2}{dt} + \frac{1}{C_2} \int_0^t i_2 \, dt + v_{C2}(0) + \frac{1}{C_1} \int_0^t (i_2 - i_1) \, dt - v_{C1}(0), \tag{2.6}$$

1.2 Examples of Initial Value Problems

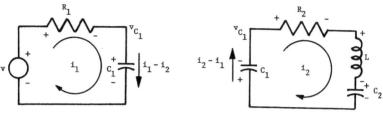

FIGURE 1.14

where $v_{C_1}(0)$ and $v_{C_2}(0)$ denote the initial voltages on capacitors C_1 and C_2, respectively.

Equations (2.5) and (2.6) can be expressed equivalently by ordinary differential equations in terms of charge q, by

$$q_1' + \frac{1}{R_1 C_1} q_1 - \frac{1}{R_1 C_1} q_2 = \frac{v}{R_1}, \qquad (2.7)$$

$$q_2'' + \frac{R_2}{L} q_2' + \left(\frac{1}{C_1 L} + \frac{1}{C_2 L}\right) q_2 - \frac{1}{L C_2} q_1 = 0. \qquad (2.8)$$

We can also describe this circuit by means of a system of first order ordinary differential equations. For example, if we let $x_1 = v_{C_1}$ (the voltage across capacitor C_1), $x_2 = v_{C_2}$ (the voltage across capacitor C_2), and $x_3 = i_2$ (the current through inductor L), then Eqs. (2.7) and (2.8) yield the system of equations

$$\begin{bmatrix} x_1' \\ x_2' \\ x_3' \end{bmatrix} = \begin{bmatrix} -1/(R_1 C_1) & 0 & -1/C_1 \\ 0 & 0 & 1/C_2 \\ 1/L & -1/L & -R_2/L \end{bmatrix} \begin{bmatrix} x_1 \\ x_2 \\ x_3 \end{bmatrix} + \begin{bmatrix} v/(R_1 C_1) \\ 0 \\ 0 \end{bmatrix}. \qquad (2.9)$$

The complete description of this circuit requires the specification of the initial data $v_{C_1}(0)$, $v_{C_2}(0)$, and $i_2(0)$.

The *nodal analysis method* is based on Kirchhoff's current law and involves the following steps:

1. Assume potentials (i.e., voltages) at all nodes in the circuit, and choose one point of node in the circuit as being at ground potential (i.e., at zero volts). The node voltages v_i measured above ground are the dependent variables in the node equations, just as the currents are dependent variables in the loop equations.

2. Utilize Kirchhoff's current law to write the appropriate equations to describe the circuit. This results in the same number of integro-

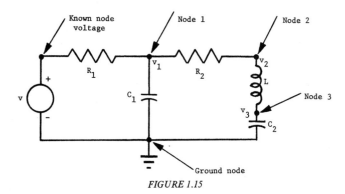
FIGURE 1.15

differential equations as there are assumed node potentials measured above ground potential. No equation is written for the node chosen at ground potential.

Example 2.4. Let us reconsider the circuit of the preceding example which is given in Fig. 1.15 with appropriately labeled node voltages. Note that the voltages of two of the five nodes are known, and therefore, node equations are required only for nodes 1–3. Using Kirchhoff's current law, we now obtain the integrodifferential equations

$$\frac{v_1 - v}{R_1} + C_1 v_1' + \frac{v_1 - v_2}{R_2} = 0, \qquad (2.10)$$

$$\frac{v_2 - v_1}{R_2} + \frac{1}{L} \int_0^t (v_2 - v_3)\, dt + i_L(0) = 0, \qquad (2.11)$$

$$\frac{1}{L} \int_0^t (v_3 - v_2)\, dt - i_L(0) + C_2 v_3' = 0, \qquad (2.12)$$

where $i_L(0)$ denotes the initial current through the inductor L. These equations can be rewritten as a system of three first order ordinary differential equations given by

$$\begin{bmatrix} v_1' \\ v_2' \\ v_3' \end{bmatrix} = \begin{bmatrix} \frac{-1}{C_1}\left(\frac{1}{R_1} + \frac{1}{R_2}\right) & \frac{1}{R_2 C_1} & 0 \\ \frac{-1}{C_1}\left(\frac{1}{R_1} + \frac{1}{R_2}\right) & -\left(\frac{R_2}{L} - \frac{1}{R_2 C_1}\right) & \frac{R_2}{L} \\ \frac{1}{R_2 C_2} & \frac{-1}{R_2 C_2} & 0 \end{bmatrix} \begin{bmatrix} v_1 \\ v_2 \\ v_3 \end{bmatrix} + \begin{bmatrix} \frac{v}{R_1 C_1} \\ \frac{v}{R_1 C_1} \\ 0 \end{bmatrix}. \qquad (2.13)$$

1.2 Examples of Initial Value Problems

In order to complete the description of this circuit, we need to specify the initial data $v_1(0)$, $v_2(0)$, and $v_3(0)$.

Since the system of equations (2.9) (obtained by Kirchhoff's voltage law) describes the same circuit as Eq. (2.13) (obtained by Kirchhoff's current law), one would expect that it would be possible to obtain Eq. (2.13) from (2.9), and vice versa, by means of an appropriate transformation. This is indeed the case. An inspection of Figs. 1.13 and 1.15 reveals that

$$x_1 = v_1, \quad x_2 = v_3, \quad x_3 = \frac{-v_2}{R_2} + \frac{v_1}{R_2}. \tag{2.14}$$

If we combine (2.9) with (2.14) we obtain (2.13), and if we combine (2.9) with (2.13) we obtain (2.14).

In Chapter 3 we shall obtain general results for linear equations which will show that the systems of equations (2.9) and (2.13) are representations of the same circuit with respect to two different sets of coordinates.

D. Some Examples of Nonlinear Systems

We now give several examples of systems which are described by some rather well-known differential equations which are not necessarily linear equations, as were the preceding cases. To simplify our discussion and to limit it to a manageable scope, we concentrate on second order differential equations of the form

$$\frac{dx^2}{dt^2} + p(t, x, x') = q(t), \quad t \geq 0, \tag{2.15}$$

where $x(0)$ and $x'(0)$ are specified, and where the functions $p(\cdot)$ and $q(\cdot)$ are specified. If we let $x_1 = x$ and $x_1' = x_2$, then Eq. (2.15) can equivalently be represented by

$$\begin{bmatrix} x_1' \\ x_2' \end{bmatrix} = \begin{bmatrix} x_2 \\ -p(t, x_1, x_2) \end{bmatrix} + \begin{bmatrix} 0 \\ q(t) \end{bmatrix} \tag{2.16}$$

with $[x_1(0) \; x_2(0)]^T = [x(0) \; x'(0)]^T$.

Example 2.5. An important special case of (2.15) is the **Lienard equation** given by

$$\frac{d^2x}{dt^2} + f(x)\frac{dx}{dt} + g(x) = 0, \tag{2.17}$$

where $f: R \to R$ and $g: R \to R$ are continuously differentiable functions with $f(x) \geq 0$ for all $x \in R$ and with $xg(x) > 0$ for all $x \neq 0$. This equation can be used to represent, for example, RLC circuits with nonlinear circuit elements (R, L, C).

An important special case of the Lienard equation is the **van der Pol equation** given by

$$\frac{d^2x}{dt^2} - \varepsilon(1 - x^2)\frac{dx}{dt} + x = 0, \tag{2.18}$$

where $\varepsilon > 0$ is a parameter. This equation represents rather well certain electronic oscillators.

Example 2.6. Another special case of Eq. (2.15) arising in applications is

$$\frac{d^2x}{dt^2} + h\frac{x'}{|x'|} + \omega_0^2 x = 0, \tag{2.19}$$

where $h > 0$ and $\omega_0^2 > 0$ are parameters. If we define the **sign function** by

$$\operatorname{sgn} \theta = \begin{cases} +1, & \theta > 0, \\ 0, & \theta = 0, \\ -1, & \theta < 0, \end{cases} \tag{2.20}$$

then Eq. (2.19) can be written as

$$\frac{d^2x}{dt^2} + h \operatorname{sgn} x' + \omega_0^2 x = 0. \tag{2.21}$$

Equation (2.21) has been used to represent a mass sliding on a surface and attached to a linear spring as shown in Fig. 1.16. The nonlinear term $h \operatorname{sgn} x'$ represents the *dry friction* force caused by the sliding of the mass on a dry surface. The magnitudes of h and ω_0^2 are determined by M, K, and the nature of the sliding surfaces. As usual, x represents the displacement of the mass.

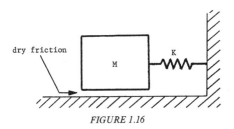

FIGURE 1.16

1.2 Examples of Initial Value Problems

Example 2.7. Another special case of Eq. (2.15) encountered in the literature is **Rayleigh's equation**, given by

$$\frac{d^2x}{dt^2} - \varepsilon\left[1 - \frac{1}{3}\left(\frac{dx}{dt}\right)^2\right]\frac{dx}{dt} + x = 0. \tag{2.22}$$

Here $\varepsilon > 0$ is a parameter.

Example 2.8. Another important special case of Eq. (2.15) is given by

$$\frac{d^2x}{dt^2} + g(x) = 0, \tag{2.23}$$

where $g(x)$ is continuous on R and where $xg(x) > 0$ for all $x \neq 0$. This equation can be used to represent a system consisting of a mass and a nonlinear spring, as shown in Fig. 1.17. Hence, we call this system a "mass on a nonlinear spring." Here, x denotes displacement and $g(x)$ denotes the restoring force due to the spring. We shall now identify several special cases that have been considered in the literature.

If $g(x) = k(1 + a^2x^2)x$, where $k > 0$ and $a^2 > 0$ are parameters, then Eq. (2.23) assumes the form

$$\frac{d^2x}{dt^2} + k(1 + a^2x^2)x = 0. \tag{2.24}$$

This system is called a *mass on a hard spring*. [More generally, one may assume only that $g'(x)$ and $g''(x)$ are positive.]

If $g(x) = k(1 - a^2x^2)x$, where $k > 0$ and $a^2 > 0$ are parameters, then Eq. (2.23) assumes the form

$$\frac{d^2x}{dt^2} + k(1 - a^2x^2)x = 0 \qquad (|x| < a^{-1}). \tag{2.25}$$

This system is referred to as a *mass on a soft spring*. [Again, this can be generalized to the requirement that $g'(x) > 0$ and $g''(x) < 0$.]

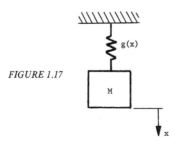

FIGURE 1.17

Equation (2.23) includes, of course, the case of a *mass on a linear spring*, also called a *harmonic oscillator*, given by

$$\frac{d^2x}{dt^2} + kx = 0, \tag{2.26}$$

where $k > 0$ is a parameter.

The motivation for the preceding terms (hard, soft, and linear spring), is made clear in Fig. 1.18, where the plots of the spring restoring forces versus displacement are given.

If $g(x) = k^2 x |x|$, where $k^2 > 0$ is a parameter, then Eq. (2.23) assumes the form

$$\frac{d^2x}{dt^2} + k^2 x |x| = 0. \tag{2.27}$$

This system is often called a *mass on a square-law spring*.

Example 2.9. An important special case of (2.23) is the equation given by

$$m\frac{d^2x}{dt^2} + k \sin x = 0, \tag{2.28}$$

where $k > 0$ is a parameter. This equation describes the motion of a constant mass moving in a circular path about the axis of rotation normal to a

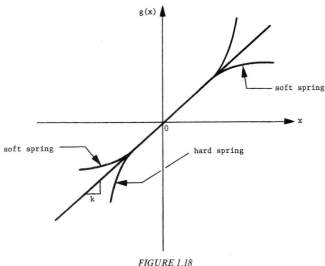

FIGURE 1.18

1.2 Examples of Initial Value Problems

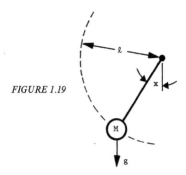

FIGURE 1.19

constant gravitational field, as shown in Fig. 1.19. The parameter k depends upon the radius l of the circular path, the gravitational acceleration g, and the mass. Here x denotes the angle of deflection measured from the vertical.

Example 2.10. Our last special case of Eq. (2.15) which we consider is the **forced Duffing's equation** (without damping), given by

$$\frac{d^2x}{dt^2} + \omega_0^2 x + hx^3 = G\cos\omega_1 t, \qquad (2.29)$$

where $\omega_0^2 > 0, h > 0, G > 0$, and $\omega_1 > 0$. This equation has been investigated extensively in the study of nonlinear resonance (ferroresonance) and can be used to represent an externally forced system consisting of a mass and nonlinear spring, as well as nonlinear circuits of the type shown in Fig. 1.20. Here the underlying variable x denotes the total instantaneous flux in the core of the inductor.

In the examples just considered, the equations are obtained by the use of physical laws, such as Newton's second law and Kirchhoff's voltage and current laws. There are many types of systems, such as models encountered in economics, ecology, biology, which are not based on laws of physics. For purposes of illustration, we consider now some examples of

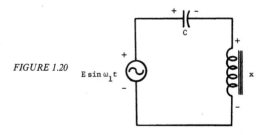

FIGURE 1.20

Volterra's population equations which attempt to model biological growth mathematically.

Example 2.11. A simple model representing the spreading of a disease in a given population is represented by the equations

$$\begin{aligned} x_1' &= -ax_1 + bx_1x_2, \\ x_2' &= -bx_1x_2, \end{aligned} \tag{2.30}$$

where x_1 denotes the density of infected individuals, x_2 denotes the density of noninfected individuals, and $a > 0$ and $b > 0$ are parameters. These equations are valid only for the case $x_1 \geq 0$ and $x_2 \geq 0$.

The second equation in (2.30) states that the noninfected individuals become infected at a rate proportional to x_1x_2. This term is a measure of the interaction between the two groups. The first equation in (2.30) consists of two terms: $-ax_1$ which is the rate at which individuals die from the disease or survive and become forever immune, and bx_1x_2 which is the rate at which previously noninfected individuals become infected.

To complete the initial value problem, it is necessary to specify nonnegative initial data $x_1(0)$ and $x_2(0)$.

Example 2.12. A simple predator–prey model is given by the equations

$$\begin{aligned} x_1' &= -ax_1 + bx_1x_2, \\ x_2' &= cx_2 - dx_1x_2, \end{aligned} \tag{2.31}$$

where $x_1 \geq 0$ denotes the density of predators (e.g., foxes), $x_2 \geq 0$ denotes the density of prey (e.g., rabbits), and $a > 0$, $b > 0$, $c > 0$, and $d > 0$ are parameters.

Note that if $x_2 = 0$, then the first equation in (2.31) reduces to $x_1' = -ax_1$, which implies that in the absence of prey, the density of predators will diminish exponentially to zero. On the other hand, if $x_2 \neq 0$, then the first equation in (2.31) indicates that x_1' contains a growth term proportional to x_2. Note also that if $x_1 = 0$, then the second equation reduces to $x_2' = cx_2$ and x_2 will grow exponentially while when $x_1 \neq 0$, x_2' contains a decay term proportional to x_1.

Once more, we need to specify nonnegative initial data, $x_1(0) = x_{10}$ and $x_2(0) = x_{20}$.

Example 2.13. A model for the growth of a (well-stirred and homogeneous) population with unlimited resources is

$$x' = cx, \quad c > 0,$$

1.2 Examples of Initial Value Problems

where x denotes population density and c is a constant. If the resources for growth are limited, then $c = c(x)$ should be a decreasing function of x instead of a constant. In the "linear" case, this function assumes the form $a - bx$ where $a, b > 0$ are constants, and one obtains the **Verhulst–Pearl equation**

$$x' = ax - bx^2.$$

Similar reasoning can be applied to population growth for two competing species. For example, consider a set of equations which describe two kinds of species (e.g., small fish) that prey on each other, i.e., the adult members of species A prey on young members of species B, and vice versa. In this case we have equations of the form

$$\begin{aligned} x_1' &= ax_1 - bx_1x_2 - cx_1^2, \\ x_2' &= dx_2 - ex_1x_2 - fx_2^2, \end{aligned} \quad (2.32)$$

where a, b, c, d, e, and f are positive parameters, where $x_1 \geq 0$ and $x_2 \geq 0$, and where nonnegative initial data $x_1(0) = x_{10}$ and $x_2(0) = x_{20}$ must be specified.

E. Hamiltonian Systems

Conservative dynamical systems are those systems which contain no energy dissipating elements. Such systems, with n degrees of freedom, can be characterized by means of a **Hamiltonian function** $H(p, q)$, where $q^T = (q_1, \ldots, q_n)$ denotes n generalized position coordinates and $p^T = (p_1, \ldots, p_n)$ denotes n generalized momentum coordinates. We assume $H(p, q)$ is of the form

$$H(p, q) = T(q, q') + W(q), \quad (2.33)$$

where T denotes the kinetic energy and W denotes the potential energy of the system. These energy terms are obtained from the path independent line integrals

$$T(q, q') = \int_0^{q'} p(q, \xi)^T d\xi = \int_0^{q'} \sum_{i=1}^n p_i(q, \xi) d\xi_i, \quad (2.34)$$

$$W(q) = \int_0^q f(\eta)^T d\eta = \int_0^q \sum_{i=1}^n f_i(\eta) d\eta_i, \quad (2.35)$$

where $f_i, i = 1, \ldots, n$, denote generalized potential forces.

In order that the integral in (2.34) be path independent, it is necessary and sufficient that

$$\frac{\partial p_i(q,q')}{\partial q'_j} = \frac{\partial p_j(q,q')}{\partial q'_i}, \quad i,j = 1,\ldots,n. \tag{2.36}$$

A similar statement can be made about Eq. (2.35).

Conservative dynamical systems are described by the system of $2n$ ordinary differential equations

$$\begin{aligned} q'_i &= \frac{\partial H}{\partial p_i}(p,q), & i = 1,\ldots,n \\ p'_i &= -\frac{\partial H}{\partial q_i}(p,q), & i = 1,\ldots,n. \end{aligned} \tag{2.37}$$

Note that if we compute the derivative of $H(p,q)$ with respect to t for (2.37) [i.e., along the solutions $q_i(t), p_i(t), i = 1, \ldots, n$] then we obtain, by the chain rule,

$$\begin{aligned}\frac{dH}{dt}(p(t),q(t)) &= \sum_{i=1}^n \frac{\partial H}{\partial p_i}(p,q)p'_i + \sum_{i=1}^n \frac{\partial H}{\partial q_i}(p,q)q'_i \\ &= \sum_{i=1}^n -\frac{\partial H}{\partial p_i}(p,q)\frac{\partial H}{\partial q_i}(p,q) + \sum_{i=1}^n \frac{\partial H}{\partial q_i}(p,q)\frac{\partial H}{\partial p_i}(p,q) \\ &= -\sum_{i=1}^n \frac{\partial H}{\partial p_i}(p,q)\frac{\partial H}{\partial q_i}(p,q) + \sum_{i=1}^n \frac{\partial H}{\partial p_i}(p,q)\frac{\partial H}{\partial q_i}(p,q) \equiv 0.\end{aligned}$$

In other words, in a conservative system (2.37) the Hamiltonian, i.e., the total energy, will be constant along the solutions of (2.37). This constant is determined by the initial data $(p(0), q(0))$.

Example 2.14. Consider the system depicted in Fig. 1.21. The kinetic energy terms for masses M_1 and M_2 are

$$T_1(x'_1) = \tfrac{1}{2}M_1(x'_1)^2, \qquad T_2(x'_2) = \tfrac{1}{2}M_2(x'_2)^2,$$

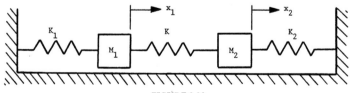

FIGURE 1.21

1.2 Examples of Initial Value Problems

respectively, the potential energy terms for springs K_1, K_2, K are
$$W_1(x_1) = \tfrac{1}{2}K_1 x_1^2, \qquad W_2(x_2) = \tfrac{1}{2}K_2 x_2^2, \qquad W(x_1,x_2) = \tfrac{1}{2}K(x_1 - x_2)^2,$$
respectively, and the Hamiltonian function for the system is given by
$$H(x_1,x_2,x_1',x_2') = \tfrac{1}{2}[M_1(x_1')^2 + M_2(x_2')^2 + K_1 x_1^2 + K_2 x_2^2 + K(x_1 - x_2)^2].$$
From (2.37) we now obtain the two second order ordinary differential equations
$$M_1 x_1'' = -K_1 x_1 - K(x_1 - x_2),$$
$$M_2 x_2'' = -K_2 x_2 - K(x_1 - x_2)(-1),$$
or
$$\begin{aligned} M_1 x_1'' + K_1 x_1 + K(x_1 - x_2) &= 0, \\ M_2 x_2'' + K_2 x_2 + K(x_2 - x_1) &= 0. \end{aligned} \qquad (2.38)$$

If we let $x_1 = y_1$, $x_1' = y_2$, $x_2 = y_3$, $x_2' = y_4$, then Eqs. (2.38) can equivalently be expressed as
$$\begin{bmatrix} y_1' \\ y_2' \\ y_3' \\ y_4' \end{bmatrix} = \begin{bmatrix} 0 & 1 & 0 & 0 \\ -(K_1 + K)/M_1 & 0 & K/M_1 & 0 \\ 0 & 0 & 0 & 1 \\ K/M_2 & 0 & -(K_2 + K)/M_2 & 0 \end{bmatrix} \begin{bmatrix} y_1 \\ y_2 \\ y_3 \\ y_4 \end{bmatrix}. \qquad (2.39)$$

Note that if in Fig. 1.5 we let $B_1 = B_2 = B = 0$, then Eq. (2.39) reduces to Eq. (2.2).

In order to complete the description of the system of Fig. 1.21 we must specify the initial data $x_1(0) = y_1(0)$, $x_1'(0) = y_2(0)$, $x_2(0) = y_3(0)$, $x_2'(0) = y_4(0)$.

Example 2.15. Let us consider the nonlinear spring-mass system shown in Fig. 1.22, where $g(x)$ denotes the potential force of the

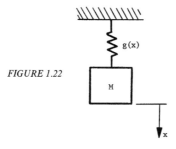

FIGURE 1.22

spring. The potential energy for this system is given as

$$W(x) = \int_0^x g(\eta)\,d\eta,$$

the kinetic energy for this system is given by

$$T(x') = \tfrac{1}{2}M(x')^2,$$

and the Hamiltonian function is given by

$$H(x,x') = \tfrac{1}{2}M(x')^2 + \int_0^x g(\eta)\,d\eta. \tag{2.40}$$

In view of Eqs. (2.37) and (2.40) we obtain the second order ordinary differential equation

$$\frac{d}{dt}(Mx') = -g(x)$$

or

$$Mx'' + g(x) = 0. \tag{2.41}$$

Equation (2.41) along with the initial data $x(0) = x_{10}$ and $x'(0) = x_{20}$ describe completely the system of Fig. 1.22. By letting $x_1 = x$ and $x_2 = x'$, this initial value problem can be described equivalently by the system of equations

$$\begin{aligned} x_1' &= x_2, \\ x_2' &= -\frac{1}{M}(g(x_1)), \end{aligned} \tag{2.42}$$

with the initial data given by $x_1(0) = x_{10}$, $x_2(0) = x_{20}$.

It should be noted that along the solutions of (2.42) we have

$$\frac{dH}{dt}(x_1,x_2) = g(x_1)x_2 + Mx_2\left(-\frac{1}{M}g(x_1)\right) = 0,$$

as expected.

The Hamiltonian formulation is of course also applicable to conservative rotational mechanical systems, electric circuits, electromechanical systems, and the like.

F. Lagrange's Equation

If a dynamical system contains elements which dissipate energy, such as viscous friction elements in mechanical systems, and resistors in electric circuits, then we can use Lagrange's equation to describe such

1.2 Examples of Initial Value Problems

systems. For a system with n degrees of freedom, this equation is given by

$$\frac{d}{dt}\left(\frac{\partial L}{\partial q'_i}(q,q')\right) - \frac{\partial L}{\partial q_i}(q,q') + \frac{\partial D}{\partial q'_i}(q') = F_i, \qquad i = 1,\ldots,n, \qquad (2.43)$$

where $q^T = (q_1,\ldots,q_n)$ denotes the generalized position vector. The function $L(q,q')$ is called the **Lagrangian** and is defined as

$$L(q,q') = T(q,q') - W(q),$$

i.e., it is the difference between the kinetic energy T and the potential energy W.

The function $D(q')$ denotes **Rayleigh's dissipation function** which we shall assume to be of the form

$$D(q') = \tfrac{1}{2}\sum_{i=1}^{n}\sum_{j=1}^{n}\beta_{ij}q'_i q'_j,$$

where $[\beta_{ij}]$ is a positive semidefinite matrix. The dissipation function D represents one-half the rate at which energy is dissipated as heat; it is produced by friction in mechanical systems and by resistance in electric circuits.

Finally, F_i in Eq. (2.43) denotes an applied force and includes all external forces which are associated with the q_i coordinate. The force F_i is defined as being positive when it acts so as to increase the value of the coordinate q_i.

Example 2.16. Consider the system depicted in Fig. 1.23 which is clearly identical to the system given in Fig. 1.5. For this system we have

$$T(q,q') = \tfrac{1}{2}M_1(x'_1)^2 + \tfrac{1}{2}M_2(x'_2)^2,$$
$$W(q) = \tfrac{1}{2}K_1 x_1^2 + \tfrac{1}{2}K_2 x_2^2 + \tfrac{1}{2}K(x_1 - x_2)^2,$$
$$D(q') = \tfrac{1}{2}B_1(x'_1)^2 + \tfrac{1}{2}B_2(x'_2)^2 + \tfrac{1}{2}B(x'_1 - x'_2)^2,$$

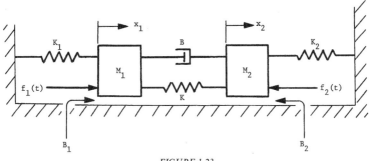

FIGURE 1.23

and
$$F_1(t) = f_1(t), \qquad F_2(t) = -f_2(t).$$
The Lagrangian assumes the form
$$L(q, q') = \tfrac{1}{2}M_1(x'_1)^2 + \tfrac{1}{2}M_2(x'_2)^2 - \tfrac{1}{2}K_1 x_1^2 - \tfrac{1}{2}K_2 x_2^2 - \tfrac{1}{2}K(x_1 - x_2)^2.$$
We now have

$$\frac{\partial L}{\partial x'_1} = M_1 x'_1, \qquad\qquad \frac{\partial L}{\partial x'_2} = M_2 x'_2,$$

$$\frac{d}{dt}\left(\frac{\partial L}{\partial x'_1}\right) = M_1 x''_1, \qquad \frac{d}{dt}\left(\frac{\partial L}{\partial x'_2}\right) = M_2 x''_2,$$

$$\frac{\partial L}{\partial x_1} = -K_1 x_1 - K(x_1 - x_2), \qquad \frac{\partial L}{\partial x_2} = -K_2 x_2 - K(x_2 - x_1),$$

$$\frac{\partial D}{\partial x'_1} = B_1 x'_1 + B(x'_1 - x'_2), \qquad \frac{\partial D}{\partial x'_2} = B_2 x'_2 + B(x'_2 - x'_1).$$

In view of Lagrange's equation we now obtain the two second order ordinary differential equations
$$\begin{aligned} M_1 x''_1 + (B + B_1)x'_1 + (K + K_1)x_1 - Bx'_2 - Kx_2 &= f_1(t), \\ M_2 x''_2 + (B + B_2)x'_2 + (K + K_2)x_2 - Bx'_1 - Kx_1 &= -f_2(t). \end{aligned} \qquad (2.44)$$

These equations are clearly in agreement with Eq. (2.1), which was obtained by using Newton's second law.

If we let $y_1 = x_1$, $y_2 = x'_1$, $y_3 = x_2$, $y_4 = x'_2$, then we can express (2.44) by the system of four first order ordinary differential equations given in (2.2).

Example 2.17. Consider the mass-linear dashpot/nonlinear spring system shown in Fig. 1.24, where $g(x)$ denotes the potential force due to the spring and $f(t)$ is an externally applied force.

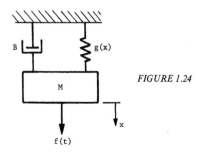

FIGURE 1.24

1.2 Examples of Initial Value Problems

The Lagrangian is given by

$$L(x, x') = \tfrac{1}{2}M(x')^2 - \int_0^x g(\eta)\,d\eta$$

and Rayleigh's dissipation function is given by

$$D(x') = \tfrac{1}{2}B(x')^2.$$

Now

$$\frac{\partial L}{\partial x'} = Mx', \qquad \frac{d}{dt}\left(\frac{\partial L}{\partial x'}\right) = Mx'',$$

$$\frac{\partial L}{\partial x} = -g(x), \qquad \frac{\partial D}{\partial x'} = Bx'.$$

Invoking Lagrange's equation, we obtain the equation

$$Mx'' + Bx' + g(x) = f(t). \tag{2.45}$$

The complete description of this initial value problem includes the initial data $x(0) = x_{10}$, $x'(0) = x_{20}$.

Lagrange's equation can be applied equally as well to rotational mechanical systems, electric circuits, and so forth. This will be demonstrated further in Section G.

G. Electromechanical Systems

In describing electromechanical systems, we can make use of Newton's second law and Kirchhoff's voltage and current laws, or we can invoke Lagrange's equation. We demonstrate these two approaches by means of two specific examples.

Example 2.18. The schematic of Fig. 1.25 represents a simplified model of an armature voltage-controlled dc servomotor. This motor consists of a stationary field and a rotating armature and load. We assume that all effects of the field are negligible in the description of this system. We now identify the indicated parameters and variables: e_a, externally applied armature voltage; i_a, armature current; R_a, resistance of armature winding; L_a, inductance of armature winding; e_m, back emf voltage induced by the rotating armature winding; B, viscous damping due to friction in bearings, due to windage, etc.; J, moment of inertia of armature and load; and θ, shaft position.

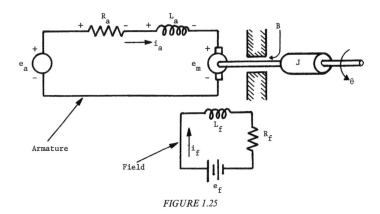

FIGURE 1.25

The back emf voltage (with polarity as shown) is given by

$$e_m = K\theta', \tag{2.46}$$

where θ' denotes the angular velocity of the shaft and $K > 0$ is a constant. The torque T generated by the motor is given by

$$T = K_T i_a \tag{2.47}$$

where $K_T > 0$ is a constant. This torque will cause an angular acceleration θ'' of the load and armature which we can determine from Newton's second law by the equation

$$J\theta'' + B\theta' = T(t). \tag{2.48}$$

Also, using Kirchhoff's voltage law we obtain for the armature circuit the equation

$$L_a \frac{di_a}{dt} + i_a R_a + e_m = e_a. \tag{2.49}$$

Combining Eqs. (2.46) and (2.49) and Eqs. (2.47) and (2.48), we obtain the differential equations

$$\frac{di_a}{dt} + \frac{R_a}{L_a} i_a + \frac{K}{L_a} \frac{d\theta}{dt} = \frac{e_a}{L_a} \quad \text{and} \quad \frac{d^2\theta}{dt^2} + \frac{B}{J} \frac{d\theta}{dt} = \frac{K_T}{J} i_a.$$

To complete the description of this initial value problem we need to specify the initial data $\theta(0) = \theta_0$, $\theta'(0) = \theta'_0$ and $i_a(0) = i_{a0}$.

Letting $x_1 = \theta$, $x_2 = \theta'$, $x_3 = i_a$, we can represent this system equivalently by the system of first order ordinary differential equations given

1.2 Examples of Initial Value Problems

by

$$\begin{bmatrix} x'_1 \\ x'_2 \\ x'_3 \end{bmatrix} = \begin{bmatrix} 0 & 1 & 0 \\ 0 & -B/J & K_T/J \\ 0 & -K/L_a & -R_a/L_a \end{bmatrix} \begin{bmatrix} x_1 \\ x_2 \\ x_3 \end{bmatrix} + \begin{bmatrix} 0 \\ 0 \\ e_a/L_a \end{bmatrix}$$

with the initial data given by $(x_1(0)\ x_2(0)\ x_3(0))^T = (\theta_0\ \theta'_0\ i_{a0})^T$.

Example 2.19. Consider the capacitor microphone depicted in Fig. 1.26. Here we have a capacitor constructed from a fixed plate and a moving plate with mass M, as shown. The moving plate is suspended from the fixed frame by a spring which has a spring constant K and which also has some damping expressed by the damping constant B. Sound waves exert an external force $f(t)$ on the moving plate. The output voltage v_s, which appears across the resistor R, will reproduce electrically the sound-wave patterns which strike the moving plate.

When $f(t) \equiv 0$ there is a charge q_0 on the capacitor. This produces a force of attraction between the plates so that the spring is stretched by an amount x_1 and the space between the plates is x_0. When sound waves exert a force on the moving plate, there will be a resulting motion displacement x which is measured from the equilibrium position. The distance between the plates will then be $x_0 - x$ and the charge on the plates will be $q_0 + q$.

The expression for the capacitance C is rather complex, but when displacements are small, it is approximately given by

$$C = \varepsilon A/(x_0 - x)$$

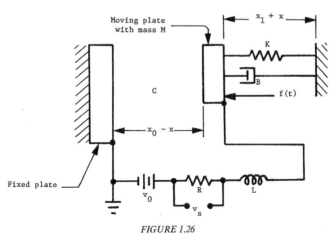

FIGURE 1.26

with $C_0 = \varepsilon A/x_0$, where $\varepsilon > 0$ is the dielectric constant for air and A is the area of the plate.

By inspection of Fig. 1.26 we now have

$$T = \tfrac{1}{2}L(q')^2 + \tfrac{1}{2}M(x')^2,$$

$$W = \frac{1}{2C}(q_0 + q)^2 + \tfrac{1}{2}K(x_1 + x)^2 = \frac{1}{2\varepsilon A}(x_0 - x)(q_0 + q)^2 + \tfrac{1}{2}K(x_1 + x)^2,$$

$$L = \tfrac{1}{2}L(q')^2 + \tfrac{1}{2}M(x')^2 - \frac{1}{2\varepsilon A}(x_0 - x)(q_0 + q)^2 - \tfrac{1}{2}K(x_1 + x)^2,$$

and

$$D = \tfrac{1}{2}R(q')^2 + \tfrac{1}{2}B(x')^2.$$

This is a two-degree-of-freedom system, where one of the degrees of freedom is displacement x of the moving plate and the other degree of freedom is the current $i = q'$. From Lagrange's equation we obtain

$$Mx'' + Bx' - \frac{1}{2\varepsilon A}(q_0 + q)^2 + K(x_1 + x) = f(t),$$

$$Lq'' + Rq' + \frac{1}{\varepsilon A}(x_0 - x)(q_0 + q) = v_0,$$

or

$$Mx'' + Bx' + Kx - c_1 q - c_2 q^2 = F(t),$$
$$Lq'' + Rq' + [x_0/(\varepsilon A)]q - c_3 x - c_4 xq = V, \tag{2.50}$$

where $c_1 = q_0/(\varepsilon A)$, $c_2 = 1/(2\varepsilon A)$, $c_3 = q_0/(\varepsilon A)$, $c_4 = 1/(\varepsilon A)$, $F(t) = f(t) - Kx_1 + [1/(2\varepsilon A)]q_0$, and $V = v_0 - [1/(\varepsilon A)]q_0$.

If we let $y_1 = x$, $y_2 = x'$, $y_3 = q$, and $y_4 = q'$, we can represent Eqs. (2.50) equivalently by the system of equations

$$\begin{bmatrix} y_1' \\ y_2' \\ y_3' \\ y_4' \end{bmatrix} = \begin{bmatrix} 0 & 1 & 0 & 0 \\ -K/M & -B/M & c_1/M & 0 \\ 0 & 0 & 0 & 1 \\ c_3/L & 0 & -y_0/(\varepsilon AL) & -R/L \end{bmatrix} \begin{bmatrix} y_1 \\ y_2 \\ y_3 \\ y_4 \end{bmatrix}$$
$$+ \begin{bmatrix} 0 \\ (c_2/M)y_3^2 \\ 0 \\ (c_4/L)y_1 y_3 \end{bmatrix} + \begin{bmatrix} 0 \\ (1/M)F(t) \\ 0 \\ (1/L)V \end{bmatrix}.$$

To complete the description of this initial value problem, we need to specify the initial data $x(0) = y_1(0)$, $x'(0) = y_2(0)$, $q(0) = y_3(0)$, and $q'(0) = i(0) = y_4(0)$.

PROBLEMS

1. Given the second order equation $y'' + f(y)y' + g(y) = 0$, write an equivalent system using the transformations
  (a) $x_1 = y, x_2 = y'$, and
  (b) $x_1 = y, x_2 = y' + \int_0^y f(s)\,ds$.

In how many different ways can this second order equation be written as an equivalent system of two first order equations?

2. Write
$$y'' + 3\sin(zy) + z' = \cos t, \qquad z''' + z'' + 3y' + z'y = t$$
as an equivalent system of first order equations. What initial conditions must be given in order to specify an initial value problem?

3. Suppose ϕ_1 and ϕ_2 solve the initial value problem
$$x'_1 = 3x_1 + x_2, \qquad x_1(0) = 1$$
$$x'_2 = -x_1 + x_2, \qquad x_2(0) = -1.$$
Find a second order differential equation which ϕ_1 will solve. Compute $\phi'_1(0)$. Do the same for ϕ_2.

4. Solve the following problems.
  (a) $x' = x^3, x(0) = 1$;
  (b) $x'' + x = 0, x(0) = 1, x'(0) = -1$;
  (c) $x'' - x = 0, x(0) = 1, x'(0) = -1$;
  (d) $x' = h(t)x, x(\tau) = \xi$;
  (e) $x' = h(t)x + k(t), x(\tau) = \xi$;
  (f) $x'_1 = -2x_2, x'_2 = -3x_1$;
  (g) $x'' + x' + x = 0$.

5. Let $x = (q^T, p^T)^T \in R^{2n}$ where $p, q \in R^n$ and let $H(q, p) = \tfrac{1}{2}x^T S x$ where S is a real, symmetric $2n \times 2n$ matrix.

  (a) Show that the corresponding Hamiltonian differential equation has the form $x' = JSx$, where $J = \begin{bmatrix} 0 & E_n \\ -E_n & 0 \end{bmatrix}$ and E_n is the $n \times n$ identity matrix.

  (b) Show that if $y = Tx$ where T is a $2n \times 2n$ matrix which satisfies the relation $T^*JT = J$ (where T^* is the adjoint of T) and det $T \neq 0$, then y will satisfy a linear Hamiltonian differential equation. Compute the Hamiltonian for this new equation.

6. Write the differential equations and the initial conditions needed to completely describe the linear mechanical translational system depicted in Fig. 1.27. Compute the Langrangian function for this mechanical system.

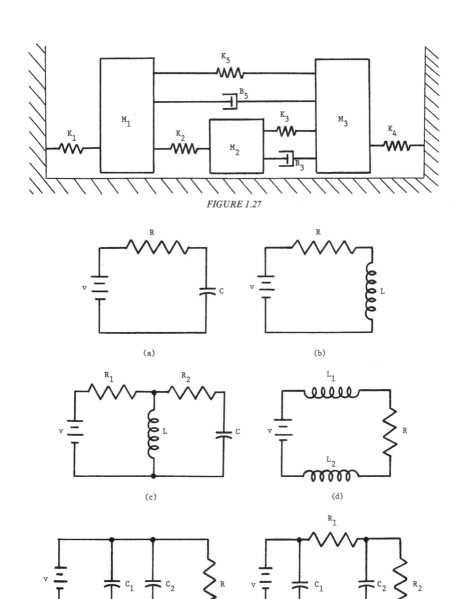

FIGURE 1.27

FIGURE 1.28

If the damping coefficients B_3 and B_5 are zero, this system is a Hamiltonian system. In this case, compute the Hamiltonian function.

7. Write differential equations which describe the linear circuits depicted in Fig. 1.28. Choose coordinates and write each differential equation as a system of first order equations.

8. Write differential equations which describe the linear circuit depicted in Fig. 1.29. Use the Maxwell mesh current method and then use the nodal analysis method.

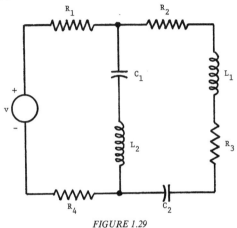

FIGURE 1.29

If $v = 0$ and $R_i = 0$ for $i = 1, \ldots, 4$, then the resulting system is a Hamiltonian system. Find the Hamiltonian.

9. A block of mass M is free to slide on a frictionless rod as indicated in the accompanying Fig. 1.30. The attached spring is linear. At equilibrium, the spring is not under tension or compression. Find the equation governing the motion of the block.

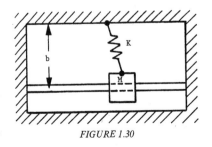

FIGURE 1.30

10. A thin inelastic cable connects a point mass M to a linear spring–linear damper over a frictionless pulley with moment of inertia J (see Fig. 1.31). Find the equation governing the motion of this mass. Assume that the cable does not slip over the pulley.

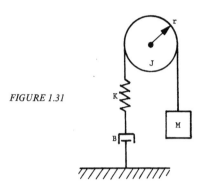

FIGURE 1.31

11. A mass, linear spring, and linear damper are connected under the lever arrangement depicted in Fig. 1.32. Write the equation governing the motion of the mass M.

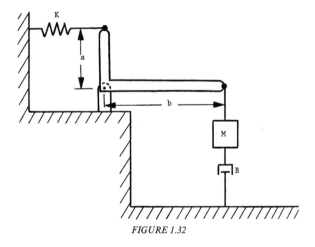

FIGURE 1.32

FUNDAMENTAL THEORY

2

The purpose of this chapter is to present some basic results on existence, uniqueness, continuation, and continuity with respect to parameters for the initial value problem

$$x' = f(t,x), \qquad x(\tau) = \xi. \tag{I}$$

Related material on comparison theory and on invariance theorems will also be given.

This chapter consists of nine parts. In Section 1 we establish some notation, we recall some well-known background material, and we establish some preliminary results which will be required later. In Section 2 we concern ourselves with the existence of solutions of initial value problems, in Section 3 we consider the continuation of solutions of initial value problems, in Section 4 we address the question of uniqueness of solutions of such problems, and in Section 5 we consider continuity of solutions of initial value problems with respect to parameters.

In order not to get bogged down with too many details at the same time, we develop all of the results in Sections 2–5 for the initial value problem (I') characterized by the scalar ordinary differential equation (E'). In Section 6 we first recall some additional facts from linear algebra. This background material makes it possible to extend all of the results of Sections 2–5 in a straightforward manner to initial value problems (I) characterized by the system of equations (E). To demonstrate this, we state and prove some sample results for linear nonhomogeneous systems (LN) and we ask the reader to do the same for initial value problems (I).

In Section 7 we consider the differentiability of solutions of (I) with respect to parameters, and in Section 8 we present our first comparison and invariance results. (We shall provide further comparison and invariance results in Chapter 5.) This chapter is concluded in Section 9 with a brief discussion of existence and uniqueness of holomorphic solutions to initial value problems characterized by systems of complex valued ordinary differential equations (C).

2.1 PRELIMINARIES

In the present section we establish some notation which will be used throughout the remainder of this book and we recall and summarize some background material which will be required in our presentation. This section consists of four parts. In Section A we consider continuous functions, in Section B we present certain inequalities, in Section C we discuss the notion of lim sup, and in Section D we present Zorn's lemma.

A. Continuous Functions

Let J be an interval of real numbers with nonempty interior. We shall use the notation

$$C(J) = \{f : f \text{ maps } J \text{ into } R \text{ and } f \text{ is continuous}\}.$$

When J contains one or both endpoints, then the continuity is one-sided at these points. Also, with k any positive integer, we shall use the notation

$$C^k(J) = \{f : \text{the derivatives } f^{(j)} \text{ exist and } f^{(j)} \in C(J) \\ \text{for } j = 0, 1, \ldots, k, \text{ where } f^{(0)} = f\}.$$

If f maps J into C, the complex numbers, then $f \in C^k(J)$ will mean that the real and complex parts of f satisfy the preceding property. Furthermore, if f is a real or complex vector valued function, then $f \in C^k(J)$ will mean that each component of f satisfies the preceding condition. Finally, for any subset D of the n space R^n with nonempty interior, we can define $C(D)$ and $C^k(D)$ similarly.

2.1 Preliminaries

A function $f \in C(J)$ is called piecewise-C^k if $f \in C^{k-1}(J)$ and $f^{(k-1)}$ has a continuous derivative for all t in J with the possible exception of a finite set of points where $f^{(k)}$ may have jump discontinuities.

Definition 1.1. Let $\{f_m\}$ be a sequence of real valued functions defined on a set $D \subset R^n$.

(i) The sequence $\{f_m\}$ is called a **uniform Cauchy sequence** if for any positive ε there exists an integer $M(\varepsilon)$ such that when $m > k \geq M$ one has $|f_k(x) - f_m(x)| < \varepsilon$ for all x in D.

(ii) The sequence $\{f_m\}$ is said to **converge uniformly** on D to a function f if for any $\varepsilon > 0$ there exists $M(\varepsilon)$ such that when $m > M$ one has $|f_m(x) - f(x)| < \varepsilon$ uniformly for all x in D.

We now recall the following well-known results which we state without proof.

Theorem 1.2. Let $\{f_m\} \subset C(K)$ where K is a compact (i.e., a closed and bounded) subset of R^n. Then $\{f_m\}$ is a uniform Cauchy sequence on K if and only if there exists a function f in $C(K)$ such that $\{f_m\}$ converges to f uniformly on K.

Theorem 1.3. (Weierstrass). Let u_k, $k = 1, 2, \ldots$, be given real valued functions defined on a set $D \subset R^n$. Suppose there exist nonnegative constants M_k such that $|u_k(x)| \leq M_k$ for all x in D and

$$\sum_{k=1}^{\infty} M_k < \infty.$$

Then the sum $\sum_{k=1}^{\infty} u_k(x)$ converges uniformly on D.

In the next definition we introduce the concept of equicontinuity which will be crucial in the development of this chapter.

Definition 1.4. Let $\mathscr{F}$ be a family of real valued functions defined on a set $D \subset R^n$. Then

(i) $\mathscr{F}$ is called **uniformly bounded** if there is a nonnegative constant M such that $|f(x)| \leq M$ for all x in D and for all f in $\mathscr{F}$.

(ii) $\mathscr{F}$ is called **equicontinuous** on D if for any $\varepsilon > 0$ there is a $\delta > 0$ (independent of x, y and f) such that $|f(x) - f(y)| < \varepsilon$ whenever $|x - y| < \delta$ for all x and y in D and for all f in $\mathscr{F}$.

We now state and prove the Ascoli–Arzela lemma which identifies an important property of equicontinuous families of functions.

Theorem 1.5. Let D be a closed, bounded subset of R^n and let $\{f_m\}$ be a real valued sequence of functions in $C(D)$. If $\{f_m\}$ is equicontinuous and uniformly bounded on D, then there is a subsequence $\{m_k\}$ and a function f in $C(D)$ such that $\{f_{m_k}\}$ converges to f uniformly on D.

Proof. Let $\{r_j\}$ be a dense subset of D. The sequence of real numbers $\{f_m(r_1)\}$ is bounded since $\{f_m\}$ is uniformly bounded on D. Hence, a subsequence will converge. Label this convergent subsequence $\{f_{1m}(r_1)\}$ and label the point to which it converges $f(r_1)$. Now the sequence $\{f_{1m}(r_2)\}$ is also a bounded sequence. Thus, there is a subsequence $\{f_{2m}\}$ of $\{f_{1m}\}$ which converges at r_2 to a point which we shall call $f(r_2)$. Continuing in this manner, one obtains subsequences $\{f_{km}\}$ of $\{f_{k-1,m}\}$ and numbers $f(r_k)$ such that $f_{km}(r_k) \to f(r_k)$ as $m \to \infty$ for $k = 1, 2, 3, \ldots$. Since the sequence $\{f_{km}\}$ is a subsequence of all previous sequences $\{f_{jm}\}$ for $1 \leq j \leq k-1$, it will converge at each point r_j with $1 \leq j \leq k$.

We now obtain a subsequence by "diagonalizing" the foregoing infinite collection of sequences. In doing so, we set $g_m = f_{mm}$ for all m. If the terms f_{km} are written as the elements of a semi-infinite matrix, as shown in Fig. 2.1, then the elements g_m are the diagonal elements of this matrix.

$$\begin{matrix} f_{11} & f_{12} & f_{13} & f_{14} & \cdots \\ f_{21} & f_{22} & f_{23} & f_{24} & \cdots \\ f_{31} & f_{32} & f_{33} & f_{34} & \cdots \\ f_{41} & f_{42} & f_{43} & f_{44} & \cdots \\ \vdots & \vdots & \vdots & \vdots & \ddots \end{matrix}$$

FIGURE 2.1 *Diagonalizing a collection of sequences.*

Since $\{g_m\}$ is eventually a subsequence of every sequence $\{f_{km}\}$, then $g_m(r_k) \to f(r_k)$ as $m \to \infty$ for $k = 1, 2, 3, \ldots$. To see that g_m converges uniformly on D, fix $\varepsilon > 0$. For any rational r_j there exists $M_j(\varepsilon)$ such that $|g_m(r_j) - g_i(r_j)| < \varepsilon$ for all $m, i \geq M_j(\varepsilon)$. By equicontinuity, there is a $\delta > 0$ such that $|g_i(x) - g_i(y)| < \varepsilon$ for all i when $x, y \in D$ and $|x - y| < \delta$. Thus for $|x - r_j| < \delta$ and $m, i \geq M_j(\varepsilon)$, we have

$$|g_m(x) - g_i(x)| \leq |g_m(x) - g_m(r_j)| + |g_m(r_j) - g_i(r_j)| + |g_i(r_j) - g_i(x)| < 3\varepsilon.$$

The collection of neighborhoods $B(r_j, \delta) = \{z \in R : |r_j - z| < \delta\}$ covers D. Since D is a closed and bounded subset of real n space R^n (i.e., since D is **compact**), by the Heine–Borel theorem a finite subset $B(r_{j1}, \delta), \ldots, B(r_{jL}, \delta)$ will cover D. Let $M(\varepsilon) = \max\{M_{j1}(\varepsilon), \ldots, M_{jL}(\varepsilon)\}$. If m and i are larger than $M(\varepsilon)$ and if x is any point of D, then $x \in B(r_{jl}, \delta)$ for some l between 1 and L.

2.1 Preliminaries

So $|g_m(x) - g_i(x)| < 3\varepsilon$ as above. This shows that $\{g_m\}$ is a uniformly Cauchy sequence on D. Apply now Theorem 1.2 to complete the proof. ∎

B. Inequalities

The following version of the **Gronwall inequality** will be required later.

Theorem 1.6. Let r be a continuous, nonnegative function on an interval $J = [a, b] \subset R$ and δ and k be nonnegative constants such that

$$r(t) \leq \delta + \int_a^t kr(s)\,ds \qquad (t \in J).$$

Then

$$r(t) \leq \delta \exp[k(t-a)] \qquad \text{for all} \quad t \text{ in } J.$$

Proof. Define $R(t) = \delta + \int_a^t kr(s)\,ds$. Then $R(a) = \delta$, $R(t) \geq r(t)$, and

$$R'(t) - kR(t) \leq R'(t) - kr(t) = 0.$$

Multiply this inequality by $K(t) = \exp[k(a - t)]$ to obtain

$$K(t)R'(t) - kK(t)R(t) \leq 0 \qquad \text{or} \qquad [K(t)R(t)]' \leq 0.$$

Since the integral from a to t of this nonpositive function is again nonpositive, then

$$K(s)R(s)\big|_a^t = K(t)R(t) - \delta \leq 0.$$

Thus $R(t) \leq \delta/K(t)$ or $r(t) \leq R(t) \leq \delta/K(t)$ as desired. ∎

C. Lim Sup

We let ∂D denote the **boundary** of a set $D \subset R^n$ and we let $\bar{D} = D \cup \partial D$ denote the **closure** of D.

Given a sequence of real numbers $\{a_m\}$, we define

$$\limsup_{m \to \infty} a_m \triangleq \overline{\lim_{m \to \infty}} a_m = \inf_{m \geq 1}\left(\sup_{k \geq m} a_k\right)$$

and

$$\liminf_{m \to \infty} a_m \triangleq \underline{\lim_{m \to \infty}} a_m = \sup_{m \geq 1}\left(\inf_{k \geq m} a_k\right).$$

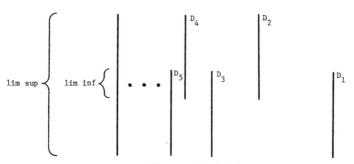

FIGURE 2.2 Lim sup *and* lim inf *of sets*.

It is easily checked that $-\infty \leq \liminf a_m \leq \limsup a_m \leq +\infty$ and that the lim sup and lim inf of a_m are, respectively, the largest and smallest limit points of the sequence $\{a_m\}$. Also, the limit of a_m exists if and only if the lim sup and lim inf are equal. In this case the limit is their common value.

In the same vein as above, if f is an extended real valued function on D, then for any $b \in \bar{D}$,

$$\limsup_{x \to b} f(x) = \inf_{\varepsilon > 0} (\sup\{f(y): y \in D, 0 < |y - b| \leq \varepsilon\}).$$

The lim inf is similarly defined. We call f **upper semicontinuous** if for each x in D,

$$f(x) \geq \limsup_{y \to x} f(y).$$

Also, we call f **lower semicontinuous** if for each x in D,

$$f(x) \leq \liminf_{y \to x} f(y).$$

Finally, if $\{D_m\}$ is a sequence of subsets of R^n, then

$$\limsup_{m \to \infty} D_m = \bigcap_{m=1}^{\infty} \left(\overline{\bigcup_{k \geq m} D_k} \right), \quad \text{and} \quad \liminf_{m \to \infty} D_m = \bigcap_{m=1}^{\infty} \bigcup_{k \geq m} \bar{D}_k.$$

In Fig. 2.2 an example of lim sup and lim inf is depicted when the D_m are intervals.

D. Zorn's Lemma

Before we can present Zorn's lemma, we need to introduce several concepts.

2.2 Existence of Solutions

A **partially ordered set**, $(A, \leq)$, consists of a set A and a relation $\leq$ on A such that for any a, b, and c in A,

(i) $a \leq a$,
(ii) $a \leq b$ and $b \leq c$ implies that $a \leq c$, and
(iii) $a \leq b$ and $b \leq a$ implies that $a = b$.

A **chain** is a subset A_0 of A such that for all a and b in A_0, either $a \leq b$ or $b \leq a$. An **upper bound** for a chain A_0 is an element $a_0 \in A$ such that $b \leq a_0$ for all b in A_0. A **maximal element** for A, if it exists, is an element a_1 of A such that for all b in A, $a_1 \leq b$ implies $a_1 = b$.

The next result, which we give without proof, is called **Zorn's lemma**.

Theorem 1.7. If each chain in a partially ordered set $(A, \leq)$ has an upper bound, then A has a maximal element.

2.2 EXISTENCE OF SOLUTIONS

In the present section we develop conditions for the existence of solutions of initial value problems characterized by scalar first order ordinary differential equations. In section 6 we give existence results for initial value problems involving systems of first order ordinary differential equations. The results of the present section do not ensure that solutions to initial value problems are unique.

Let $D \subset R^2$ be a **domain**, that is, let D be an open, connected, nonempty set in the (t, x) plane. Let $f \in C(D)$. Given (τ, ξ) in D, we seek a solution ϕ of the initial value problem

$$x' = f(t, x), \qquad x(\tau) = \xi. \tag{I'}$$

The reader may find it instructive to refer to Fig. 1.1. Recall that in order to find a solution of (I'), it suffices to find a solution of the equivalent integral equation

$$\phi(t) = \xi + \int_\tau^t f(s, \phi(s))\, ds. \tag{V}$$

This will be done in the following where we shall assume only that f is continuous on D. Later on, when we consider uniqueness of solutions, we shall need more assumptions on f.

We shall arrive at the main existence result in several steps. The first of these involves an existence result for a certain type of approximate solution which we introduce next.

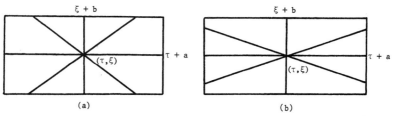

FIGURE 2.3 (a) Case $c = b/M$. (b) Case $c = a$.

Definition 2.1 An ε-**approximate** solution of (I') on an interval J containing τ is a real valued function ϕ which is piecewise C^1 on J and satisfies $\phi(\tau) = \xi, (t, \phi(t)) \in D$ for all t in J and which satisfies

$$|\phi'(t) - f(t, \phi(t))| < \varepsilon$$

at all points t of J where $\phi'(t)$ exists.

Now let $S = \{(t, x): |t - \tau| \le a, |x - \xi| \le b\}$ be a fixed rectangle in D containing (τ, ξ). Since $f \in C(D)$, it is bounded on S and there is an $M > 0$ such that $|f(t, x)| \le M$ for all (t, x) in S. Define

$$c = \min\{a, b/M\}. \tag{2.1}$$

A pictorial demonstration of (2.1) is given in Fig. 2.3. We are now in a position to prove the following existence result.

Theorem 2.2. If $f \in C(D)$ and if c is as defined in (2.1), then any $\varepsilon > 0$ there is an ε-approximate solution of (I') on the interval $|t - \tau| \le c$.

Proof. Given $\varepsilon > 0$, we shall show that there is an ε-approximate solution on $[\tau, \tau + c]$. The proof for the interval $[\tau - c, \tau]$ is similar. The approximate solution will be made up of a finite number of straight line segments joined at their ends to achieve continuity.

Since f is continuous and S is a closed and bounded set, then f is uniformly continuous on S. Hence, there is a $\delta > 0$ such that $|f(t, x) - f(s, y)| < \varepsilon$ whenever (t, x) and (s, y) are in S with $|t - s| \le \delta$ and $|x - y| \le \delta$. Now subdivide the interval $[\tau, \tau + c]$ into m equal subintervals by a partition $\tau = t_0 < t_1 < t_2 < \cdots < t_m = \tau + c$, where $t_{j+1} - t_j < \min\{\delta, \delta/M\}$ and where M is the bound for f given above. On the interval $t_0 \le t \le t_1$, let $\phi(t)$ be the line segment issuing from (τ, ξ) with slope $f(\tau, \xi)$. On $t_1 \le t \le t_2$, let $\phi(t)$ be the line segment starting at $(t_1, \phi(t_1))$ with slope $f(t_1, \phi(t_1))$. Continue in this manner to define ϕ over $t_0 \le t \le t_m$. A typical situation is as shown in Fig. 2.4. The resulting ϕ is piecewise linear and hence

2.2 Existence of Solutions

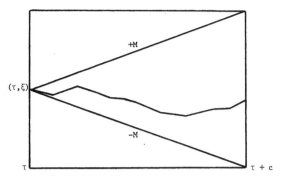

FIGURE 2.4 *Typical ε-approximate solution.*

piecewise C^1 and $\phi(\tau) = \xi$. Indeed, on $t_j \leq t \leq t_{j+1}$ we have

$$\phi(t) = \phi(t_j) + f(t_j, \phi(t_j))(t - t_j). \tag{2.2}$$

Since the slopes of the linear segments in (2.2) are bounded between $\pm M$, then $(t, \phi(t))$ cannot leave S before time $t_m = \tau + c$ (see Fig. 2.4).

To see that ϕ is an ε-approximate solution, we use (2.2) to compute

$$|\phi'(t) - f(t, \phi(t))| = |f(t_j, \phi(t_j)) - f(t, \phi(t))| < \varepsilon.$$

This inequality is true by the choice of δ, since $|t_j - t| \leq |t_j - t_{j+1}| < \delta$ and

$$|\phi(t) - \phi(t_j)| \leq M|t - t_j| \leq M(\delta/M) = \delta.$$

This completes the proof. ∎

The approximations defined in the proof of Theorem 2.2 are called **Euler polygons** and (2.2) with $t = t_{j+1}$ is called **Euler's method**. This technique and more sophisticated piecewise polynomial approximations are common in determining numerical approximations to solutions of (I′) via computer simulations.

Theorem 2.3. *If $f \in C(D)$ and $(\tau, \xi) \in D$, then (I′) has a solution defined on $|t - \tau| \leq c$.*

Proof. Let ε_m be a monotone decreasing sequence of real numbers with limit zero, e.g., $\varepsilon_m = 1/m$. Let c be given by (2.1) and let ϕ_m be the ε_m-approximate solution given by Theorem 2.2. Then $|\phi_m(t) - \phi_m(s)| \leq M|t - s|$ for all t, s in $[\tau - c, \tau + c]$ and for all $m \geq 1$. This means that $\{\phi_m\}$ is an equicontinuous sequence. The sequence is also uniformly bounded since

$$|\phi_m(t)| \leq |\phi_m(\tau)| + |\phi_m(t) - \phi_m(\tau)| \leq |\xi| + Mc.$$

By the Ascoli–Arzela lemma (Theorem 1.5) there is a subsequence $\{\phi_{m_k}\}$ which converges uniformly on $J = [\tau - c, \tau + c]$ to a continuous function ϕ.

Now define
$$E_m(t) = \phi'_m(t) - f(t, \phi_m(t))$$
so that E_m is piecewise continuous and $|E_m(t)| \leq \varepsilon_m$ on J. Rearranging this equation and integrating, we see that

$$\phi_m(t) = \xi + \int_a^t [f(s, \phi_m(s)) + E_m(s)]\, ds. \tag{2.3}$$

Now since ϕ_{m_k} tends to ϕ uniformly on J and since f is uniformly continuous on S, it follows that $f(t, \phi_{m_k}(t))$ tends to $f(t, \phi(t))$ uniformly on J, say,

$$\sup_{t \in J} |f(t, \phi_{m_k}(t)) - f(t, \phi(t))| = \alpha_k \to 0 \quad \text{as} \quad k \to \infty.$$

Thus, on J we have

$$\left| \int_\tau^t [f(s, \phi_{m_k}(s)) + E_{m_k}(s)]\, ds - \int_\tau^t f(s, \phi(s))\, ds \right|$$
$$\leq \left| \int_\tau^t |f(s, \phi_{m_k}(s)) - f(s, \phi(s))|\, ds \right| + \left| \int_\tau^t |E_{m_k}(s)|\, ds \right|$$
$$\leq \left| \int_\tau^t \alpha_k\, ds \right| + \left| \int_\tau^t \varepsilon_{m_k}\, ds \right| \leq (\alpha_k + \varepsilon_{m_k})c \to 0$$

as $k \to \infty$. Hence, we can take the limit as $k \to \infty$ in (2.3) with $m = m_k$ to obtain (V). ∎

As an example, consider the problem
$$x' = x^{1/3}, \qquad x(\tau) = 0.$$

Since $x^{1/3}$ is continuous, there is a solution (which can be obtained by separating variables). Indeed it is easy to verify that $\phi(t) = [2(t - \tau)/3]^{3/2}$ is a solution. This solution is not unique since $\psi(t) \equiv 0$ is also clearly a solution. Conditions which ensure uniqueness of solutions of (I') are given in Section 2.4.

Theorem 2.3 asserts the existence of a solution of (I') "locally," i.e., only on a sufficiently short time interval. In general, this assertion cannot be changed to existence of a solution for all $t \geq \tau$ (or for all $t \leq \tau$) as the following example shows. Consider the problem

$$x' = x^2, \qquad x(\tau) = \xi.$$

2.3 Continuation of Solutions

By separation of variables we can compute that the solution is

$$\phi(t) = \xi[1 - \xi(t - \tau)]^{-1}.$$

This solution exists forward in time for $\xi > 0$ only until $t = \tau + \xi^{-1}$.

Finally, we note that when f is discontinuous, a solution in the sense of Section 1.1 may or may not exist. For example, if $s(x) = 1$ for $x \geq 0$ and $s(x) = -1$ for $x < 0$, then the equation

$$x' = -s(x), \qquad x(\tau) = 0, \qquad t \geq \tau,$$

has no C^1 solution. Furthermore, there is no elementary way to generalize the idea of solution to include this example. On the other hand, the equation

$$x' = s(x), \qquad x(\tau) = 0$$

has the unique solution $\phi(t) = t - \tau$ for $t \geq \tau$.

2.3 CONTINUATION OF SOLUTIONS

Once the existence of a solution of an initial value problem has been ascertained over some time interval, it is reasonable to ask whether or not this solution can be extended to a larger time interval in the sense explained below. We call this process continuation of solutions. In the present section we address this problem for the scalar initial value problem (I'). We shall consider the continuation of solutions of an initial value problem (I), characterized by a system of equations, in Section 2.6 and in the problems at the end of this chapter.

To be more specific, let ϕ be a solution of (E') on an interval J. By a **continuation** of ϕ we mean an extension ϕ_0 of ϕ to a larger interval J_0 in such a way that the extension solves (E') on J_0. Then ϕ is said to be **continued** or **extended** to the larger interval J_0. When no such continuation is possible (or is not possible by making J bigger on its left or on its right), then ϕ is called **noncontinuable** (or, respectively, noncontinuable to the left or to the right).

The examples from the last section illustrate these ideas nicely.

For $x' = x^2$, the solution

$$\phi(t) = (1 - t)^{-1} \qquad \text{on} \quad -1 < t < 1$$

is continuable forever to the left but it is noncontinuable to the right.

For $x' = x^{1/3}$, the solution

$$\psi(t) \equiv 0 \quad \text{on} \quad -1 < t < 0$$

is continuable to the right from zero in more than one way. For example, both $\psi_0(t) \equiv 0$ or $\psi_0(t) = (2t/3)^{3/2}$ for $t > 0$ will work. The solution ψ is also continuable to the left using $\psi_0(t) = 0$ for all $t < -1$.

Theorem 3.1. Let $f \in C(D)$ with f bounded on D. Suppose ϕ is a solution of

$$x' = f(t, x) \tag{E'}$$

on the interval $J = (a, b)$. Then

(i) the two limits

$$\lim_{t \to a^+} \phi(t) = \phi(a^+) \quad \text{and} \quad \lim_{t \to b^-} \phi(t) = \phi(b^-)$$

exist; and

(ii) if $(a, \phi(a^+))$ (respectively, $(b, \phi(b^-))$ is in D, then the solution ϕ can be continued to the left past the point $t = a$ (respectively, to the right past the point $t = b$).

Proof. We consider the endpoint b. The proof for the endpoint a is similar.

Let M be a bound for $|f(t,x)|$ on D, fix $\tau \in J$, and define $\xi = \phi(\tau)$. Then for $\tau < t < u < b$ the solution ϕ satisfies (V) so that

$$|\phi(u) - \phi(t)| = \left| \int_t^u f(s, \phi(s))\, ds \right|$$

$$\leq \int_t^u |f(s, \phi(s))|\, ds \leq \int_\tau^u M\, ds = M(u - t). \tag{3.1}$$

Given any sequence $\{t_m\} \subset (\tau, b)$ with t_m tending monotonically to b, we see from the estimate (3.1) that $\{\phi(t_m)\}$ is a Cauchy sequence. Thus, the limit $\phi(b^-)$ exists.

If $(b, \phi(b^-))$ is in D, then by the local existence theorem (Theorem 2.3) there is a solution ϕ_0 of (E') which satisfies $\phi_0(b) = \phi(b^-)$. The solution $\phi_0(t)$ will be defined on some interval $b \leq t \leq b + c$ for some $c > 0$. Define $\phi_0(t) = \phi(t)$ on $a < t < b$. Then ϕ_0 is continuous on $a < t < b + c$ and satisfies

$$\phi_0(t) = \xi + \int_\tau^t f(s, \phi_0(s))\, ds, \qquad a < t < b, \tag{3.2}$$

2.3 Continuation of Solutions

and

$$\phi_0(t) = \phi(b^-) + \int_b^t f(s, \phi_0(s)) \, ds, \qquad b \le t < b + c.$$

The limit of (3.2) as t tends to b is seen to be

$$\phi(b^-) = \xi + \int_\tau^b f(s, \phi_0(s)) \, ds.$$

Thus

$$\phi_0(t) = \xi + \int_\tau^b f(s, \phi_0(s)) \, ds + \int_b^t f(s, \phi_0(s)) \, ds$$
$$= \xi + \int_\tau^t f(s, \phi_0(s)) \, ds$$

on $b \le t < b + c$. We see that ϕ_0 solves (V) on $a < t < b + c$. Therefore, ϕ_0 solves (I') on $a < t < b + c$. ∎

As a consequence of Theorem 3.1, we have the following result.

Corollary 3.2. If f is in $C(D)$ and if ϕ is a solution of (E') on an interval J, then ϕ can be continued to a maximal interval $J^* \supset J$ in such a way that $(t, \phi(t))$ tends to ∂D as $t \to \partial J^*$ (and $|t| + |\phi(t)| \to \infty$ if ∂D is empty). The extended solution ϕ^* on J^* is noncontinuable.

Proof. Let ϕ be given on J. The **graph** of ϕ is the set

$$\mathrm{Gr}(\phi) = \{(t, \phi(t)) : t \in J\}.$$

Given any two solutions ϕ_1 and ϕ_2 of (E') which extend ϕ we define $\phi_1 \le \phi_2$ if and only if $\mathrm{Gr}(\phi_1) \subset \mathrm{Gr}(\phi_2)$, i.e., if and only if ϕ_2 is an extension of ϕ_1. The relation $\le$ determines a partial ordering on continuations of ϕ over open intervals. If $\{\phi_\alpha : \alpha \in A\}$ is any chain of such extensions, then $\bigcup \{\mathrm{Gr}(\phi_\alpha) : \alpha \in A\}$ is the graph of a continuation which we can call ϕ_A. This ϕ_A is an upper bound for the chain. By Zorn's lemma there is a maximal element ϕ^*. Clearly ϕ^* is a noncontinuable extension of the original solution ϕ.

Let J^* be the domain of ϕ^*. By Theorem 3.1 the interval $J^* = (a, b)$ must be open—otherwise ϕ^* could not be maximal. Assume that $b < \infty$ and suppose that $(t, \phi^*(t))$ does not approach ∂D on any sequence $t_m \to b^-$. Then $(t, \phi^*(t))$ remains in a compact subset K of D when t runs over the interval $[c, b)$ for any $c \in (a, b)$. Since f must be bounded on K, then by Theorem 3.1 we can continue ϕ^* past b. But this is impossible since ϕ^* is noncontinuable. Thus $(t, \phi^*(t))$ must approach ∂D on some sequence $t_m \to b^-$.

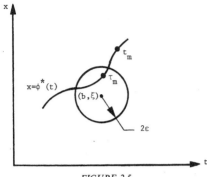

FIGURE 2.5

We claim that $(t, \phi^*(t)) \to \partial D$ as $t \to b^-$. For if this is not the case, there will be a sequence $\tau_m \to b^-$ and a point $(b, \xi) \in D$ such that $\phi^*(\tau_m) \to \xi$. Let ε be one third of the distance from (b, ξ) to ∂D or let $\varepsilon = 1$ if $\partial D = \phi$. Without loss of generality we can assume that $\tau_m < t_m < \tau_{m+1}$, $(\tau_m, \phi^*(\tau_m)) \in B((b, \xi), \varepsilon)$, and $(t_m, \phi^*(t_m)) \notin N \triangleq B((b, \xi), 2\varepsilon)$ for all $m \geq 1$ (see Fig. 2.5). Let M be a bound for $|f(t, x)|$ over N. Then from (E') we see that

$$\varepsilon \leq |\phi^*(t_m) - \phi^*(\tau_m)| = \left| \int_{\tau_m}^{t_m} f(u, \phi^*(u)) \, du \right| \leq M(t_m - \tau_m).$$

Thus $t_m - \tau_m \geq \varepsilon/M$ for all m. But this is impossible since $t_m \to b^-$ and $b < \infty$. Hence, we see that $(t, \phi^*(t)) \to \partial D$ as $t \to b^-$. A similar argument applies to the endpoint $t = a$. ∎

We now consider a simple situation where the foregoing result can be applied.

Theorem 3.3. Let $h(t)$ and $g(x)$ be positive, continuous functions on $t_0 \leq t < \infty$ and $0 < x < \infty$ such that for any $A > 0$

$$\lim_{B \to \infty} \int_A^B \frac{dx}{g(x)} = +\infty. \tag{3.3}$$

Then all solutions of

$$x' = h(t)g(x), \qquad x(\tau) = \xi \tag{3.4}$$

with $\tau \geq t_0$ and $\xi > 0$ can be continued to the right over the entire interval $\tau \leq t < \infty$.

Proof. If the result is not true, then there is a solution $\phi(t)$ and a $T > \tau$ such that $\phi(\tau) = \xi$ and such that $\phi(t)$ exists on $\tau \leq t < T$ but cannot

2.4 Uniqueness of Solutions

be continued to T. Since ϕ solves (3.4), $\phi'(t) > 0$ on $\tau \le t < T$ and ϕ is increasing. Hence by Corollary 3.2 it follows that $\phi(t) \to +\infty$ as $t \to T^-$. By separation of variables it follows that

$$\int_\tau^t h(s)\,ds = \int_\tau^t \frac{\phi'(t)\,dt}{g(\phi(t))}\,dt = \int_\xi^{\phi(t)} \frac{dx}{g(x)}.$$

Taking the limit as $t \to T$ and using (3.3), we see that

$$\infty = \lim_{t \to T} \int_\xi^{\phi(t)} \frac{dx}{g(x)} = \int_\tau^T h(s)\,ds < \infty.$$

This contradiction completes the proof. ∎

As a specific example, consider the equation

$$x' = h(t)x^\alpha, \qquad x(\tau) = \xi, \tag{3.5}$$

where α is a fixed positive real number. If $0 < \alpha \le 1$, then for any real number τ and any $\xi > 0$ the solution of (3.5) can be continued to the right for all $t \ge \tau$.

From this point on, when we speak of a **solution** without qualification, we shall mean a noncontinuable solution. In all other circumstances we shall speak of a "local solution" or we shall state the interval where we assume the solution exists.

2.4 UNIQUENESS OF SOLUTIONS

We now develop conditions for the uniqueness of solutions of initial value problems involving scalar first order ordinary differential equations. Later, in Section 6 and in the problems, we consider the uniqueness of solutions of initial value problems characterized by systems of first order ordinary differential equations.

We shall require the following concept.

Definition 4.1. A function $f \in C(D)$ is said to satisfy a **Lipschitz condition** in D with **Lipschitz constant** L if

$$|f(t,x) - f(t,y)| \le L|x - y|$$

for all points (t, x) and (t, y) in D. In this case $f(t, x)$ is also said to be **Lipschitz continuous** in x.

For example, if $f \in C(D)$ and if $\partial f/\partial x$ exists and is continuous in D, then f is Lipschitz continuous on any compact and convex subset D_0 of D. To see this, let L_0 be a bound for $|\partial f/\partial x|$ on D_0. If (t, x) and (t, y) are in D_0, then by the mean value theorem there is a z on the line between x and y such that

$$|f(t,x) - f(t,y)| = \left|\frac{\partial f}{\partial x}(t,z)(x-y)\right| \le L_0|x-y|.$$

We are now in a position to state and prove our first uniqueness result.

Theorem 4.2. If $f \in C(D)$ and if f satisfies a Lipschitz condition in D with Lipschitz constant L, then the initial value problem (I') has at most one solution on any interval $|t - \tau| \le d$.

Proof. Suppose for some $d > 0$ there are two solutions ϕ_1 and ϕ_2 on $|t - \tau| \le d$. Since both solutions solve the integral equation (V), we have on $\tau \le t \le \tau + d$,
$$\phi_1(t) - \phi_2(t) = \int_\tau^t [f(s, \phi_1(s)) - f(s, \phi_2(s))]\,ds$$
and
$$|\phi_1(t) - \phi_2(t)| \le \int_\tau^t |f(s, \phi_1(s)) - f(s, \phi_2(s))|\,ds$$
$$\le \int_\tau^t L|\phi_1(s) - \phi_2(s)|\,ds.$$
Apply the Gronwall inequality (Theorem 1.6) with $\delta = 0$ and $k = L$ to see that $|\phi_1(t) - \phi_2(t)| \le 0$ on the given interval. Thus, $\phi_1(t) = \phi_2(t)$ on this interval. A similar argument works on $\tau - d \le t \le \tau$. ∎

Corollary 4.3. If f and $\partial f/\partial x$ are both in $C(D)$, then for any (τ, ξ) in D and any interval J containing τ, if a solution of (I') exists on J, it must be unique.

The proof of this result follows from the comments given after Definition 4.1 and from Theorem 4.2. We leave the details to the reader.

The next result gives an indication of how solutions of (I') vary with ξ and f.

Theorem 4.4. Let f be in $C(D)$ and let f satisfy a Lipschitz condition in D with Lipschitz constant L. If ϕ and ψ solve (E') on an interval $|t - \tau| \le d$ with $\psi(\tau) = \xi_0$ and $\phi(\tau) = \xi$, then

$$|\phi(t) - \psi(t)| \le |\xi - \xi_0|\exp(L|t - \tau|).$$

2.4 Uniqueness of Solutions

Proof. Consider first $t \geq \tau$. Subtract the integral equations satisfied by ϕ and ψ and then estimate as follows:

$$|\phi(t) - \psi(t)| \leq |\xi - \xi_0| + \int_\tau^t |f(s, \phi(s)) - f(s, \psi(s))|\, ds$$

$$\leq |\xi - \xi_0| + \int_\tau^t L|\phi(s) - \psi(s)|\, ds.$$

Apply the Gronwall inequality (Theorem 1.6) to obtain the conclusion for $0 \leq t - \tau \leq d$.

Next, define $\phi_0(t) = \phi(-t)$, $\psi_0(t) = \psi(-t)$, and $\tau_0 = -\tau$, so that

$$\phi_0(t) = \xi - \int_{\tau_0}^t f(-s, \phi_0(s))\, ds, \qquad \tau_0 \leq t \leq \tau_0 + d,$$

and

$$\psi_0(t) = \xi - \int_{\tau_0}^t f(-s, \psi_0(s))\, ds, \qquad \tau_0 \leq t \leq \tau_0 + d.$$

Using the estimate established in the preceding paragraph, we have

$$|\phi(-t) - \psi(-t)| \leq |\xi - \xi_0| \exp(L(\tau + t))$$

on $-\tau \leq t \leq -\tau + d$. ∎

The preceding theorem can now be used to prove the following continuation result.

Theorem 4.5. *Let $f \in C(J \times R)$ for some open interval $J \subset R$ and let f satisfy a Lipschitz condition in $J \times R$. Then for any (τ, ξ) in $J \times R$, the solution of (I') exists on the entirety of J.*

Proof. The local existence and uniqueness of solutions $\phi(t, \tau, \xi)$ of (I') are clear from earlier results. If $\phi(t) = \phi(t, \tau, \xi)$ is a solution defined on $\tau \leq t < c$, then ϕ satisfies (V) so that

$$\phi(t) - \xi = \int_\tau^t [f(s, \phi(s)) - f(s, \xi)]\, ds + \int_\tau^t f(s, \xi)\, ds$$

and

$$|\phi(t) - \xi| \leq \int_\tau^t L|\phi(s) - \xi|\, ds + \delta,$$

where $\delta = \max\{|f(s, \xi)| : \tau \leq s < c\}(c - \tau)$. By the Gronwall inequality, $|\phi(t) - \xi| \leq \delta \exp[L(c - \tau)]$ on $\tau \leq t < c$. Hence $|\phi(t)|$ is bounded on $[\tau, c)$ for all $c > \tau$, $c \in J$. By Corollary 3.2, $\phi(t)$ can be continued for all $t \in J$, $t \geq \tau$. The same argument can be applied when $t \leq \tau$. ∎

If the solution $\phi(t, \tau, \xi)$ of (I') is unique, then the ε-approximate solutions constructed in the proof of Theorem 2.2 will tend to ϕ as $\varepsilon \to 0^+$ (cf. the problems at the end of this chapter). This is the basis for justifying

Euler's method—a numerical method of constructing approximations to ϕ. (Much more efficient numerical approximations are available when f is very smooth.) Assuming that f satisfies a Lipschitz condition, an alternate classical method of approximation, related to the contraction mapping theorem, is the **method of successive approximations**. Such approximations will now be studied in detail.

Let f be in $C(D)$, let S be a rectangle in D centered at (τ, ξ), and let M and c be as defined in (2.1). Successive approximations for (I'), or equivalently for (V), are defined as follows:

$$\phi_0(t) = \xi, \qquad (4.1)$$
$$\phi_{m+1}(t) = \xi + \int_\tau^t f(s, \phi_m(s))\,ds \qquad (m = 0, 1, 2, \ldots)$$

for $|t - \tau| \leq c$. It must be shown that this sequence $\{\phi_m\}$ is well defined on the given interval.

Theorem 4.6. *If f is in $C(D)$ and if f is Lipschitz continuous on S with constant L, then the successive approximations $\phi_m, m = 0, 1, \ldots,$ given by (4.1) exist on $|t - \tau| \leq c$, are continuous there, and converge uniformly, as $m \to \infty$, to the unique solution of (I').*

Proof. The proof will be given on the interval $\tau \leq t \leq \tau + c$. The proof for the interval $[\tau - c, \tau]$ can then be accomplished be reversing time as in the proof of Theorem 4.4.

First we need to prove the following statements:

(i) ϕ_m exists on $[\tau, \tau + c]$,
(ii) $\phi_m \in C^1[\tau, \tau + c]$, and
(iii) $|\phi_m(t) - \xi| \leq M(t - \tau)$ on $[\tau, \tau + c]$

for all $m \geq 0$. We shall prove these items together using induction on the integer m.

Each statement is clear when $m = 0$. Assume that each statement is true for a fixed integer $m \geq 0$. By (iii) and the choice of c, it follows that $(t, \phi_m(t)) \in S \subset D$ for all $t \in [\tau, \tau + c]$. Thus $f(t, \phi_m(t))$ exists and is continuous in t while $|f(t, \phi_m(t))| \leq M$ on the interval. This means that the integral

$$\phi_{m+1}(t) = \xi + \int_\tau^t f(s, \phi_m(s))\,ds$$

exists, $\phi_{m+1} \in C^1[\tau, \tau + c]$, and

$$|\phi_{m+1}(t) - \xi| = \left|\int_\tau^t f(s, \phi_m(s))\,ds\right| \leq M(t - \tau).$$

This completes the induction.

2.4 Uniqueness of Solutions

Now define $\Phi_m(t) = \phi_{m+1}(t) - \phi_m(t)$ so that

$$|\Phi_m(t)| \leq \int_\tau^t |f(s, \phi_m(s)) - f(s, \phi_{m-1}(s))|\, ds$$
$$\leq \int_\tau^t L|\phi_m(s) - \phi_{m-1}(s)|\, ds = L \int_\tau^t \Phi_{m-1}(s)\, ds.$$

Notice that

$$|\Phi_0(t)| = \left|\int_\tau^t f(s, \xi)\, ds\right| \leq M(t - \tau).$$

These two estimates can be combined to see that

$$|\Phi_1(t)| \leq L \int_\tau^t M(s - \tau)\, ds = LM(t - \tau)^2/2!,$$

that

$$|\Phi_2(t)| \leq L \int_\tau^t [LM(s - \tau)^2/2!]\, ds \leq L^2 M(t - \tau)^3/3!,$$

and by induction that

$$|\Phi_m(t)| \leq ML^m(t - \tau)^{m+1}/(m + 1)!.$$

Hence, the mth term of the series

$$\phi_0(t) + \sum_{k=0}^{\infty} [\phi_{k+1}(t) - \phi_k(t)] \tag{4.2}$$

is bounded on $[\tau, \tau + c]$ by

$$\frac{M}{L} \frac{(Lc)^{m+1}}{(m + 1)!}.$$

Since

$$e^{Lc} = \sum_{k=0}^{\infty} (Lc)^k/k! < \infty,$$

it follows from the Weierstrass comparison test (Theorem 1.3) that the series (4.2) converges uniformly to a continuous function ϕ. This means that the sequence of partial sums

$$\phi_0 + \sum_{k=0}^{m} (\phi_{k+1} - \phi_k) = \phi_0 + (\phi_1 - \phi_0) + \cdots + (\phi_{m+1} - \phi_m) = \phi_{m+1}$$

tends uniformly to ϕ as $m \to \infty$. Since the bound (iii) is true for all ϕ_m, it is also true in the limit, i.e.,

$$|\phi(t) - \xi| \leq M(t - \tau).$$

Thus, $f(t, \phi(t))$ exists and is a continuous function of t. As in the proof of Theorem 2.3, it now follows that

$$\phi(t) = \lim_{m \to \infty} \phi_{m+1}(t) = \xi + \lim_{m \to \infty} \int_\tau^t f(s, \phi_m(s)) \, ds$$

$$= \xi + \int_\tau^t f(s, \phi(s)) \, ds, \qquad \tau \leq t \leq \tau + c.$$

Hence, ϕ solves (V). ∎

As a specific example, consider the initial value problem

$$x' = x + t, \qquad x(0) = \xi.$$

The first three successive approximations are

$$\phi_0(t) = \xi,$$

$$\phi_1(t) = \xi + \int_0^t [\xi + s] \, ds = \xi(1 + t) + t^2/2,$$

$$\phi_2(t) = \xi + \int_0^t [\xi(1 + s) + s^2/2 + s] \, ds$$

$$= \xi(1 + t + t^2/2!) + (t^2/2 + t^3/3!).$$

The reader should try to find the formula for the general term $\phi_m(t)$ and the form of the limit of ϕ_m as $m \to \infty$.

2.5 CONTINUITY OF SOLUTIONS WITH RESPECT TO PARAMETERS

The discussion in Chapter 1 clearly shows that in most applications one would expect that (I′) has a unique solution. Moreover, one would expect that this solution should vary continuously with respect to (τ, ξ) and with respect to any parameters of the function f. This continuity is clearly impossible at points where the solution of (I′) is not unique. We shall see that uniqueness of solutions is not only a necessary but also a sufficient condition for continuity. The exact statements of such results must be made with care since different (noncontinuable) solutions will generally be defined on different intervals.

In the present section we concern ourselves with scalar initial value problems (I′). We shall consider the continuity of solutions with respect to parameters for the vector equation (I) in Section 2.6 and in the problems given at the end of this chapter.

2.5 Continuity of Solutions with Respect to Parameters

In order to begin our discussion of the questions raised previously, we need a preliminary result which we establish next. Consider a sequence of initial value problems

$$x(t) = \xi_m + \int_\tau^t f_m(s, x(s))\, ds \qquad (5.1)$$

with noncontinuable solutions $\phi_m(t)$ defined on intervals J_m. Assume that f and $f_m \in C(D)$, that $\xi_m \to \xi$ as $m \to \infty$ and that $f_m \to f$ uniformly on compact subsets of D.

Lemma 5.1. Let D be bounded. Suppose a solution ϕ of (I') exists on an interval $J = [\tau, b)$, or on $[\tau, b]$, or on the "degenerate interval" $[\tau, \tau]$, and suppose that $(t, \phi(t))$ does not approach ∂D as $t \to b^-$, i.e.,

$$\operatorname{dist}((t, \phi(t)), \partial D) \triangleq \inf\{|t - s| + |\phi(t) - x| : (s, x) \notin D\} \geq \eta > 0$$

for all $t \in J$. Suppose that $\{b_m\} \subset J$ is a sequence which tends to b while the solutions $\phi_m(t)$ of (5.1) are defined on $[\tau, b_m] \subset J_m$ and satisfy

$$\Phi_m = \sup\{|\phi_m(t) - \phi(t)| : \tau \leq t \leq b_m\} \to 0$$

as $m \to \infty$. Then there is a number $b' > b$, where b' depends only on η, and there is a subsequence $\{\phi_{m_j}\}$ such that ϕ_{m_j} and ϕ are defined on $[\tau, b']$ and $\phi_{m_j} \to \phi$ as $j \to \infty$ uniformly on $[\tau, b']$.

Proof. Define $G = \{(t, \phi(t)) : t \in J\}$, the graph of ϕ over J. By hypothesis, the distance from G to ∂D is at least $\eta = 3A > 0$. Define

$$D(b) = \{(t, x) \in D : \operatorname{dist}((t, x), G) \leq b\}.$$

Then $D(2A)$ is a compact subset of D and f is bounded there, say $|f(t, x)| \leq M$ on $D(2A)$. Since $f_m \to f$ uniformly on $D(2A)$, it may be assumed (by increasing the size of M) that $|f_m(t, x)| \leq M$ on $D(2A)$ for all $m \geq 1$. Choose m_0 such that for $m \geq m_0$, $\Phi_m < A$. This means that $(t, \phi_m(t)) \in D(A)$ for all $m \geq m_0$ and $t \in [\tau, b_m]$. Choose $m_1 \geq m_0$ so that if $m \geq m_1$, then $b - b_m < A/(2M)$. Define $b' = b + A/(2M)$.

Fix $m \geq m_1$. Since $(t, \phi_m(t)) \in D(A)$ on $[\tau, b_m]$, then $|\phi'_m(t)| \leq M$ on $[\tau, b_m]$ and until such time as $(t, \phi_m(t))$ leaves $D(2A)$. Hence

$$|\phi_m(t) - \phi_m(b_m)| \leq M|t - b_m| \leq MA/M = A$$

for so long as both $(t, \phi_m(t)) \in D(2A)$ and $|t - b_m| \leq A/M$. Thus $(t, \phi_m(t)) \in D(2A)$ on $\tau \leq t \leq b_m + A/M$. Moreover $b_m + A/M > b'$ when m is large.

Thus, it has been shown that $\{\phi_m : m \geq m_1\}$ is a uniformly bounded family of functions and each is Lipschitz continuous with Lipschitz constant M on $[\tau, b']$. By Ascoli's lemma (Theorem 1.5), a subsequence $\{\phi_{m_j}\}$ will converge uniformly to a limit ϕ. The arguments used at the end of the proof of Theorem 2.3 show that

$$\lim_{j \to \infty} \int_\tau^t f(s, \phi_{m_j}(s))\, ds = \int_\tau^t f(s, \phi(s))\, ds.$$

Thus, the limit of

$$\phi_{m_j}(t) = \xi_{m_j} + \int_\tau^t f(s, \phi_{m_j}(s))\, ds + \int_\tau^t [f_{m_j}(s, \phi_{m_j}(s)) - f(s, \phi_{m_j}(s))]\, ds$$

as $j \to \infty$, is

$$\phi(t) = \xi + \int_\tau^t f(s, \phi(s))\, ds. \quad \blacksquare$$

We are now in a position to prove the following result.

Theorem 5.2. Let $f, f_m \in C(D)$, let $\xi_m \to \xi$, and let $f_m \to f$ uniformly on compact subsets of D. If $\{\phi_m\}$ is a sequence of noncontinuable solutions of (5.1) defined on intervals J_m, then there is a subsequence $\{m_j\}$ and a noncontinuable solution ϕ of (I') defined on an interval J_0 containing τ such that

(i) $\liminf_{j \to \infty} J_{m_j} \supset J_0$, and

(ii) $\phi_{m_j} \to \phi$ uniformly on compact subsets of J_0 as $j \to \infty$.

If in addition the solution of (I') is unique, then the entire sequence $\{\phi_m\}$ tends to ϕ uniformly for t on compact subsets of J_0.

Proof. With $J = [\tau, \tau]$ (a single point) and $b_m = \tau$ for all $m \geq 1$ apply Lemma 5.1. (If D is not bounded, use a subdomain.) Thus, there is a subsequence of $\{\phi_m\}$ which converges uniformly to a limit function ϕ on some interval $[\tau, b']$, $b' > \tau$. Let B_1 be the supremum of these numbers b'. If $B_1 = +\infty$, choose b_1 to be any fixed b'. If $B_1 < \infty$, let b_1 be a number $b' > \tau$ such that $B_1 - b' < 1$. Let $\{\phi_{1m}\}$ be a subsequence of $\{\phi_m\}$ which converges uniformly on $[\tau, b_1]$.

Suppose for induction that we are given $\{\phi_{km}\}$, b_k, and $B_k > b_k$ with $\phi_{km} \to \phi$ uniformly on $[\tau, b_k]$ as $m \to \infty$. Define B_{k+1} as the supremum of all numbers $b' > b_k$ such that a subsequence of $\{\phi_{km}\}$ will converge uniformly on $[\tau, b']$. Clearly $b_k < B_{k+1} \leq B_k$. If $B_{k+1} = +\infty$, pick $b_{k+1} > b_k + 1$ and if $B_{k+1} < \infty$, pick b_{k+1} so that $b_k < b_{k+1} < B_{k+1}$ and $b_{k+1} > B_{k+1} - 1/(k+1)$. Let $\{\phi_{k+1,m}\}$ be a subsequence of $\{\phi_{km}\}$ which converges uniformly on $[\tau, b_{k+1}]$ to a limit ϕ. Clearly, by possibly deleting finitely many terms of the new subsequence, we can assume without loss of generality that $|\phi_{k+1,m}(t) - \phi(t)| < 1/(k+1)$ for $t \in [\tau, b_{k+1}]$ and $m \geq k + 1$.

Since $\{b_k\}$ is monotonically increasing, it has a limit $b \leq +\infty$. Define $J_0 = [\tau, b)$. The diagonal sequence $\{\phi_{mm}\}$ will eventually become a subsequence of each sequence $\{\phi_{km}\}$. Hence $\phi_{mm} \to \phi$ as $m \to \infty$ with conver-

2.5 Continuity of Solutions with Respect to Parameters

gence uniform on compact subsets of J_0. By the argument used at the end of the proof of Lemma 5.1, the limit ϕ must be a solution of (I′).

If $b = \infty$, then ϕ is clearly noncontinuable. If $b < \infty$, then this means that B_k tends to b from above. If ϕ could be continued to the right past b, i.e., if $(t, \phi(t))$ stays in a compact subset of D as $t \to b^-$, then by Lemma 5.1 there would be a number $b' > b$, a continuation of ϕ, and a subsequence of $\{\phi_{mm}\}$ which would converge uniformly on $[\tau, b']$ to ϕ. Since $b' > b$ and $B_k \to b^+$, then for sufficiently large k (i.e., when $b' > B_k$), this would contradict the definition of B_k. Hence, ϕ must be noncontinuable. Since a similar argument works for $t < \tau$, parts (i) and (ii) are proved.

Now assume that the solution of (I′) is unique. If the entire sequence $\{\phi_m\}$ does not converge to ϕ uniformly on compact subsets of J_0, then there is a compact set $K \subset J_0$, an $\varepsilon > 0$, a sequence $\{t_k\} \subset K$, and a subsequence $\{\phi_{m_k}\}$ such that

$$|\phi_{m_k}(t_k) - \phi(t_k)| \geq \varepsilon. \tag{5.2}$$

By the part of the present theorem which has already been proved, there is a subsequence, we shall still call it $\{\phi_{m_k}\}$ in order to avoid a proliferation of subscripts, which converges uniformly on compact subsets of an interval J' to a solution ψ of (I′). By uniqueness $J' = J_0$ and $\phi = \psi$. Thus $\phi_{m_k} \to \phi$ as $k \to \infty$ uniformly on $K \subset J_0$ which contradicts (5.2). ∎

In Theorem 5.2, conclusion (i) cannot be strengthened from "contained in" to "equality," as can be seen from the following example. Define

$$f(t, x) = x^2 \qquad \text{for} \quad t < 1$$

and

$$f(t, x) = x^2[1 + (t-1)x^2]^{-1} \qquad \text{for} \quad t \geq 1.$$

Clearly f is continuous on R^2 and Lipschitz continuous in x on each compact subset of R^2. Consider the solution $\phi(t, \xi)$ of (I′) for $\tau = 0$ and $0 < \xi < 1$. Clearly

$$\phi(t, \xi) = \xi(1 - \xi t)^{-1} \qquad \text{on} \quad -\infty < t \leq 1.$$

By Theorem 2.3 the solution can be continued over a small interval $1 \leq t \leq 1 + c$. By Theorem 4.5 the solution $\phi(t, \xi)$ can be continued for all $t \geq 1 + c$. Thus, for $0 < \xi < 1$ the maximum interval of existence of $\phi(t, \xi)$ is $R = (-\infty, \infty)$. However, for

$$x' = f(t, x), \qquad x(0) = 1$$

the solution $\phi(t, 1) = (1 - t)^{-1}$ exists only for $-\infty < t < 1$.

As an application of the Theorem 5.2 we consider an autonomous equation

$$x' = g(x) \tag{5.3}$$

and we assume that $f(t, x)$ tends to $g(x)$ as $t \to \infty$. We now prove the following result.

Corollary 5.3. Let $g(x)$ be continuous on D_0, let $f \in C(R \times D_0)$, and let $f(t, x) \to g(x)$ uniformly for x on compact subsets of D_0 as $t \to \infty$. Suppose there is a solution $\phi(t)$ of (I') and a compact set $D_1 \subset D_0$ such that $\phi(t) \in D_1$ for all $t \geq \tau$. Then given any sequence $t_m \to \infty$ there will exist a subsequence $\{t_{m_j}\}$ and a solution ψ of (5.3) such that

$$\phi(t + t_{m_j}) \to \psi(t) \quad \text{as } j \to \infty \tag{5.4}$$

with convergence uniform for t in compact subsets of R.

Proof. Define $\phi_m(t) = \phi(t + t_m)$ for $m = 1, 2, 3, \ldots$ and for $t \geq \tau - t_m$. Then ϕ_m is a solution of

$$x' = f(t + t_m, x), \qquad x(0) = \phi(t_m).$$

Since $\xi_m = \phi(t_m) \in D_1$ and since D_1 is compact, then a subsequence $\{\xi_m\}$ will converge to some point ξ of D_1. Theorem 5.2 asserts that by possibly taking a further subsequence, we can assume that $\phi_{m_j}(t) \to \psi(t)$ as $j \to \infty$ uniformly for t on compact subsets of J_0. Here ψ is a solution of (5.3) defined on J_0 which satisfies $\psi(0) = \xi$. Since $\phi(t) \in D_1$ for all $t \geq \tau$, it follows from (5.4) that $\psi(t) \in D_1$ for $t \in R$. Since D_1 is a compact subset of the open set D_0, this means that $\psi(t)$ does not approach the boundary of D_0 and, hence, can be continued for all t, i.e., $J_0 = R$. ∎

Given a solution ϕ of (I') defined on a half line $[\tau, \infty)$, the **positive limit set** of ϕ is defined as

$$\Omega(\phi) = \{\xi: \text{there is a sequence } t_m \to \infty \text{ such that } \phi(t_m) \to \xi\}.$$

[If ϕ is defined for $t \leq \tau$, then the negative limit set $A(\phi)$ is defined similarly.] A set M is called **invariant with respect to** (5.3) if for any $\xi \in M$ and any $\tau \in R$, there is a solution ψ of (5.3) satisfying $\psi(\tau) = \xi$ and satisfying $\psi(t) \in M$ for all $t \in R$. The conclusion of Corollary 5.3 implies that $\Omega(\phi)$ is invariant with respect to (5.3). This conclusion will prove very useful later (e.g., in Chapter 5).

Now consider a family of initial value problems

$$x' = f(t, x, \lambda), \qquad x(\tau) = \xi \tag{I$_\lambda$}$$

where f maps a set $D \times D_\lambda$ into R continuously and D_λ is an open set in R^l space. We assume that solutions of (I$_\lambda$) are unique. Let $\phi(t, \tau, \xi, \lambda)$ denote the (unique and noncontinuable) solution of (I$_\lambda$) for $(\tau, \xi) \in D$ and $\lambda \in D_\lambda$ on the interval $\alpha(\tau, \xi, \lambda) < t < \beta(\tau, \xi, \lambda)$. We are now in a position to prove the following result.

2.6 Systems of Equations

Corollary 5.4. Under the foregoing assumptions, define

$$\mathscr{S} = \{(t, \tau, \xi, \lambda) : (\tau, \xi) \in D, \lambda \in D_\lambda, \alpha(\tau, \xi, \lambda) < t < \beta(\tau, \xi, \lambda)\}.$$

Then $\phi(t, \tau, \xi, \lambda)$ is continuous on $\mathscr{S}$, α is upper semicontinuous in (τ, ξ, λ), and β is lower semicontinuous in $(\tau, \xi, \lambda) \in D \times D_\lambda$.

Proof. Define $\psi(t, \tau, \xi, \lambda) = \phi(t + \tau, \tau, \xi, \lambda)$ so that ψ solves

$$y' = f(t + \tau, y, \lambda), \qquad y(0) = \xi. \qquad (J'_\lambda)$$

Let $(t_m, \tau_m, \xi_m, \lambda_m)$ be a sequence in $\mathscr{S}$ which tends to a limit $(t_0, \tau_0, \xi_0, \lambda_0)$ in $\mathscr{S}$. By Theorem 5.2 it follows that

$$\psi(t, \tau_m, \xi_m, \lambda_m) \to \psi(t, \tau_0, \xi_0, \lambda_0) \qquad \text{as} \quad m \to \infty$$

uniformly for t in compact subsets of $\alpha(\tau_0, \xi_0, \lambda_0) - \tau_0 < t < \beta(\tau_0, \xi_0, \lambda_0) - \tau_0$ and in particular uniformly in m for $t = t_m$. Therefore, we see that

$$|\phi(t_m, \tau_m, \xi_m, \lambda_m) - \phi(t_0, \tau_0, \xi_0, \lambda_0)| \leq |\phi(t_m, \tau_m, \xi_m, \lambda_m) - \phi(t_m, \tau_0, \xi_0, \lambda_0)|$$
$$+ |\phi(t_m, \tau_0, \xi_0, \lambda_0) - \phi(t_0, \tau_0, \xi_0, \lambda_0)| \to 0 \qquad \text{as} \quad m \to \infty.$$

This proves that ϕ is continuous on $\mathscr{S}$.

To prove the remainder of the conclusions, we note that by Theorem 5.2(i), if J_m is the interval $(\alpha(\tau_m, \xi_m, \lambda_m), \beta(\tau_m, \xi_m, \lambda_m))$, then

$$\liminf_{m \to \infty} J_m \supset J_0.$$

The remaining assertions follow immediately. ∎

As a particular example, note that the solutions of the initial value problem

$$x' = \sum_{j=1}^{l-2} \lambda_j x^j + \sin(\lambda_{l-1} t + \lambda_l), \qquad x(\tau) = \xi$$

depend continuously on the parameters $(\lambda_1, \lambda_2, \ldots, \lambda_l, \tau, \xi)$.

2.6 SYSTEMS OF EQUATIONS

In Section 1.1D it was shown that an nth order ordinary differential equation can be reduced to a system of first order ordinary differential equations. In Section 1.1B it was also shown that arbitrary systems of n first order differential equations can be written as a single vector equation

$$x' = f(t, x) \tag{E}$$

while the initial value problem for (E) can be written as

$$x' = f(t, x), \qquad x(\tau) = \xi. \tag{I}$$

The purpose of this section is to show that the results of Sections 2–5 can be extended from the scalar case [i.e., from (E') and (I')] to the vector case [i.e., to (E) and (I)] with no essential changes in the proofs.

A. Preliminaries

In our subsequent development we require some additional concepts from linear algebra which we recall next.

Let X be a vector space over a field $\mathscr{F}$. We will require that $\mathscr{F}$ be either the real numbers R or the complex numbers C. A function $|\cdot|: X \to R^+ = [0, \infty)$ is said to be a **norm** if

(i) $|x| \geq 0$ for every vector $x \in X$ and $|x| = 0$ if and only if x is the null vector (i.e., $x = 0$);

(ii) for every scalar $\alpha \in \mathscr{F}$ and for every vector $x \in X$, $|\alpha x| = |\alpha| |x|$ where $|\alpha|$ denotes the absolute value of α when $\mathscr{F} = R$ and $|\alpha|$ denotes the modulus of α when $\mathscr{F} = C$; and

(iii) for every x and y in X, $|x + y| \leq |x| + |y|$.

In the present chapter as well as in the remainder of this book, we shall be concerned primarily with the vector space R^n over R and with the vector space C^n over C. We now define an important class of norms on R^n. A similar class of norms can be defined on C^n in the obvious way. Thus, given a vector $x = (x_1, x_2, \ldots, x_n)^T \in R^n$, let

$$|x|_p = \left(\sum_{i=1}^n |x_i|^p \right)^{1/p}, \qquad 1 \leq p < \infty$$

and let

$$|x|_\infty = \max\{|x_i| : 1 \leq i \leq n\}.$$

It is an easy matter to show that for every p, $1 \leq p < \infty$, $|\cdot|_p$ is a norm on R^n and also, that $|\cdot|_\infty$ is a norm on R^n. Of particular interest to us will be the

2.6 Systems of Equations

cases $p = 1$ and 2, i.e., the cases

$$|x|_1 = \sum_{i=1}^{n} |x_i| \quad \text{and} \quad |x|_2 = \left(\sum_{i=1}^{n} |x_i|^2 \right)^{1/2}.$$

The latter is called the **Euclidean norm**.

The foregoing norms on R^n (or on C^n) are related by various inequalities, including the relations

$$|x|_\infty \le |x|_1 \le n|x|_\infty,$$
$$|x|_\infty \le |x|_2 \le \sqrt{n}|x|_\infty,$$
$$|x|_2 \le |x|_1 \le \sqrt{n}|x|_2.$$

The reader is asked to verify the validity of these formulas. These inequalities show that from the point of view of convergence properties, the foregoing norms are equivalent (i.e., one norm yields no different results than the others). Thus, when the particular norm being used is obvious from context, or when it is unimportant which particular norm is being used, we shall write $|x|$ in place of $|x|_p$ or $|x|_\infty$.

Using the concept of norm, we can define the **distance** between two vectors x and y in R^n (or in C^n, or more generally, in X) as $d(x, y) = |x - y|$. The following three fundamental properties of distance are true:

(i) $|x - y| \ge 0$ for all vectors x, y and $|x - y| = 0$ if and only if $x = y$;

(ii) $|x - y| = |y - x|$ for all vectors x, y; and

(iii) $|x - z| \le |x - y| + |y - z|$ for all vectors x, y, z.

We can now define a **spherical neighborhood** in R^n (in C^n) with center at x_0 and with radius $h > 0$ as

$$B(x_0, h) = \{x \in R^n : |x - x_0| < h\}.$$

If in particular the center of a spherical neighborhood with radius h is the origin, then

$$B(h) = \{x \in R^n : |x| < h\}.$$

We shall also use the notation $\overline{B(x_0, h)} = \{x \in R^n : |x - x_0| \le h\}$ and $\overline{B(h)} = \{x \in R^n : |x| \le h\}$.

We can also define the **norm of a matrix**. Let $R^{m \times n}$ denote the set of all real valued $m \times n$ matrices and let $C^{m \times n}$ denote the set of all complex $m \times n$ matrices. We define the norm of a matrix $A \in R^{m \times n}$ by the function $|\cdot|$, where

$$|A| = \sup\{|Ax| : x \in R^n \text{ with } |x| = 1\}.$$

The norm of $A \in C^{m \times n}$ is defined similarly. It is easily verified that

(M1) $|Ax| \leq |A||x|$ for any $x \in R^n$ (for any $x \in C^n$),
(M2) $|A + B| \leq |A| + |B|$,
(M3) $|\alpha A| = |\alpha||A|$ for all scalars α,
(M4) $|A| \geq 0$ and $|A| = 0$ if and only if A is the zero matrix,
(M5) $|AG| \leq |A||G|$, and
(M6) $|A| \leq \sum_{i=1}^{m} \left(\max_{1 \leq j \leq m} |a_{ij}| \right) \leq \sum_{i=1}^{m} \sum_{j=1}^{n} |a_{ij}|$

where A, B are any matrices in $R^{m \times n}$ (in $C^{m \times n}$) and G is any matrix in $R^{n \times k}$. Properties (M2)–(M4) clearly justify the use of the term matrix norm.

We can now also consider the convergence of vectors in R^n (in C^n). Thus, a sequence of vectors $\{x_m\} = \{(x_{1m}, x_{2m}, \ldots, x_{nm})^T\}$ is said to **converge** to a vector x, i.e., $x_m \to x$, if

$$\lim_{m \to \infty} |x_m - x| = 0.$$

Equivalently, $x_m \to x$ if and only if for each coordinate $k = 1, \ldots, n$ one has $x_{km} \to x_k$.

Next, let $g(t) = (g_1(t), \ldots, g_n(t))^T$ be a vector valued function defined on some interval J. Assume that each component of g is differentiable and integrable on J. Then differentiation and integration of g are defined componentwise, i.e.,

$$g'(t) = (g'_1(t), \ldots, g'_n(t))^T$$

and

$$\int_a^b g(t)\,dt = \left(\int_a^b g_1(t)\,dt, \ldots, \int_a^b g_n(t)\,dt \right)^T.$$

It is easily verified that for $b > a$,

$$\left| \int_a^b g(t)\,dt \right| \leq \int_a^b |g(t)|\,dt.$$

Finally, if D is an open connected nonempty set in the (t, x) space $R \times R^n$ and if $f: D \to R^n$, then f is said to satisfy a **Lipschitz condition** with Lipschitz constant L if and only if for all (t, x) and (t, y) in D,

$$|f(t, x) - f(t, y)| \leq L|x - y|.$$

This is an obvious extension of the scalar notion of a Lipschitz condition.

2.6 Systems of Equations

B. Systems of Equations

Every result given in Sections 2–5 can now be stated in vector form and proved, using the same methods as in the scalar case and invoking obvious modifications (such as the replacement of absolute values of scalars by the norms of vectors). We shall ask the reader to verify some of these results for the vector case in the problem section at the end of this chapter.

In the following result we demonstrate how systems of equations are treated. Instead of presenting one of the results from Sections 2–5 for the vector case, we state and prove a new result for linear non-homogeneous systems

$$x' = A(t)x + g(t), \tag{LN}$$

where $x \in R^n$, $A(t) = [a_{ij}(t)]$ is an $n \times n$ matrix, and $g(t)$ is an n vector valued function.

Theorem 6.1. Suppose that $A(t)$ and $g(t)$ in (LN) are defined and continuous on an interval J. [That is, suppose that each component $a_{ij}(t)$ of $A(t)$ and each component $g_k(t)$ of $g(t)$ is defined and continuous on an interval J.] Then for any τ in J and any $\xi \in R^n$, Eq. (LN) has a unique solution satisfying $x(\tau) = \xi$. This solution exists on the *entire* interval J and is continous in (t, τ, ξ). If A and g depend continuously on parameters $\lambda \in R^l$, then the solution will also vary continuously with λ.

Proof. First note that $f(t, x) \triangleq A(t)x + g(t)$ is continuous in (t, x). Moreover, for t on any compact subinterval J_0 of J there will be an $L_0 \geq 0$ such that

$$|f(t, x) - f(t, y)| = |A(t)(x - y)| \leq |A(t)||x - y|$$
$$\leq \left(\sum_{i=1}^{n} \max_{1 \leq j \leq n} |a_{ij}(t)|\right)|x - y| \leq L_0|x - y|.$$

Thus f satisfies a Lipschitz condition on $J_0 \times R^n$. The continuity implies existence (Theorem 2.3) while the Lipschitz condition implies uniqueness (Theorem 4.2) and continuity with respect to parameters (Corollary 5.4). To prove continuation over the interval J_0, let K be a bound for $\int_\tau^t |g(s)|\, ds$ over J_0. Then

$$|x(t)| \leq |\xi| + \int_\tau^t (|A(s)||x(s)| + |g(s)|)\, ds \leq (|\xi| + K) + \int_\tau^t L_0|x(s)|\, ds.$$

By the Gronwall inequality, $|x(t)| \leq (|\xi| + K)\exp(L_0|t - \tau|)$ for as long as $x(t)$ exists on J_0. By Corollary 3.2, the solution exists over all of J_0. ∎

For example, consider the mechanical system depicted in Fig. 1.5 whose governing equations are given in (1.2.2). Given any continuous functions $f_i(t)$, $i = 1, 2$, and initial data $(x_{10}, x'_{10}, x_{20}, x'_{20})^T$ at $\tau \in R$, there is according to Theorem 6.1 a unique solution on $-\infty < t < \infty$. This solution varies continuously with respect to the initial data and with respect to all parameters K, K_i, B, B_i, and M_i ($i = 1, 2$). Similar statements can be made for the rotational system depicted in Fig. 1.7 and for the circuits of Fig. 1.13 [see Eqs. (1.2.9) and also (1.2.13)].

For a nonlinear system such as the van der Pol equation (1.2.18), we can predict that unique solutions exist at least on small intervals and that these solutions vary continuously with respect to parameters. We also know that solutions can be continued both backwards and forwards either for all time or until such time as the solution becomes unbounded. The question of exactly how far a given solution of a nonlinear system can be continued has not been satisfactorily settled. It must be argued separately in each given case.

That the fundamental questions of existence, uniqueness, and so forth, have not yet been dealt with in a completely satisfactory way can be seen from Example 1.2.6, the Lienard equation with dry friction,

$$x'' + h \operatorname{sgn}(x') + \omega_0^2 x = 0,$$

where $h > 0$ and $\omega_0 > 0$. Since one coefficient of this equation has a locus of discontinuities at $x' = 0$, the theory already given will not apply on this curve. The existence and the behavior of solutions on a domain containing this curve of discontinuity must be studied by different and much more complex methods.

2.7. DIFFERENTIABILITY WITH RESPECT TO PARAMETERS

In the present section we consider systems of equations (E) and initial value problems (I). Given $f \in C(D)$ with f differentiable with respect to x, we definine the **Jacobian matrix** $f_x = \partial f / \partial x$ as the $n \times n$ matrix whose (i, j)th element is $\partial f_i / \partial x_j$, i.e.,

$$f_x = \partial f / \partial x = [\partial f_i / \partial x_j].$$

In this section, and throughout the remainder of this book, E will denote the **identity matrix**. When the dimension of E is to be emphasized, we shall write E_n to denote an $n \times n$ identity matrix.

2.7 Differentiability with Respect to Parameters

In the present section we show that when f_x exists and is continuous, then the solution ϕ of (I) depends smoothly on the parameters of the problem.

Theorem 7.1. Let $f \in C(D)$, let f_x exist and let $f_x \in C(D)$. If $\phi(t,\tau,\xi)$ is the solution of (E) such that $\phi(\tau,\tau,\xi) = \xi$, then ϕ is of class C^1 in (t,τ,ξ). Each vector valued function $\partial\phi/\partial\xi_i$ or $\partial\phi/\partial\tau$ will solve

$$y' = f_x(t,\phi(t,\tau,\xi))y \tag{7.1}$$

as a function of t while

$$\frac{\partial \phi}{\partial \tau}(\tau,\tau,\xi) = -f(\tau,\xi) \quad \text{and} \quad \frac{\partial \phi}{\partial \xi}(\tau,\tau,\xi) = E_n.$$

Proof. In any small spherical neighborhood of any point $(\tau,\xi) \in D$, the function f is Lipschitz continuous in x. Hence $\phi(t,\tau,\xi)$ exists locally, is unique, is continuable while it remains in D, and is continuous in (t,τ,ξ). Note also that (7.1) is a linear equation with continuous coefficient matrix. Thus by Theorem 6.1 solutions of (7.1) exist for as long as $\phi(t,\tau,\xi)$ is defined.

Fix a point (t,τ,ξ) and define $\xi(h) = (\xi_1 + h, \xi_2, \ldots, \xi_n)^T$ for all h with $|h|$ so small that $(\tau,\xi(h)) \in D$. Define

$$z(t,\tau,\xi,h) = (\phi(t,\tau,\xi(h)) - \phi(t,\tau,\xi))/h, \quad h \neq 0.$$

Differentiate z with respect to t and then apply the mean value theorem to each component z_i, $1 \leq i \leq n$, to obtain

$$z_i'(t,\tau,\xi,h) = [f_i(t,\phi(t,\tau,\xi(h))) - f_i(t,\phi(t,\tau,\xi))]/h$$
$$= \sum_{j=1}^n \left[\frac{\partial f_i}{\partial x_j}(t,\phi(t,\tau,\xi)) + P_{ij}(t,\tau,\xi,h)\right] z_j(t,\tau,\xi,h),$$

where

$$P_{ij}(t,\tau,\xi,h) = \frac{\partial f_i}{\partial x_j}(t,\bar{\phi}_i) - \frac{\partial f_i}{\partial x_j}(t,\phi(t,\tau,\xi))$$

and $\bar{\phi}_i$ is a point on the line segment between $\phi(t,\tau,\xi(h))$ and $\phi(t,\tau,\xi)$. The elements P_{ij} of the matrix P are continuous in (t,τ,ξ) and as $h \to 0$ $P_{ij}(t,\tau,\xi,h) \to 0$. Hence by continuity with respect to parameters, it follows that for any sequence $h_k \to 0$ we have

$$\lim_{k \to 0} z(t,\tau,\xi,h_k) = y_1(t,\tau,\xi),$$

where y_1 is that solution of (7.1) which satisfies the initial condition $y_1(\tau,\tau,\xi) = (1, 0, \ldots, 0)^T$. A similar argument applies to $\partial\phi/\partial\xi_k$ for $k = 2, 3, \ldots, n$ and for the existence of $\partial\phi/\partial\tau$. To obtain the initial condition for $\partial\phi/\partial\tau$, we note that

$$[\phi(\tau, \tau + h, \xi) - \phi(\tau, \tau, \xi)]/h = [\phi(\tau, \tau + h, \xi) - \xi]/h$$

$$= h^{-1} \int_{\tau+h}^{\tau} f(s, \phi(s, \tau + h, \xi)) \, ds$$

$$= -h^{-1} \int_{\tau}^{\tau+h} f(s, \phi(s, \tau + h, \xi)) \, ds \to -f(\tau, \xi)$$

as $h \to 0$. ∎

A similar analysis can be applied to (I_λ) to prove the next result.

Theorem 7.2. Let $f(t, x, \lambda)$ be continuous on $D \times D_\lambda$ and let f_x and $\partial f/\partial \lambda_k$, $1 \leq k \leq l$ exist and be continuous on $D \times D_\lambda$. Then the solution $\phi(t, \tau, \xi, \lambda)$ of (I_λ) is of class C^1 in (t, τ, ξ, λ). Moreover, $\partial\phi/\partial\lambda_k$ solves the initial value problem

$$y' = f_x(t, \phi(t, \tau, \xi, \lambda), \lambda)y + f_{\lambda_k}(t, \phi(t, \tau, \xi, \lambda), \lambda), \qquad y(\tau) = 0.$$

Proof. This result follows immediately by applying Theorem 7.1 to the $(n + l)$-dimensional system

$$x' = f(t, x, \lambda), \qquad \lambda' = 0. \qquad \blacksquare$$

The reader is invited to interpret the meaning of these results for some of the specific examples given in Chapter 1.

2.8 COMPARISON THEORY

This is the only section of the present chapter where it is crucial in our treatment of some results that the differential equation in question be a scalar equation. We point out that the results below on maximal solutions could be generalized to vector systems, however, only under the strong assumption that the system of equations is quasimonotone (see the problems at the end of the chapter).

Consider the scalar initial value problem (I') where $f \in C(D)$ and D is a domain in the (t, x) space. Any solution of (I') can be bracketed

2.8 Comparison Theory

between the two special solutions called the maximal solution and the minimal solution. More precisely, we define the **maximal solution** ϕ_M of (I') to be that noncontinuable solution of (I') such that if ϕ is any other solution of (I'), then $\phi_M(t) \geq \phi(t)$ for as long as both solutions are defined. The **minimal solution** ϕ_m of (I') is defined to be that noncontinuable solution of (I') such that if ϕ is any other solution of (I'), then $\phi_m(t) \leq \phi(t)$ for as long as both solutions are defined. Clearly, when ϕ_M and ϕ_m exist, they are unique. Their existence will be proved below.

Given $\varepsilon \geq 0$, consider the family of initial value problems

$$X' = f(t, X) + \varepsilon, \qquad X(\tau) = \xi + \varepsilon. \tag{8.ε}$$

Let $X(t, \varepsilon)$ be any fixed solution of (8.ε) which is noncontinuable to the right. We are now in a position to prove the following result.

Theorem 8.1. Let $f \in C(D)$ and let $\varepsilon \geq 0$.

(i) If $\varepsilon_1 > \varepsilon_2$, then $X(t, \varepsilon_1) > X(t, \varepsilon_2)$ for as long as both solutions exist and $t \geq \tau$.

(ii) There exist β as well as a solution X^* of (I') defined on $[\tau, \beta)$ and noncontinuable to the right such that

$$\lim_{\varepsilon \to 0^+} X(t, \varepsilon) = X^*(t)$$

with convergence uniform for t on compact subsets of $[\tau, \beta)$.

(iii) X^* is the maximal solution of (I'), i.e., $X^* = \phi_M$.

Proof. Since $X(\tau, \varepsilon_1) = \xi + \varepsilon_1 > \xi + \varepsilon_2 = X(\tau, \varepsilon_2)$, then by continuity $X(t, \varepsilon_1) > X(t, \varepsilon_2)$ for t near τ. Hence if (i) is not true, then there is a first time $t > \tau$ where the two solutions become equal. At that time,

$$X'(t, \varepsilon_1) = f(t, X(t, \varepsilon_1)) + \varepsilon_1 = f(t, X(t, \varepsilon_2)) + \varepsilon_1$$
$$> f(t, X(t, \varepsilon_2)) + \varepsilon_2 = X'(t, \varepsilon_2).$$

This is impossible since $X(s, \varepsilon_1) > X(s, \varepsilon_2)$ on $\tau < s < t$. Hence (i) is true.

To prove (ii), pick any sequence $\{\varepsilon_m\}$ which decreases to zero and let $X_m(t) = X(t, \varepsilon_m)$ be defined on the maximal intervals $[\tau, \beta_m)$. By Theorem 5.2 there is a subsequence of $\{X_m\}$ (which we again label by $\{X_m\}$ in order to avoid double subscripts) and there is a noncontinuable solution X^* of (I') defined on an interval $[\tau, \beta)$ such that

$$[\tau, \beta) \subset \liminf_{m \to \infty} [\tau, \beta_m), \qquad X^*(t) = \lim_{m \to \infty} X_m(t)$$

with the last limit uniform for t on compact subsets of $[\tau, \beta)$.

For any compact set $J \subset [\tau, \beta)$, J will be a subset of $[\tau, \beta_m)$, where m is sufficiently large. If $\varepsilon_{m+1} < \varepsilon < \varepsilon_m$, then by the monotonicity proved in part (i), $X_{m+1}(t) < X(t, \varepsilon) < X_m(t)$ for t in J. Thus, $X(t, \varepsilon) \to X^*(t)$ uniformly on J as $m \to \infty$. This proves (ii).

To prove that $X^* = \phi_M$ let ϕ be any solution of (I'). Then ϕ solves (8.ε) with $\varepsilon = 0$. By part (i), $X(t, \varepsilon) > \phi(t)$ when $\varepsilon > 0$ and both solutions exist. Take the limit as $\varepsilon \to 0^+$ to obtain $X^*(t) = \lim X(t, \varepsilon) \geq \phi(t)$. Hence $X^* = \phi_M$. ∎

The minimal solution of (I') is the maximal solution of the problem

$$y' = -f(t, -y), \qquad y(\tau) = -\xi \qquad (y = -x).$$

Hence, the minimal solution will exist whenever f is continuous. The maximal solution for $t < \tau$ can be obtained from

$$y' = -f(-s, y), \qquad y(-\tau) = \xi, \qquad s \geq -\tau (s = -t).$$

Given a function $x \in C(\alpha, \beta)$, the **upper right Dini derivative** $D^+ x$ is defined by

$$D^+ x(t) = \lim_{h \to 0^+} \sup [x(t+h) - x(t)]/h.$$

Note that $D^+ x$ is the derivative of x whenever x' exists. With this notation we now consider the differential inequality

$$D^+ x(t) \leq f(t, x(t)). \tag{8.1}$$

We call any function $x(t)$ satisfying (8.1) a **solution** of (8.1).

We are now in a position to prove the following result.

Lemma 8.2. If $x(t)$ is a continuous solution of (8.1) with $x(\tau) \leq \xi$, if $f \in C(D)$, and if ϕ_M is the maximal solution of (I'), then $x(t) \leq \phi_M(t)$ for as long as both functions exist and $t \geq \tau$.

Proof. Fix $\varepsilon > 0$ and let $X(t, \varepsilon)$ solve (8.ε). Clearly $x(t) < X(t, \varepsilon)$ at $t = \tau$ and hence in a neighborhood of τ. It is claimed that $X(t, \varepsilon) \geq x(t)$ for as long as both exist. If this were not the case, then there would be a first time t when it is not true. Thus, there would be a decreasing sequence $\{h_m\}$ with $x(t + h_m) > X(t + h_m, \varepsilon)$. Clearly $x(t) = X(t, \varepsilon)$ so that

$$D^+ x(t) = \lim_{h \to 0^+} \sup [x(t+h) - x(t)]/h \geq \lim_{m \to \infty} \sup [x(t+h_m) - x(t)]/h_m$$

$$= \lim_{m \to \infty} \sup [x(t+h_m) - X(t\ \varepsilon)]/h_m \geq \lim_{m \to \infty} [X(t+h_m, \varepsilon) - X(t, \varepsilon)]/h_m.$$

2.8 Comparison Theory

Thus
$$D^+x(t) \geq X'(t,\varepsilon) = f(t, X(t,\varepsilon)) + \varepsilon$$
$$= f(t, x(t)) + \varepsilon > f(t, x(t)) \geq D^+x(t),$$
a contradiction.

Since $X(t,\varepsilon) \geq x(t)$ for all $\varepsilon > 0$, we can let $\varepsilon \to 0^+$ and use Theorem 8.1 to obtain the conclusion. ∎

We are also in a position to prove the next result.

Lemma 8.3. Let $\phi(t)$ be n vector valued and assume that $\phi \in C^1(\alpha, \beta)$. Then $D^+|\phi(t)| \leq |\phi'(t)|$ for all $t \in (\alpha, \beta)$.

Proof. By the triangle inequality, if $h > 0$, then
$$[|\phi(t+h)| - |\phi(t)|]/h \leq |[\phi(t+h) - \phi(t)]/h|.$$
Take the lim sup as $h \to 0^+$ on both sides of the preceding inequality to complete the proof. ∎

The foregoing results can now be combined to obtain the following **comparison theorem**.

Theorem 8.4. Let $f \in C(D)$ where D is a domain in the (t, x) space $R \times R^n$ and let ϕ be a solution of (I). Let $F(t, v)$ be a continuous function such that $|f(t, x)| \leq F(t, |x|)$ for all (t, x) in D. If $\eta \geq |\phi(\tau)|$ and if v_M is the maximal solution of
$$v' = F(t, v), \qquad v(\tau) = \eta, \tag{8.2}$$
then $|\phi(t)| \leq v_M(t)$ for as long as both functions exist.

Proof. By Lemma 8.3 it follows if $v(t) = |\phi(t)|$, then
$$D^+v(t) = D^+|\phi(t)| \leq |\phi'(t)| = |f(t, \phi(t))|$$
$$\leq F(t, |\phi(t)|) = F(t, v(t)).$$
By Lemma 8.2 it follows that $v_M(t) \geq v(t) = |\phi(t)|$. ∎

For example, if $|f(t, x)| \leq A|x| + B$ for $t \in J$, $x \in R^n$, then (8.2) reduces to
$$v' = Av + B, \qquad v(\tau) = \eta.$$
Thus, $v_M(t) = \exp[A(t - \tau)](\eta - B/A) + B/A$. Since v_M exists for all $t \in J$, then so do the solutions of (E). From this example it should be clear that the comparison theory can often be useful in obtaining continuation results and certain types of solution estimates. However, we note that in Theorem

8.3 it is necessary that F be nonnegative. This severely restricts the use of the comparison theory, particularly in analyzing stability properties of solutions of (E) (see Chapter 5).

2.9 COMPLEX VALUED SYSTEMS*

Systems of n complex valued ordinary differential equations of the form

$$z' = f(t, z, \lambda) \tag{C}$$

were introduced in Section 1.1E. Since (C) can be separated into its real and imaginary parts to obtain a real $2n$-dimensional system, the theory of such systems has already been covered. What has not been addressed is the natural question of when solutions of (C) are holomorphic in t or in the parameter λ. This is the topic of the present section.

A function $F(w)$ defined on a domain D in complex n space C^n with $w = (w_1, \ldots, w_n)^T$ is called **holomorphic** in D if each component F_i of F is continuous on D and if for each $w_0 \in D$ there is a neighborhood $N = \{w : |w - w_0| < \varepsilon\}$ in which F_i is holomorphic in each w_k separately (with all other w_j, $j \neq k$, held fixed). If F is holomorphic in D, then for each $w_0 = (w_{10}, \ldots, w_{n0})^T$ there is a neighborhood N in which F can be expanded in a convergent power series in the n variables,

$$F(w_1, \ldots, w_n) = \sum_{k_1=0}^{\infty} \sum_{k_2=0}^{\infty} \cdots \sum_{k_n=0}^{\infty} A(k_1, \ldots, k_n)(w_1 - w_{10})^{k_1} \cdots (w_n - w_{n0})^{k_n}.$$

Thus, F has partial derivatives of all orders which are also holomorphic functions in D. Furthermore, recall that if $\{f_n\}$ is a sequence of functions which are holomorphic in D and if $\{f_n\}$ converges uniformly on compact subsets of D to a limit f, then f must also be holomorphic in D.

As can be seen from the examples in Chapter 1, it is a common situation for (I_λ) that $f(t, x, \lambda)$ is holomorphic in (x, λ) or even in (t, x, λ). In order to emphasize that t is allowed to become complex, we replace t by z and (C) by

$$dw/dz = f(z, w, \lambda), \tag{C_λ}$$

where f is holomorphic on a domain D in complex $(1 + n + l)$ space. Let (z_0, w_0, λ_0) be a point in D, let S be a rectangle

$$|z - z_0| \leq a, \qquad |w - w_0| \leq b, \qquad |\lambda - \lambda_0| \leq c$$

in D with $|f| \leq M$ on S, and let d be the minimum of a and b/M. Then one can construct the successive approximations

$$w_0(z) \equiv w_0 \qquad \text{and} \qquad w_{m+1}(z) = w_0 + \int_{z_0}^{z} f(u, w_m(u), \lambda) \, du$$

where the integral is the complex contour integral taken along the straight line from z_0 to z. By arguments similar to those used to prove Theorem 4.6, the following can be proved.

Theorem 9.1. If f, S, and d are defined as above, then the successive approximations $w_m(z)$ given above are well defined on $|z - z_0| \leq d$ and converge to a solution $w(z, z_0, w_0, \lambda)$ as $m \to \infty$. This solution of (C_λ) is unique once initial conditions z_0 and w_0 are fixed. Moreover, $w(z, z_0, w_0, \lambda)$ is holomorphic in (z, z_0, w_0, λ).

If in (C) f is only continuous in t for t real but is holomorphic in (z, λ), then the solution will be holomorphic in (ξ, λ) (see the problems at the end of this chapter).

PROBLEMS

1. Let J be a compact subinterval of R and let $\mathscr{F}$ be a subset of $C(J)$. Show that if each sequence in $\mathscr{F}$ contains a uniformly convergent subsequence, then $\mathscr{F}$ is both equicontinuous and uniformly bounded.

2. (*Gronwall inequality*) Let r, k, and f be real and continuous functions which satisfy $r(t) \geq 0$, $k(t) \geq 0$, and

$$r(t) \leq f(t) + \int_a^t k(s) r(s) \, ds, \qquad a \leq t \leq b.$$

Show that

$$r(t) \leq \int_a^t f(s) k(s) \exp\left[\int_s^t k(u) \, du \right] ds + f(t), \qquad a \leq t \leq b.$$

3. Show that the initial value problem $x' = x^{1/3}$, $x(0) = 0$, has infinitely many different solutions. Find the maximal and the minimal solutions. (Remember to consider $t < 0$.)

4. (a) Carefully restate and prove Theorems 2.2 and 2.3 for a system (I) of n real equations.

(b) In the same way, restate and prove Theorems 3.1 and 4.2.

5. Show that if $g \in C^1(R)$ and $f \in C(R)$, then the solution $y(t, \tau, A, B)$ of

$$y'' + f(y)y' + g(y) = 0, \qquad y(\tau) = A, \quad y'(\tau) = B$$

exists locally, is unique, and can be continued so long as y and y' remain bounded. *Hint*: Use a transformation.

6. Let $f \in C(D)$, let $(\tau, \xi) \in D$, and let (I') have a unique solution ϕ. For each ε with $0 < \varepsilon < 1$, assume that (I') has an ε-approximate solution ϕ_ε defined on $|t - \tau| \leq c$, where c is given as in (2.1). Show that

$$\lim_{\varepsilon \to 0^+} \phi_\varepsilon(t) = \phi(t)$$

uniformly in t on compact subsets of $|t - \tau| < c$.

7. Let $f(t, x, \lambda)$ and $F(t, v, \lambda)$ be continuous functions on $R \times R^n \times R^l$ and $R \times R \times R^l$ with $F(t, 0, \lambda) = 0$ and assume that $|f(t, x, \lambda) - f(t, y, \lambda)| \leq F(t, |x - y|, \lambda)$. Assume that for any (τ, c) with $c \geq 0$ the solution of

$$v' = F(t, v, \lambda), \qquad v(\tau) = c$$

is unique. Show that solutions $\phi(t, \tau, \xi, \lambda)$ of (I_λ) vary continuously with (t, τ, ξ, λ).

8. Show that for any $\phi \in C[\tau, \infty)$ the positive limit set satisfies

$$\Omega(\phi) = \bigcap_{t > \tau} \overline{\{\phi(s): s \geq t\}}.$$

Show that if the range of ϕ is contained in a compact set $K \subset R^n$, then $\Omega(\phi)$ is nonempty, compact, connected, and $\phi(t) \to \Omega(\phi)$ as $t \to \infty$.

9. (a) Show that for $x \in R^n$ (or $x \in C^n$),

$$|x|_\infty \leq |x|_1 \leq n|x|_\infty,$$
$$|x|_\infty \leq |x|_2 \leq \sqrt{n}|x|_\infty,$$
$$|x|_2 \leq |x|_1 \leq \sqrt{n}|x|_2.$$

(b) Given $x = (x_1, \ldots, x_n)^T$ and $y = (y_1, \ldots, y_n)^T$ in R^n, show that $\sum_{j=1}^n x_j y_j \leq |x|_2 |y|_2$.

(c) Show that $|\cdot|_1, |\cdot|_2,$ and $|\cdot|_\infty$ each define a norm.

10. Let $f \in C(D)$ and let f have continuous partial derivatives with respect to x up to and including order $k \geq 1$. Show that the solution $\phi(t, \tau, \xi)$ of (E) has continuous partial derivatives in t, τ, and ξ through order k.

11. Let h and g be positive, continuous, and real valued functions on $0 \leq t \leq \infty$ and $0 < x < \infty$, respectively. Suppose that (3.3) is true.

(i) If $\int_0^\infty h(t)\,dt < \infty$, show that any solution of (3.4) has a finite limit as $t \to \infty$.

(ii) If
$$\lim_{\varepsilon \to 0^+} \int_\varepsilon^1 \frac{dx}{g(x)} = +\infty, \qquad (10.1)$$
show that all solutions of (3.4) with $x(\tau) > 0$ can be continued to the left until $t = 0$.

12. Let $f \in C(R \times R^n)$ with $|f(t,x)| \le h(t)g(|x|)$. Assume that h and g are positive continuous functions such that (3.3) and (10.1) are true. If $\int_0^\infty h(t)\,dt < \infty$, then any solution ϕ of (E) with $\tau > 0$ exists over the interval $0 < t < \infty$ and has a finite limit at $t = 0$ and at $t = \infty$.

13. Consider a $2n$-dimensional Hamiltonian system with Hamiltonian function $H \in C^2(R^n \times R^n)$. Suppose that for some fixed k the surface S defined by $H(x,y) = k$ is bounded. Show that all solutions starting on the surface S can be continued for all t in R.

14. Show that any solution of
$$x'' + x + x^3 = 0$$
exists for all $t \in R$. Can the same be said about the equation
$$x'' + x' + x + x^3 = 0?$$

15. Let $x(t)$ and $y(t)$ denote the density at time t of a wolf and a moose population, respectively, say on a certain island in Lake Superior. (Wolves eat moose.) Assume that the animal populations are "well stirred" and that there are no other predators or prey on the island. Under these conditions, a simple model of this predator–prey system is
$$x' = x(-a + by),$$
$$y' = y(c - dx),$$
where a, b, c, and d are positive constants and where $x(0) > 0$ and $y(0) > 0$. Show that:

(a) Solutions are defined for all $t \ge 0$.
(b) Neither the wolf nor the moose population can die out within a *finite* period of time.

16. Let $f: R \to R$ with f Lipschitz continuous on any compact interval $K \subset R$. Show that $x' = f(x)$ has no nonconstant solution ϕ which is periodic.

17. A function $f \in C(D)$ is said to be **quasimonotone in** x if each component $f_i(t, x_1, \ldots, x_n)$ is nondecreasing in x_j for $j = 1, \ldots, i-1, i+1, \ldots, n$. We define the **maximal solution** ϕ_M of (I) to be that noncontinuable solution

which has the following property: if ϕ is any other solution of (I) and if ϕ_{Mi} is the ith component of ϕ_M, then $\phi_{Mi}(t) \geq \phi_i(t)$ for all t such that both solutions exist, $i = 1, \ldots, n$.

Show that if $f \in C(D)$ and if f is quasimonotone in x, then there exists a maximal solution ϕ_M for (I).

18. Let $f \in C(D)$ and let f be quasimonotone in x (see Problem 17). Let $x(t)$ be a continuous function which satisfies (8.1) componentwise. If $x_i(\tau) \leq \phi_{Mi}(\tau)$ for $i = 1, \ldots, n$, then $x_i(t) \leq \phi_{Mi}(t)$ for as long as $t \geq \tau$ and both solutions exist.

19. Let $f: R \times D \to R^n$ where D is an open subset of R^n. Suppose for each compact subset $K \subset D$, f is uniformly continuous and bounded on $R \times K$. Let ϕ be a solution of (I) which remains in a compact subset $K_1 \subset D$ for all $t \geq \tau$. Given any sequence $t_m \to \infty$, show that there is a subsequence $t_{m_k} \to \infty$, a continuous function $g \in C(R \times D)$, and a solution ψ such that $\psi(t) \in K_1$ and
$$\psi'(t) = g(t, \psi(t)), \qquad -\infty < t < \infty.$$
Moreover, as $k \to \infty$, $f(t + t_{m_k}, x) \to g(t, x)$ uniformly for $(t\,x)$ in compact subsets of $R \times D$ and $\phi(t + t_{m_k}) \to \psi(t)$ uniformly for t on compact subsets of R.

20.* Prove Theorem 9.1

21.* Suppose $f(t, x, \lambda)$ is continuous for (t, x, λ) in D and is holomorphic in (x, λ) for each fixed t. Let $\phi(t, \tau, \xi, \lambda)$ be the solution of (I_λ) for (τ, ξ, λ) in D. Then for each fixed t and τ, prove that ϕ is holomorphic in (ξ, λ).

22.* Suppose that ϕ is the solution of (1.2.29) which satisfies $\phi(\tau) = \xi$ and $\phi'(\tau) = \eta$. In which of the variables $t, \tau, \xi, \eta, \omega_0, \omega_1, h$, and G does ϕ vary holomorphically?

23. Let $f \in C(D_0)$, $D_0 \subset R^n$, and let f be smooth enough so that solutions $\phi(t, \tau, \xi)$ of
$$x' = f(x), \qquad x(\tau) = \xi \tag{A}$$
are unique. Show that $\phi(t, \tau, \xi) = \phi(t - \tau, 0, \xi)$ for all $\xi \in D_0$, all $\tau \in R$, and all t such that ϕ is defined.

24. Let $f \in C(D)$, let f be periodic with period T in t, and let f be smooth enough so that solutions ϕ of (I) are unique. Show that for any integer m,
$$\phi(t, \tau, \xi) = \phi(t + mT, \tau + mT, \xi)$$
for all $(\tau, \xi) \in D$ and for all t where ϕ is defined.

The next four problems require the notion of **complete metric space** which should be recalled or learned by the reader at this time.

25.* (*Banach fixed point theorem*) Given a metric space (X, ρ) (where ρ denotes a metric defined on a set X), a **contraction mapping** T is a function $T: X \to X$ such that for some constant k, with $0 < k < 1$, $\rho(T(x), T(y)) \le k\rho(x, y)$ for all x and y in X. A **fixed point** of T is a point x in X such that $T(x) = x$. Use successive approximations to prove the following: If T is a contraction mapping on a complete metric space X, then T has a unique fixed point in X.

26.* Show that the following metric spaces are all complete. (Here α is some fixed real number.)

   (a) $X = C[a, b]$ and $\rho(f, g) = \max\{|f(t) - g(t)|e^{\alpha(t-a)} : a \le t \le b\}$.

   (b) $X = \{f \in C[a, \infty) : e^{\alpha t} f(t) \text{ is bounded on } [a, \infty)\}$ and $\rho(f, g) = \sup\{|f(t) - g(t)|e^{\alpha t} : a \le t < \infty\}$.

27.* Let $f \in C(R^+ \times R^n)$ and let f be Lipschitz continuous in x with Lipschitz constant L. In Problem 26(a), let $a = \tau$, $\alpha = L$, and choose b in the interval $\tau < b < \infty$. Show that

$$(T\phi)(t) = \xi + \int_\tau^t f(s, \phi(s))\, ds, \qquad \tau \le t \le b,$$

is a contraction mapping on (X, ρ).

28.* Prove Theorem 4.1 using a contraction mapping argument.

3 | LINEAR SYSTEMS

Both in the theory of differential equations and in their applications, linear systems of ordinary differential equations are extremely important. This can be seen from the examples in Chapter 1 which include linear translational and rotational mechanical systems and linear circuits. Linear systems of ordinary differential equations are also frequently used as a "first approximation" to nonlinear problems. Moreover, the theory of linear ordinary differential equations is often useful as an integral part of the analysis of many nonlinear problems. This will become clearer in some of the subsequent chapters (Chapters 5 and 6).

In this chapter, we first study the general properties of linear systems. We then turn our attention to the special cases of linear systems of ordinary differential equations with constant coefficients and linear systems of ordinary differential equations with periodic coefficients. The chapter is concluded with a discussion of special properties of nth order linear ordinary differential equations.

3.1 PRELIMINARIES

In this section, we establish some notation and we summarize certain facts from linear algebra which we shall use throughout this book.

3.1 Preliminaries

A. Linear Independence

Let X be a vector space over the real or over the complex numbers. A set of vectors $\{v_1, v_2, \ldots, v_n\}$ is said to be **linearly dependent** if there exist scalars $\alpha_1, \alpha_2, \ldots, \alpha_n$, not all zero, such that

$$\alpha_1 v_1 + \alpha_2 v_2 + \cdots + \alpha_n v_n = 0.$$

If this inequality is true only for $\alpha_1 = \alpha_2 = \cdots = \alpha_n = 0$, then the set $\{v_1, v_2, \ldots, v_n\}$ is said to be **linearly independent**.

If $v_k = [x_{1k}, x_{2k}, \ldots, x_{nk}]^T$ is a real or complex n vector, then $[v_1, v_2, \ldots, v_n]$ denotes the matrix whose ith column is v_i, i.e.,

$$[v_1, v_2, \ldots, v_n] = \begin{bmatrix} x_{11} & x_{12} & \cdots & x_{1n} \\ \vdots & \vdots & & \vdots \\ x_{n1} & x_{n2} & \cdots & x_{nn} \end{bmatrix}.$$

In this case, the set $\{v_1, v_2, \ldots, v_n\}$ is linearly independent if and only if the determinant of the above matrix is not zero, i.e.,

$$\det[v_1, v_2, \ldots, v_n] \neq 0.$$

A **basis** for a vector space X is a linearly independent set of vectors such that every vector in X can be expressed as a linear combination of these vectors. In R^n or C^n, the set

$$e_1 = \begin{bmatrix} 1 \\ 0 \\ \vdots \\ 0 \\ 0 \end{bmatrix}, \quad e_2 = \begin{bmatrix} 0 \\ 1 \\ 0 \\ \vdots \\ 0 \end{bmatrix}, \ldots, e_n = \begin{bmatrix} 0 \\ \cdot \\ \vdots \\ 0 \\ 1 \end{bmatrix} \quad (1.1)$$

is a basis called the **natural basis**.

B. Matrices

Given an $m \times n$ matrix $A = [a_{ij}]$, we denote the **rank** of A by $\rho(A)$ and the (complex) **conjugate matrix** by $\bar{A} = [\bar{a}_{ij}]$. The **transpose** of A is $A^T = [a_{ji}]$ and the **adjoint** is $A^* = \bar{A}^T$. A matrix A is **symmetric** if $A = A^T$ and **self-adjoint** when $A = A^*$.

C. Jordan Canonical Form

Two $n \times n$ matrices A and B are said to be **similar** if there is a nonsingular matrix P such that $A = P^{-1}BP$. The polynomial $p(\lambda) = \det(A - \lambda E_n)$ is called the **characteristic polynomial** of A. (Here E_n denotes the $n \times n$ identity matrix and λ is a scalar.) The roots of $p(\lambda)$ are called the **eigenvalues** of A. By an **eigenvector** (or **right eigenvector**) of A associated with the eigenvalue λ, we mean a nonzero $x \in C^n$ such that $Ax = \lambda x$.

Now let A be an $n \times n$ matrix. We may regard A as a mapping of C^n with the natural basis into itself, i.e., we may regard $A: C^n \to C^n$ as a linear operator. To begin with, let us assume that A has *distinct eigenvalues* $\lambda_1, \ldots, \lambda_n$. Let v_i be an eigenvector of A corresponding to λ_i, $i = 1, \ldots, n$. Then it can be easily shown that the set of vectors $\{v_1, \ldots, v_n\}$ is linearly independent over C, and as such, it can be used as a basis for C^n. Now let $\tilde{A}$ be the representation of A with respect to the basis $\{v_1, \ldots, v_n\}$. Since the ith column of $\tilde{A}$ is the representation of $Av_i = \lambda_i v_i$ with respect to the basis $\{v_1, \ldots, v_n\}$, it follows that

$$\tilde{A} = \begin{bmatrix} \lambda_1 & & & 0 \\ & \lambda_2 & & \\ & & \ddots & \\ 0 & & & \lambda_n \end{bmatrix}.$$

Since A and $\tilde{A}$ are matrix representations of the same linear transformation, it follows that A and $\tilde{A}$ are similar matrices. Indeed, this can be checked by computing

$$\tilde{A} = P^{-1}AP,$$

where $P = [v_1, \ldots, v_n]$ and where the v_i are eigenvectors corresponding to λ_i, $i = 1, \ldots, n$.

When a matrix $\tilde{A}$ is obtained from a matrix A via a similarity transformation P, we say that matrix A **has been diagonalized**. Now if the matrix A has repeated eigenvalues, then it is not always possible to diagonalize it. In generating a "convenient" basis for C^n in this case, we introduce the concept of **generalized eigenvector**. Specifically, a vector v is called a **generalized eigenvector of rank** k of A, associated with an eigenvalue λ if and only if

$$(A - \lambda E_n)^k v = 0 \quad \text{and} \quad (A - \lambda E_n)^{k-1} v \neq 0.$$

Note that when $k = 1$, this definition reduces to the preceding definition of eigenvector.

3.1 Preliminaries

Now let v be a generalized eigenvector of rank k associated with the eigenvalue λ. Define

$$\begin{aligned}
v_k &= v, \\
v_{k-1} &= (A - \lambda E_n)v = (A - \lambda E_n)v_k, \\
v_{k-2} &= (A - \lambda E_n)^2 v = (A - \lambda E_n)v_{k-1}, \\
&\vdots \\
v_1 &= (A - \lambda E_n)^{k-1} v = (A - \lambda E_n)v_2.
\end{aligned} \quad (1.2)$$

Then for each i, $1 \leq i \leq k$, v_i is a generalized eigenvector of rank i. We call the set of vectors $\{v_1, \ldots, v_k\}$ a **chain of generalized eigenvectors**.

For generalized eigenvectors, we have the following important results:

(i) The generalized eigenvectors $v_1, \ldots, v_k$ defined in (1.2) are linearly independent.

(ii) The generalized eigenvectors of A associated with different eigenvalues are linearly independent.

(iii) If u and v are generalized eigenvectors of rank k and l, respectively, associated with the same eigenvalue λ, and if u_i and v_j are defined by

$$\begin{aligned}
u_i &= (A - \lambda E_n)^{k-i} u, \quad i = 1, \ldots, k, \\
v_j &= (A - \lambda E_n)^{l-j} v, \quad j = 1, \ldots, l,
\end{aligned}$$

and if u_1 and v_1 are linearly independent, then the generalized eigenvectors $u_1, \ldots, u_k, v_1, \ldots, v_l$ are linearly independent.

These results can be used to construct a new basis for C^n such that the matrix representation of A with respect to this new basis, is in the **Jordan canonical form** J. We characterize J in the following result: For every complex $n \times n$ matrix A there exists a nonsingular matrix P such that the matrix

$$J = P^{-1} A P$$

is in the canonical form

$$J = \begin{bmatrix} J_0 & & & 0 \\ & J_1 & & \\ & & \ddots & \\ 0 & & & J_s \end{bmatrix}, \quad (1.3)$$

where J_0 is a diagonal matrix with diagonal elements $\lambda_1, \ldots, \lambda_k$ (not necessarily distinct), i.e.,

$$J_0 = \begin{bmatrix} \lambda_1 & & 0 \\ & \ddots & \\ 0 & & \lambda_k \end{bmatrix},$$

and each J_p is an $n_p \times n_p$ matrix of the form

$$J_p = \begin{bmatrix} \lambda_{k+p} & 1 & 0 & \cdots & 0 \\ 0 & \lambda_{k+p} & 1 & \ddots & \vdots \\ \vdots & \vdots & \ddots & \ddots & 1 \\ 0 & 0 & \cdots & & \lambda_{k+p} \end{bmatrix}, \quad p = 1, \ldots, s,$$

where λ_{k+p} need not be different from λ_{k+q} if $p \neq q$ and $k + n_1 + \cdots + n_s = n$. The numbers λ_i, $i = 1, \ldots, k + s$, are the eigenvalues of A. If λ_i is a simple eigenvalue of A, it appears in the block J_0. The blocks $J_0, J_1, \ldots, J_s$ are called **Jordan blocks** and J is called the **Jordan canonical form**.

Note that a matrix may be similar to a diagonal matrix without having simple eigenvalues. The identity matrix E is such an example. Also, it can be shown that any real symmetric matrix A or complex self-adjoint matrix A has only real eigenvalues (which may be repeated) and is similar to a diagonal matrix.

We now give a procedure for computing a set of basis vectors which yield the Jordan canonical form J of an $n \times n$ matrix A and the required nonsingular transformation P which relates A to J:

1. Compute the eigenvalues of A. Let $\lambda_1, \ldots, \lambda_m$ be the distinct eigenvalues of A with multiplicities $n_1, \ldots, n_m$, respectively.

2. Compute n_1 linearly independent generalized eigenvectors of A associated with λ_1 as follows: Compute $(A - \lambda_1 E_n)^i$ for $i = 1, 2, \ldots$ until the rank of $(A - \lambda_1 E_n)^k$ is equal to the rank of $(A - \lambda_1 E_n)^{k+1}$. Find a generalized eigenvector of rank k, say u. Define $u_i = (A - \lambda_1 E_n)^{k-i} u$, $i = 1, \ldots, k$. If $k = n_1$, proceed to step 3. If $k < n_1$, find another linearly independent generalized eigenvector with the largest possible rank; i.e., try to find another generalized eigenvector with rank k. If this is not possible, try $k - 1$, and so forth, until n_1 linearly independent generalized eigenvectors are determined. Note that if $\rho(A - \lambda_1 E_n) = r$, then there are totally $(n - r)$ chains of generalized eigenvectors associated with λ_1.

3. Repeat step 2 for $\lambda_2, \ldots, \lambda_m$.

3.1 Preliminaries

4. Let $u_1, \ldots, u_k, \ldots$ be the new basis. Observe, from (1.2), that

$$Au_1 = \lambda_1 u_1 = [u_1 u_2 \cdots u_k \cdots] \begin{bmatrix} \lambda_1 \\ 0 \\ \vdots \\ 0 \end{bmatrix},$$

$$Au_2 = u_1 + \lambda_1 u_2 = [u_1 u_2 \cdots u_k \cdots] \begin{bmatrix} 1 \\ \lambda_1 \\ 0 \\ \vdots \\ 0 \end{bmatrix},$$

$$\vdots$$

$$Au_k = u_{k-1} + \lambda_1 u_k = [u_1 u_2 \cdots u_k \cdots] \begin{bmatrix} 0 \\ \vdots \\ 0 \\ 1 \\ \lambda_1 \\ 0 \\ \vdots \\ 0 \end{bmatrix} \leftarrow k\text{th position,}$$

which yields the representation of A with respect to the new basis

$$J = \begin{bmatrix} \begin{matrix} \lambda_1 & 1 & \cdots & 0 \\ 0 & \lambda_1 & \ddots & \vdots \\ \vdots & & \ddots & 1 \\ 0 & & \cdots & \lambda_1 \end{matrix} & \\ \underbrace{}_{k} & \end{bmatrix} \Big\} k .$$

Note that each chain of generalized eigenvectors generates a Jordan block whose order equals the length of the chain.

5. The similarity transformation which yields $J = Q^{-1}AQ$ is given by $Q = [u_1, \ldots, u_k, \ldots]$.

6. Rearrange the Jordan blocks in the desired order to yield (1.3) and the corresponding similarity transformation P.

Example 1.1. The characteristic equation of the matrix

$$A = \begin{bmatrix} 3 & -1 & 1 & 1 & 0 & 0 \\ 1 & 1 & -1 & -1 & 0 & 0 \\ 0 & 0 & 2 & 0 & 1 & 1 \\ 0 & 0 & 0 & 2 & -1 & -1 \\ 0 & 0 & 0 & 0 & 1 & 1 \\ 0 & 0 & 0 & 0 & 1 & 1 \end{bmatrix}$$

is given by

$$\det(A - \lambda E) = (\lambda - 2)^5 \lambda = 0.$$

Thus, A has eigenvalue $\lambda_2 = 2$ with multiplicity 5 and eigenvalue $\lambda_1 = 0$ with multiplicity 1.

Now compute $(A - \lambda_2 E)^i$, $i = 1, 2, \ldots$, as follows:

$$(A - 2E) = \begin{bmatrix} 1 & -1 & 1 & 1 & 0 & 0 \\ 1 & -1 & -1 & -1 & 0 & 0 \\ 0 & 0 & 0 & 0 & 1 & 1 \\ 0 & 0 & 0 & 0 & -1 & -1 \\ 0 & 0 & 0 & 0 & -1 & 1 \\ 0 & 0 & 0 & 0 & 1 & -1 \end{bmatrix} \quad \text{and} \quad \rho(A - 2E) = 4,$$

$$(A - 2E)^2 = \begin{bmatrix} 0 & 0 & 2 & 2 & 0 & 0 \\ 0 & 0 & 2 & 2 & 0 & 0 \\ 0 & 0 & 0 & 0 & 0 & 0 \\ 0 & 0 & 0 & 0 & 0 & 0 \\ 0 & 0 & 0 & 0 & 2 & -2 \\ 0 & 0 & 0 & 0 & -2 & 2 \end{bmatrix} \quad \text{and} \quad \rho(A - 2E)^2 = 2,$$

$$(A - 2E)^3 = \begin{bmatrix} 0 & 0 & 0 & 0 & 0 & 0 \\ 0 & 0 & 0 & 0 & 0 & 0 \\ 0 & 0 & 0 & 0 & 0 & 0 \\ 0 & 0 & 0 & 0 & 0 & 0 \\ 0 & 0 & 0 & 0 & -4 & 4 \\ 0 & 0 & 0 & 0 & 4 & -4 \end{bmatrix} \quad \text{and} \quad \rho(A - 2E)^3 = 1,$$

3.1 Preliminaries

$$(A - 2E)^4 = \begin{bmatrix} 0 & 0 & 0 & 0 & 0 & 0 \\ 0 & 0 & 0 & 0 & 0 & 0 \\ 0 & 0 & 0 & 0 & 0 & 0 \\ 0 & 0 & 0 & 0 & 0 & 0 \\ 0 & 0 & 0 & 0 & 8 & -8 \\ 0 & 0 & 0 & 0 & -8 & 8 \end{bmatrix} \quad \text{and} \quad \rho(A - 2E)^4 = 1.$$

Since $\rho(A - 2E)^3 = \rho(A - 2E)^4$, we stop at $(A - 2E)^3$. It can be easily verified that if $u = [0\ 0\ 1\ 0\ 0\ 0]^T$, then $(A - 2E)^3 u = 0$ and $(A - 2E)^2 u = [2\ 2\ 0\ 0\ 0\ 0]^T \neq 0$. Therefore, u is a generalized eigenvector of rank 3. So we define

$$u_1 \triangleq (A - 2E)^2 u = [2\ 2\ 0\ 0\ 0\ 0]^T,$$
$$u_2 \triangleq (A - 2E)u = [1\ -1\ 0\ 0\ 0\ 0]^T,$$
$$u_3 \triangleq u = [0\ 0\ 1\ 0\ 0\ 0]^T.$$

Since we have only three generalized eigenvectors for $\lambda_2 = 2$ and since the multiplicity of $\lambda_2 = 2$ is five, we have to find two more linearly independent eigenvectors for $\lambda_2 = 2$. So let us try to find a generalized eigenvector of rank 2. Let $v = [0\ 0\ 1\ -1\ 1\ 1]^T$. Then $(A - 2E)v = [0\ 0\ 2\ -2\ 0\ 0]^T \neq 0$ and $(A - 2E)^2 v = 0$. Moreover, $(A - 2E)v$ is linearly independent of u_1 and hence, we have another linearly independent generalized eigenvector of rank 2. Define

$$v_2 \triangleq v = [0\ 0\ 1\ -1\ 1\ 1]^T$$

and

$$v_1 = (A - 2E)v = [0\ 0\ 2\ -2\ 0\ 0]^T.$$

Next, we compute an eigenvector associated with $\lambda_1 = 0$. Since $w = [0\ 0\ 0\ 0\ 1\ -1]^T$ is a solution of $(A - \lambda_1 E)w = 0$, the vector w will do.

Finally, with respect to the basis $w_1, u_1, u_2, u_3, v_1, v_2$, the Jordan canonical form of A is given by

$$J = \begin{bmatrix} 0 & 0 & 0 & 0 & 0 & 0 \\ 0 & 2 & 1 & 0 & 0 & 0 \\ 0 & 0 & 2 & 1 & 0 & 0 \\ 0 & 0 & 0 & 2 & 0 & 0 \\ 0 & 0 & 0 & 0 & 2 & 1 \\ 0 & 0 & 0 & 0 & 0 & 2 \end{bmatrix} = \begin{bmatrix} \lambda_1 & 0 & 0 & 0 & 0 & 0 \\ 0 & \lambda_2 & 1 & 0 & 0 & 0 \\ 0 & 0 & \lambda_2 & 1 & 0 & 0 \\ 0 & 0 & 0 & \lambda_2 & 0 & 0 \\ 0 & 0 & 0 & 0 & \lambda_2 & 1 \\ 0 & 0 & 0 & 0 & 0 & \lambda_2 \end{bmatrix}$$

and

$$P = \begin{bmatrix} w & u_1 & u_2 & u_3 & v_1 & v_2 \end{bmatrix} = \begin{bmatrix} 0 & 2 & 1 & 0 & 0 & 0 \\ 0 & 2 & -1 & 0 & 0 & 0 \\ 0 & 0 & 0 & 1 & 2 & 1 \\ 0 & 0 & 0 & 0 & -2 & -1 \\ 1 & 0 & 0 & 0 & 0 & 1 \\ -1 & 0 & 0 & 0 & 0 & 1 \end{bmatrix}.$$

The correctness of P is easily checked by computing $PJ = AP$.

3.2 LINEAR HOMOGENEOUS AND NONHOMOGENEOUS SYSTEMS

In the present section, we consider linear homogeneous systems

$$x' = A(t)x \tag{LH}$$

and linear nonhomogeneous systems

$$x' = A(t)x + g(t). \tag{LN}$$

In Chapter 2, Theorem 6.1, it was shown that these systems, subject to initial conditions $x(\tau) = \xi$, possess unique solutions for every $(\tau, \xi) \in D$, where

$$D = \{(t, x) : t \in J = (a, b), x \in R^n \text{ (or } x \in C^n)\}.$$

These solutions exist over the entire interval $J = (a, b)$ and they depend continuously on the initial conditions. In applications, it is typical that $J = (-\infty, \infty)$. We note that $\phi(t) \equiv 0$, for all $t \in J$, is a solution of (LH), with $\phi(\tau) = 0$. We call this the **trivial solution** of (LH).

Throughout this chapter we consider matrices and vectors which will be either real valued or complex valued. In the former case, the field of scalars for the x space is the field of real numbers ($F = R$) and in the latter case, the field for the x space is the field of complex numbers ($F = C$).

Theorem 2.1. The set of solutions of (LH) on the interval J forms an n-dimensional vector space.

3.2 Linear Homogeneous and Nonhomogeneous Systems

Proof. Let V denote the set of all solutions of (LH) on J. Let $\alpha_1, \alpha_2 \in F$ and let $\phi_1, \phi_2 \in V$. Then $\alpha_1\phi_1 + \alpha_2\phi_2 \in V$ since

$$\frac{d}{dt}[\alpha_1\phi_1(t) + \alpha_2\phi_2(t)] = \alpha_1 \frac{d}{dt}\phi_1(t) + \alpha_2 \frac{d}{dt}\phi_2(t)$$

$$= \alpha_1 A(t)\phi_1(t) + \alpha_2 A(t)\phi_2(t) = A(t)[\alpha_1\phi_1(t) + \alpha_2\phi_2(t)]$$

for all $t \in J$. Hence, V is a vector space.

To complete the proof, we must show that V is of dimension n. This means that we must find n linearly independent solutions $\phi_1, \ldots, \phi_n$ which span V. To this end, we choose a set of n linearly independent vectors $\xi_1, \ldots, \xi_n$ in the n-dimensional x space (i.e., in R^n or C^n). By the existence results of Chapter 2, if $\tau \in J$, then there exist n solutions $\phi_1, \ldots, \phi_n$ of (LH) such that $\phi_1(\tau) = \xi_1, \ldots, \phi_n(\tau) = \xi_n$. We first show that these solutions are linearly independent. If on the contrary these solutions are linearly dependent, then there exist scalars $\alpha_1, \ldots, \alpha_n \in F$, not all zero, such that

$$\sum_{i=1}^{n} \alpha_i \phi_i(t) = 0$$

for all $t \in J$. This implies in particular that

$$\sum_{i=1}^{n} \alpha_i \phi_i(\tau) = \sum_{i=1}^{n} \alpha_i \xi_i = 0.$$

But this contradicts the assumption that $\{\xi_1, \ldots, \xi_n\}$ is a linearly independent set. Therefore, the solutions $\phi_1, \ldots, \phi_n$ are linearly independent.

Finally, we must show that the solutions $\phi_1, \ldots, \phi_n$ span V. Let ϕ be any solution of (LH) on the interval J such that $\phi(\tau) = \xi$. Then there exist unique scalars $\alpha_1, \ldots, \alpha_n \in F$ such that

$$\xi = \sum_{i=1}^{n} \alpha_i \xi_i,$$

since by assumption, the vectors $\xi_1, \ldots, \xi_n$ form a basis for the x space. Now

$$\psi = \sum_{i=1}^{n} \alpha_i \phi_i$$

is a solution of (LH) on J such that $\psi(\tau) = \xi$. But by the uniqueness results of Chapter 2, we have

$$\phi = \psi = \sum_{i=1}^{n} \alpha_i \phi_i.$$

Since ϕ was chosen arbitrarily, it follows that the solutions $\phi_1, \ldots, \phi_n$ span V. ∎

Theorem 2.1 enables us to make the following definition.

Definition 2.2. A set of n linearly independent solutions of (LH) on J, $\{\phi_1, \ldots, \phi_n\}$, is called a **fundamental set** of solutions of (LH) and the $n \times n$ matrix

$$\Phi = [\phi_1 \ \phi_2 \cdots \phi_n]$$

is called a **fundamental matrix** of (LH).

In the sequel, we shall find it convenient to use the notation

$$\Phi = [\phi_1 \ \phi_2 \ \cdots \ \phi_n] = \begin{bmatrix} \phi_{11} & \phi_{12} & \cdots & \phi_{1n} \\ \phi_{21} & \phi_{22} & \cdots & \phi_{2n} \\ \vdots & \vdots & & \vdots \\ \phi_{n1} & \phi_{n2} & \cdots & \phi_{nn} \end{bmatrix}$$

for a fundamental matrix. Note that there are infinitely many different fundamental sets of solutions of (LH) and hence, infinitely many different fundamental matrices for (LH). We shall first need to study some basic properties of a fundamental matrix.

In the following result, $X = [x_{ij}]$ denotes an $n \times n$ matrix and the derivative of X with respect to t is defined as $X' = [x'_{ij}]$. If $A(t)$ is the $n \times n$ matrix given in (LH), then we call the system of n^2 equations

$$X' = A(t)X \tag{2.1}$$

a **matrix** (differential) **equation**.

Theorem 2.3. A fundamental matrix Φ of (LH) satisfies the matrix equation (2.1) on the interval J.

Proof. We have

$$\Phi' = [\phi'_1, \phi'_2, \ldots, \phi'_n] = [A(t)\phi_1, A(t)\phi_2, \ldots, A(t)\phi_n]$$
$$= A(t)[\phi_1, \phi_2, \ldots, \phi_n] = A(t)\Phi. \quad \blacksquare$$

The next result is called **Abel's formula**.

Theorem 2.4. If Φ is a solution of the matrix equation (2.1) on an interval J and if τ is any point of J, then

$$\det \Phi(t) = \det \Phi(\tau) \exp\left[\int_\tau^t \operatorname{tr} A(s)\,ds\right] \quad \text{for every} \quad t \in J.$$

3.2 Linear Homogeneous and Nonhomogeneous Systems

Proof. If $\Phi = [\phi_{ij}]$ and $A(t) = [a_{ij}(t)]$, then

$$\phi'_{ij} = \sum_{k=1}^{n} a_{ik}(t)\phi_{kj}.$$

Now

$$\frac{d}{dt}[\det \Phi(t)] = \begin{vmatrix} \phi'_{11} & \phi'_{12} & \cdots & \phi'_{1n} \\ \phi_{21} & \phi_{22} & \cdots & \phi_{2n} \\ \vdots & \vdots & & \vdots \\ \phi_{n1} & \phi_{n2} & \cdots & \phi_{nn} \end{vmatrix} + \begin{vmatrix} \phi_{11} & \phi_{12} & \cdots & \phi_{1n} \\ \phi'_{21} & \phi'_{22} & \cdots & \phi'_{2n} \\ \vdots & \vdots & & \vdots \\ \phi_{n1} & \phi_{n2} & \cdots & \phi_{nn} \end{vmatrix} + \cdots$$

$$+ \begin{vmatrix} \phi_{11} & \phi_{12} & \cdots & \phi_{1n} \\ \phi_{21} & \phi_{22} & \cdots & \phi_{2n} \\ \vdots & \vdots & & \vdots \\ \phi'_{n1} & \phi'_{n2} & \cdots & \phi'_{nn} \end{vmatrix}$$

$$= \begin{vmatrix} \sum_{k=1}^{n} a_{1k}(t)\phi_{k1} & \sum_{k=1}^{n} a_{1k}(t)\phi_{k2} & \cdots & \sum_{k=1}^{n} a_{1k}(t)\phi_{kn}(t) \\ \phi_{21} & \phi_{22} & \cdots & \phi_{2n} \\ \vdots & \vdots & & \vdots \\ \phi_{n1} & \phi_{n2} & \cdots & \phi_{nn} \end{vmatrix}$$

$$+ \begin{vmatrix} \phi_{11} & \phi_{12} & \cdots & \phi_{1n} \\ \sum_{k=1}^{n} a_{2k}(t)\phi_{k1} & \sum_{k=1}^{n} a_{2k}(t)\phi_{k2} & \cdots & \sum_{k=1}^{n} a_{2k}(t)\phi_{kn} \\ \phi_{31} & \phi_{32} & \cdots & \phi_{3n} \\ \vdots & \vdots & & \vdots \\ \phi_{n1} & \phi_{n2} & \cdots & \phi_{nn} \end{vmatrix} + \cdots$$

$$+ \begin{vmatrix} \phi_{11} & \phi_{12} & \cdots & \phi_{1n} \\ \phi_{21} & \phi_{22} & \cdots & \phi_{2n} \\ \vdots & \vdots & & \vdots \\ \phi_{n-1,1} & \phi_{n-1,2} & \cdots & \phi_{n-1,n} \\ \sum_{k=1}^{n} a_{nk}(t)\phi_{k1} & \sum_{k=1}^{n} a_{nk}(t)\phi_{k2} & \cdots & \sum_{k=1}^{n} a_{nk}(t)\phi_{kn} \end{vmatrix}.$$

The first term in the foregoing sum of determinants is unchanged if we subtract from the first row

(a_{12} times the second row) + (a_{13} times the third row)
$+ \cdots +$ (a_{1n} times the nth row).

This yields

$$\begin{vmatrix} \phi'_{11} & \phi'_{12} & \cdots & \phi'_{1n} \\ \phi_{21} & \phi_{22} & \cdots & \phi_{2n} \\ \vdots & \vdots & & \vdots \\ \phi_{n1} & \phi_{n2} & \cdots & \phi_{nn} \end{vmatrix} = \begin{vmatrix} a_{11}(t)\phi_{11} & a_{11}(t)\phi_{12} & \cdots & a_{11}(t)\phi_{1n} \\ \phi_{21} & \phi_{22} & \cdots & \phi_{2n} \\ \vdots & \vdots & & \vdots \\ \phi_{n1} & \phi_{n2} & \cdots & \phi_{nn} \end{vmatrix}$$

$$= a_{11}(t) \det \Phi(t).$$

Repeating this procedure for the remaining terms in the above sum of determinants, we have

$$\frac{d}{dt}[\det \Phi(t)] = a_{11}(t) \det \Phi(t) + a_{22}(t) \det \Phi(t) + \cdots + a_{nn}(t) \det \Phi(t)$$

$$= [\text{trace } A(t)] \det \Phi(t).$$

But this implies that

$$\det \Phi(t) = \det \Phi(\tau) \exp\left[\int_{\tau}^{t} \text{trace } A(s)\, ds\right]. \qquad \blacksquare$$

It follows from Theorem 2.4, since τ is arbitrary, that either $\det \Phi(t) \neq 0$ for each $t \in J$ or that $\det \Phi(t) = 0$ for every $t \in J$.

The next result allows us to characterize a fundamental matrix as a solution of (2.1) with a nonzero determinant for all t in J.

Theorem 2 5. A solution Φ of the matrix equation (2.1) is a fundamental matrix of (LH) if and only if its determinant is nonzero for all $t \in J$.

Proof. Suppose that $\Phi = [\phi_1, \phi_2, \ldots, \phi_n]$ is a fundamental matrix for (LH). Then the columns of Φ, $\phi_1, \ldots, \phi_n$, form a linearly independent set. Let ϕ be a nontrivial solution of (LH). By Theorem 2.1 there exist unique scalars $\alpha_1, \ldots, \alpha_n \in F$, not all zero, such that

$$\phi = \sum_{j=1}^{n} \alpha_i \phi_j \quad \text{or} \quad \phi = \Phi a,$$

where $a^T = [\alpha_1, \ldots, \alpha_n]$. Let $t = \tau \in J$. Then we have

$$\phi(\tau) = \Phi(\tau) a,$$

a system of n linear (algebraic) equations. By construction, this system of equations has a unique solution for any choice of $\phi(\tau)$. Hence, $\det \Phi(\tau) \neq 0$. It now follows from Theorem 2.4 that $\det \Phi(t) \neq 0$ for any $t \in J$.

Conversely, let Φ be a solution of the matrix equation (2.1) and assume that $\det \Phi(t) \neq 0$ for all $t \in J$. Then the columns of Φ, $\phi_1, \ldots, \phi_n$,

3.2 Linear Homogeneous and Nonhomogeneous Systems

are linearly independent (for all $t \in J$). Hence, Φ is a fundamental matrix of (LH). ∎

Note that a matrix may have its determinant identically zero on some interval, even though its columns are linearly independent. For example, the columns of the matrix

$$\Phi(t) = \begin{bmatrix} 1 & t & t^2 \\ 0 & 2 & t \\ 0 & 0 & 0 \end{bmatrix}$$

are linearly independent, yet $\det \Phi(t) = 0$ for all $t \in (-\infty, \infty)$. According to Theorem 2.5, this matrix $\Phi(t)$ cannot be a solution of the matrix equation (2.1) for any continuous matrix $A(t)$.

Example 2.6. For the system of equations

$$\begin{aligned} x_1' &= 5x_1 - 2x_2, \\ x_2' &= 4x_1 - x_2, \end{aligned} \quad (2.2)$$

we have

$$A(t) \equiv A = \begin{bmatrix} 5 & -2 \\ 4 & -1 \end{bmatrix} \quad \text{for all} \quad t \in (-\infty, \infty).$$

Two linearly independent solutions of (2.2) are

$$\phi_1(t) = \begin{bmatrix} e^{3t} \\ e^{3t} \end{bmatrix}, \quad \phi_2(t) = \begin{bmatrix} e^t \\ 2e^t \end{bmatrix}.$$

The reader should verify that ϕ_1 and ϕ_2 are indeed solutions of (2.2). We now have

$$\Phi(t) = \begin{bmatrix} e^{3t} & e^t \\ e^{3t} & 2e^t \end{bmatrix}$$

which satisfies the matrix equation $\Phi' = A\Phi$. Moreover,

$$\det \Phi(t) = e^{4t} \neq 0 \quad \text{for all} \quad t \in (-\infty, \infty).$$

Therefore, Φ is a fundamental matrix of (2.2) by Theorem 2.5. Also, in view of Theorem 2.4 we have

$$\det \Phi(t) = \det \Phi(\tau) \exp\left[\int_\tau^t \text{trace } A(s)\, ds\right]$$

$$= e^{4\tau} \exp\left[\int_\tau^t 4\, ds\right] = e^{4\tau} e^{4(t-\tau)} = e^{4t}$$

for all $t \in (-\infty, \infty)$.

Example 2.7. For the system of equations
$$x_1' = x_2,$$
$$x_2' = tx_2, \qquad (2.3)$$
we have
$$A(t) = \begin{bmatrix} 0 & 1 \\ 0 & t \end{bmatrix} \quad \text{for all} \quad t \in (-\infty, \infty).$$
Two linearly independent solutions of (2.3) are
$$\phi_1(t) = \begin{bmatrix} 1 \\ 0 \end{bmatrix}, \quad \phi_2(t) = \begin{bmatrix} \int_\tau^t e^{\eta^2/2} \, d\eta \\ e^{t^2/2} \end{bmatrix}.$$
The matrix
$$\Phi(t) = \begin{bmatrix} 1 & \int_\tau^t e^{\eta^2/2} \, d\eta \\ 0 & e^{t^2/2} \end{bmatrix}$$
satisfies the matrix equation $\Phi' = A(t)\Phi$ and
$$\det \Phi(t) = e^{t^2/2} \quad \text{for all} \quad t \in (-\infty, \infty).$$
Therefore, Φ is a fundamental matrix of (2.3). Also, in view of Theorem 2.4, we have
$$\det \Phi(t) = \det \Phi(\tau) \exp\left[\int_\tau^t \text{trace } A(s) \, ds\right]$$
$$= e^{\tau^2/2} \exp\left[\int_\tau^t \eta \, d\eta\right] = e^{\tau^2/2} e^{t^2/2} e^{-\tau^2/2} = e^{t^2/2}$$
for all $t \in (-\infty, \infty)$.

Theorem 2.8. If Φ is a fundamental matrix of (LH) and if C is any nonsingular constant $n \times n$ matrix, then ΦC is also a fundamental matrix of (LH). Moreover, if Ψ is any other fundamental matrix of (LH), then there exists a constant $n \times n$ nonsingular matrix P such that $\Psi = \Phi P$.

Proof. We have
$$(\Phi C)' = \Phi' C = (A(t)\Phi)C = A(t)(\Phi C)$$
and hence, ΦC is a solution of the matrix equation (2.1). But
$$\det(\Phi C) = \det \Phi \det C \neq 0.$$
By Theorem 2.5, ΦC is a fundamental matrix.

3.2 Linear Homogeneous and Nonhomogeneous Systems

Next, let Ψ be any other fundamental matrix. Consider the product $\Phi^{-1}\Psi$. Notice that since $\det \Phi(t) \neq 0$ for all $t \in J$, then Φ^{-1} exists for all t. Also, $\Phi\Phi^{-1} = E$ so by the Leibnitz rule $\Phi'\Phi^{-1} + \Phi(\Phi^{-1})' = 0$, or $(\Phi^{-1})' = -\Phi^{-1}\Phi'\Phi^{-1}$. Thus, we can compute

$$(\Phi^{-1}\Psi)' = \Phi^{-1}\Psi' + (\Phi^{-1})'\Psi = \Phi^{-1}A(t)\Psi - (\Phi^{-1}\Phi'\Phi^{-1})\Psi$$
$$= \Phi^{-1}A(t)\Psi - (\Phi^{-1}A(t)\Phi\Phi^{-1})\Psi = \Phi^{-1}A(t)\Psi - \Phi^{-1}A(t)\Psi = 0.$$

Therefore

$$\Phi^{-1}\Psi = P \quad \text{or} \quad \Psi = \Phi P. \quad \blacksquare$$

Example 2.9. For the system of equations (2.2) given in Example 2.6, we can find the fundamental matrix Ψ which satisfies the initial condition

$$\Psi(0) = \begin{bmatrix} 1 & 0 \\ 0 & 1 \end{bmatrix} = E$$

as follows. To find C such that $\Psi = \Phi C$, we must have $\Psi(0) = E = \Phi(0)C$ or $C = \Phi^{-1}(0)$. Thus, for (2.2) take

$$C = \begin{bmatrix} 1 & 1 \\ 1 & 2 \end{bmatrix}^{-1} = \begin{bmatrix} 2 & -1 \\ -1 & 1 \end{bmatrix}$$

and

$$\Psi(t) = \begin{bmatrix} (2e^{3t} - e^t) & (-e^{3t} + e^t) \\ (2e^{3t} - 2e^t) & (-e^{3t} + 2e^t) \end{bmatrix}.$$

We are now in a position to study the structure of the solutions of (LH). In doing so, we need to introduce the concept of state transition matrix. In the following definition, we use the natural basis $\{e_1, e_2, \ldots, e_n\}$ which was defined in (1.1) in Section 3.1.

Definition 2.10. A fundamental matrix Φ of (LH) whose columns are determined by the linearly independent solutions $\phi_1, \ldots, \phi_n$ with

$$\phi_1(\tau) = e_1, \ldots, \phi_n(\tau) = e_n, \quad \tau \in J,$$

is called the **state transition matrix** Φ for (LH). Equivalently, if ψ is *any* fundamental matrix of (LH), then the matrix Φ determined by

$$\Phi(t,\tau) \triangleq \Psi(t)\Psi^{-1}(\tau) \quad \text{for all} \quad t, \tau \in J,$$

is said to be **the state transition matrix** of (LH).

Example 2.11. For system (2.2) of Example 2.6, a fundamental matrix is given by

$$\Psi(t) = \begin{bmatrix} e^{3t} & e^t \\ e^{3t} & 2e^t \end{bmatrix} \quad \text{and} \quad \Psi^{-1}(t) = \begin{bmatrix} 2e^{-3t} & -e^{-3t} \\ -e^{-t} & e^{-t} \end{bmatrix}.$$

The state transition matrix of (2.2) is now given by

$$\Phi(t,\tau) = \Psi(t)\Psi^{-1}(\tau) = \begin{bmatrix} 2e^{3(t-\tau)} - e^{t-\tau} & -e^{3(t-\tau)} + e^{t-\tau} \\ 2e^{3(t-\tau)} - 2e^{t-\tau} & -e^{3(t-\tau)} + 2e^{t-\tau} \end{bmatrix}.$$

Note that the state transition matrix of (LH) is uniquely determined by the matrix $A(t)$ and is independent of the particular choice of the fundamental matrix. For example, let Ψ_1 and Ψ_2 be two different fundamental matrices for (LH). Then by Theorem 2.8, there exists a constant $n \times n$ nonsingular matrix P such that $\Psi_2 = \Psi_1 P$. By the definition of the state transition matrix, we have

$$\Phi(t,\tau) = \Psi_2(t)[\Psi_2(\tau)]^{-1} = \Psi_1(t)PP^{-1}[\Psi_1(\tau)]^{-1} = \Psi_1(t)[\psi_1(\tau)]^{-1}.$$

This shows that $\Phi(t,\tau)$ is independent of the fundamental matrix chosen.

We now summarize some of the properties of a state transition matrix and we give an explicit expression for the solution of the initial value problem (LH).

Theorem 2.12. Let $\tau \in J$, let $\phi(\tau) = \xi$, and let $\Phi(t,\tau)$ denote the state transition matrix for (LH) for all $t \in J$. Then

(i) $\Phi(t,\tau)$ is the unique solution of the matrix equation

$$\frac{\partial}{\partial t}\Phi(t,\tau) \triangleq \Phi'(t,\tau) = A(t)\Phi(t,\tau)$$

with $\Phi(\tau,\tau) = E$, the $n \times n$ identity matrix,

(ii) $\Phi(t,\tau)$ is nonsingular for all $t \in J$,

(iii) for any $t, \sigma, \tau \in J$, we have

$$\Phi(t,\tau) = \Phi(t,\sigma)\Phi(\sigma,\tau),$$

(iv) $[\Phi(t,\tau)]^{-1} \triangleq \Phi^{-1}(t,\tau) = \Phi(\tau,t)$ for all $t, \tau \in J$,

(v) the unique solution $\phi(t,\tau,\xi)$ of (LH), with $\phi(\tau,\tau,\xi) = \xi$ specified, is given by

$$\phi(t,\tau,\xi) = \Phi(t,\tau)\xi \qquad \text{for all} \quad t \in J. \tag{2.4}$$

Proof. (i) Let Ψ be any fundamental matrix of (LH). By definition, we have $\Phi(t,\tau) = \Psi(t)\Psi^{-1}(\tau)$ (independent of the choice of Ψ)

3.2 Linear Homogeneous and Nonhomogeneous Systems

and

$$\frac{\partial \Phi(t,\tau)}{\partial t} = \Phi'(t,\tau) = \Psi'(t)\Psi^{-1}(\tau) = A(t)\Psi(t)\Psi^{-1}(\tau) = A(t)\Phi(t,\tau).$$

Furthermore, $\Phi(\tau,\tau) = \Psi(\tau)\Psi^{-1}(\tau) = E$.

(ii) Since for any fundamental matrix Ψ of (LH) we have $\det \Psi(t) \neq 0$ for all $t \in J$, it follows that

$$\det \Phi(t,\tau) = \det \Psi(t)\Psi^{-1}(\tau) = \det \Psi(t) \det \Psi^{-1}(\tau) \neq 0$$

for all $t, \tau \in J$.

(iii) For any fundamental matrix Ψ of (LH) and for the state transition matrix Φ of (LH), we have

$$\Phi(t,\tau) = \Psi(t)\Psi^{-1}(\tau) = \Psi(t)\Psi^{-1}(\sigma)\Psi(\sigma)\Psi^{-1}(\tau) = \Phi(t,\sigma)\Phi(\sigma,\tau)$$

for any $t, \sigma, \tau \in J$.

(iv) For any fundamental matrix Φ of (LH) and for the state transition matrix Φ of (LH), we have

$$[\Phi(t,\tau)]^{-1} = [\Psi(t)\Psi(\tau)^{-1}]^{-1} = \Psi(\tau)\Psi^{-1}(t) = \Phi(\tau,t)$$

for any $t, \tau \in J$.

(v) By the uniqueness results in Chapter 2, we know that for every $(\tau,\xi) \in D$, (LH) has a unique solution $\phi(t)$ for all $t \in J$ with $\phi(\tau) = \xi$. To verify that (2.4) is indeed this solution, note first that $\phi(\tau) = \Phi(\tau,\tau)\xi = \xi$. Differentiating both sides of (2.4), we have

$$\phi'(t) = \Phi(t,\tau)'\xi = A(t)\Phi(t,\tau)\xi = A(t)\phi(t)$$

which shows that (2.4) is the desired solution. ∎

In engineering and physics applications, $\phi(t)$ is interpreted as representing the "state" of a (dynamical) system represented by (LH) at time t and $\phi(\tau) = \xi$ is interpreted as representing the "state" at time τ. In (2.4), $\Phi(t,\tau)$ relates the "states" of (LH) at t and τ. This explains the name "state transition matrix."

Example 2.13. For system (2.3) of Example 2.7 a fundamental matrix is given by

$$\Phi(t) = \begin{bmatrix} 1 & \int_\tau^t e^{\eta^2/2}\,d\eta \\ 0 & e^{t^2/2} \end{bmatrix}.$$

Therefore

$$\Phi^{-1}(t) = \begin{bmatrix} 1 & -e^{-t^2/2} \int_\tau^t e^{\eta^2/2}\, d\eta \\ 0 & e^{-t^2/2} \end{bmatrix}.$$

The state-transition matrix for (2.3) is now given by

$$\Phi(t,\tau) = \Phi(t)\Phi^{-1}(\tau) = \begin{bmatrix} 1 & \int_\tau^t e^{\eta^2/2}\, d\eta \\ 0 & e^{t^2/2} \end{bmatrix} \begin{bmatrix} 1 & 0 \\ 0 & e^{-\tau^2/2} \end{bmatrix}$$

$$= \begin{bmatrix} 1 & e^{-\tau^2/2} \int_\tau^t e^{\eta^2/2}\, d\eta \\ 0 & e^{(t^2-\tau^2)/2} \end{bmatrix}.$$

Now suppose that $\phi(\tau) = \xi = [1,1]^T$. Then

$$\phi(t,\tau,\xi) = \Phi(t,\tau)\xi = \begin{bmatrix} 1 + e^{-\tau^2/2} \int_\tau^t e^{\eta^2/2}\, d\eta \\ e^{(t^2-\tau^2)/2} \end{bmatrix}.$$

Finally note that

$$\phi'(t,\tau,\xi) = \begin{bmatrix} e^{t^2/2} e^{-\tau^2/2} \\ t e^{t^2/2} e^{-\tau^2/2} \end{bmatrix} = A(t)\phi(t,\tau,\xi)$$

$$= \begin{bmatrix} 0 & 1 \\ 0 & t \end{bmatrix} \begin{bmatrix} 1 + e^{-\tau^2/2} \int_\tau^t e^{\eta^2/2}\, d\eta \\ e^{(t^2-\tau^2)/2} \end{bmatrix}.$$

In the next result, we study the structure of the solution of the linear nonhomogeneous system

$$x' = A(t)x + g(t). \tag{LN}$$

Theorem 2.14. Let $\tau \in J$, let $(\tau,\xi) \in D$, and let $\Phi(t,\tau)$ denote the state transition matrix for (LH) for all $t \in J$. Then the unique solution $\phi(t,\tau,\xi)$ of (LN) satisfying $\phi(\tau,\tau,\xi) = \xi$ is given by

$$\phi(t,\tau,\xi) = \Phi(t,\tau)\xi + \int_\tau^t \Phi(t,\eta)g(\eta)\, d\eta. \tag{2.5}$$

3.2 Linear Homogeneous and Nonhomogeneous Systems

Proof. We prove the theorem by first verifying that ϕ given in (2.5) is indeed a solution of (LN) with $\phi(\tau) = \xi$. Differentiating both sides of (2.5) with respect to t we have

$$\begin{aligned}
\phi'(t,\tau,\xi) &= \Phi'(t,\tau)\xi + \Phi(t,t)g(t) + \int_\tau^t \Phi'(t,\eta)g(\eta)\,d\eta \\
&= A(t)\Phi(t,\tau)\xi + g(t) + \int_\tau^t A(t)\Phi(t,\eta)g(\eta)\,d\eta \\
&= A(t)\left[\Phi(t,\tau)\xi + \int_\tau^t \Phi(t,\eta)g(\eta)\,d\eta\right] + g(t) \\
&= A(t)\phi(t,\tau,\xi) + g(t).
\end{aligned}$$

From (2.5) we also note that $\phi(\tau,\tau,\xi) = \xi$. Therefore, ϕ given in (2.5) is a solution of (LN) with $\phi(\tau) = \xi$. By uniqueness it follows that ϕ is in fact the unique solution. ∎

Note that when $\xi = 0$, then (2.5) reduces to

$$\phi_p(t) = \int_\tau^t \Phi(t,\eta)g(\eta)\,d\eta$$

and when $\xi \neq 0$ but $g(t) \equiv 0$, then (2.5) reduces to

$$\phi_h(t) = \Phi(t,\tau)\xi.$$

Therefore, the solution of (LN) may be viewed as consisting of a component which is due to the initial data ξ, and another component which is due to the "forcing term" $g(t)$. This type of separation is in general possible only in linear systems of differential equations. We call ϕ_p a **particular solution** of the nonhomogenous system (LN).

We conclude this section with a discussion of the adjoint equation. Let Φ be a fundamental matrix for the linear homogeneous system (LH). Then

$$(\Phi^{-1})' = -\Phi^{-1}\Phi'\Phi^{-1} = -\Phi^{-1}A(t).$$

Taking the conjugate transpose of both sides, we obtain

$$(\Phi^{*-1})' = -A^*(t)\Phi^{*-1}.$$

This implies that Φ^{*-1} is a fundamental matrix for the system

$$y' = -A^*(t)y, \qquad t \in J. \tag{2.6}$$

We call (2.6) the **adjoint to** (LH), and we call the matrix equation

$$Y' = -A^*(t)Y, \qquad t \in J,$$

the **adjoint to** the matrix equation (2.1).

Theorem 2.15. If Φ is a fundamental matrix for (LH), then Ψ is a fundamental matrix for its adjoint (2.6) if and only if

$$\Psi^*\Phi = C, \tag{2.7}$$

where C is some constant nonsingular matrix.

Proof. Let Φ be a fundamental matrix for (LH) and let Ψ be a fundamental matrix for (2.6). Since Φ^{*-1} is a fundamental matrix for (2.6), then by Theorem 2.8 there exists a constant $n \times n$ nonsingular matrix P such that

$$\Psi = \Phi^{*-1} P.$$

Therefore

$$\Psi^*\Phi = P^* \triangleq C.$$

Conversely, let Φ be a fundamental matrix for (LH) which satisfies (2.7). Then

$$\Psi^* = C\Phi^{-1} \quad \text{or} \quad \Psi = \Phi^{*-1}C^*.$$

By Theorem 2.8, Ψ is a fundamental matrix of the adjoint system (2.6). ∎

3.3 LINEAR SYSTEMS WITH CONSTANT COEFFICIENTS

For the scalar differential equation

$$x' = ax, \quad x(\tau) = \xi,$$

the solution is given by $\phi(t) = e^{a(t-\tau)}\xi$. In the present section, we show that a similar result holds for the system of linear equations with constant coefficients,

$$x' = Ax. \tag{L}$$

Specifically, we show that (L) has a solution of the form $\phi(t) = e^{A(t-\tau)}\xi$ with $\phi(\tau) = \xi$. Before we can do this, however, we need to define the matrix e^{At} and discuss some of its properties. We first require the following result.

Theorem 3.1. Let A be a constant $n \times n$ matrix which may be real or complex. Let $S_N(t)$ denote the partial sum of matrices defined by the formula

$$S_N(t) = E + \sum_{k=1}^{N} \frac{t^k}{k!} A^k.$$

3.3 Linear Systems with Constant Coefficients

Then each element of the matrix $S_N(t)$ converges absolutely and uniformly on any finite t interval $(-a, a)$, $a > 0$, as $N \to \infty$.

Proof. The properties of the norm given in Section 2.6 imply that

$$\left| E + \sum_{k=1}^{\infty} (t^k A^k)/k! \right| \leq |E| + \sum_{k=1}^{\infty} a^k |A|^k / k!$$

$$\leq n + \sum_{k=1}^{\infty} (a|A|)^k / k! = (n-1) + \exp(a|A|).$$

By the Weierstrass M test (Theorem 2.1.3), it follows that $S_N(t)$ is a Cauchy sequence uniformly on $(-a, a)$. ∎

Note that by the same proof, we obtain

$$S'_N(t) = A S_{N-1}(t) = S_{N-1}(t) A.$$

Thus, the limit of $S_N(t)$ is a C^1 function on $(-\infty, \infty)$. Moreover, this limit commutes with A.

In view of Theorem 3.1, the following definition makes sense.

Definition 3.2. Let A be a constant $n \times n$ matrix which may be real or complex. We define e^{At} to be the matrix

$$e^{At} = E + \sum_{k=1}^{\infty} \frac{t^k}{k!} A^k$$

for any $-\infty < t < \infty$.

We note in particular that $e^{At}|_{t=0} = E$.

In the special case when $A(t) \equiv A$, system (LH) reduces to system (L). Consequently, the results of Section 3.2 are applicable to (L) as well as to (LH). However, because of the special nature of (L), more detailed information can be given.

Theorem 3.3. Let $J = (-\infty, \infty)$, $\tau \in J$, and let A be a given constant $n \times n$ matrix for (L). Then

  (i) $\Phi(t) \triangleq e^{At}$ is a fundamental matrix for (L) for $t \in J$.
  (ii) The state transition matrix for (L) is given by

$$\Phi(t, \tau) = e^{A(t-\tau)} \triangleq \Phi(t - \tau), \qquad t \in J.$$

  (iii) $e^{At_1} e^{At_2} = e^{A(t_1 + t_2)}$ for all $t_1, t_2 \in J$.
  (iv) $A e^{At} = e^{At} A$ for all $t \in J$.
  (v) $(e^{At})^{-1} = e^{-At}$ for all $t \in J$.
  (vi) The unique solution ϕ of (L) with $\phi(\tau) = \xi$ is given by

$$\phi(t) = e^{A(t-\tau)} \xi. \tag{3.1}$$

Proof. By Definition 3.2, we have

$$e^{At} = E + \sum_{k=1}^{\infty} \frac{t^k}{k!} A^k$$

for any $t \in (-\infty, \infty)$. By Theorem 3.1 and the remarks following its proof, we may differentiate this series term by term to obtain

$$\frac{d}{dt}[e^{At}] = \lim_{N \to \infty} AS_N(t) = \lim_{N \to \infty} S_N(t)A = Ae^{At} = e^{At}A.$$

Thus, $\Phi(t) \triangleq e^{At}$ is a solution of the matrix equation

$$\Phi' = A\Phi.$$

Next, observe that $\Phi(0) = E$. It follows from Theorem 2.4 that

$$\det[e^{At}] = e^{\operatorname{trace}(At)} \neq 0 \qquad \text{for all} \quad t \in (-\infty, \infty).$$

Therefore, $\Phi(t) = e^{At}$ is a fundamental matrix for (L). This proves parts (i) and (iv).

In proving (iii), note that for any $t_1, t_2 \in R$ we have, in view of Theorem 2.12 (iii) that

$$\Phi(t_1, t_2) = \Phi(t_1, 0)\Phi(0, t_2).$$

By Theorem 2.12(i) we see that $\Phi(t, \tau)$ solves (L) with $\Phi(\tau, \tau) = E$. It was just proved that $\Psi(t) = e^{A(t-\tau)}$ is also a solution. By uniqueness, it follows that $\Phi(t, \tau) = e^{A(t-\tau)}$. For $t = t_1$, $\tau = -t_2$, we now obtain

$$e^{A(t_1 + t_2)} = \Phi(t_1, -t_2) = \Phi(t_1)\Phi(-t_2)^{-1},$$

and for $t = t_1$, $\tau = 0$, we have

$$\Phi(t_1, 0) = e^{At_1} = \Phi(t_1).$$

Also, for $t = 0$, $\tau = -t_2$, we get

$$\Phi(0, -t_2) = e^{t_2 A} = \Phi(-t_2)^{-1}.$$

Thus

$$e^{A(t_1 + t_2)} = e^{At_1} e^{At_2} \qquad \text{for all} \quad t_1, t_2 \in R.$$

To prove the second part, note that by (iii)

$$\Phi(t, \tau) \triangleq e^{A(t-\tau)} = E + \sum_{k=1}^{\infty} \frac{(t-\tau)^k}{k!} A^k = \Phi(t - \tau)$$

is a fundamental matrix for (L) with $\Phi(\tau, \tau) \triangleq \Phi(0) = E$. As such, it is a state transition matrix for (L).

3.3 Linear Systems with Constant Coefficients

Part (v) follows immediately by observing that

$$e^{At}e^{A(-t)} = e^{A(t-t)} = E.$$

To verify (vi), differentiate both sides of (3.1). Then

$$\phi'(t) = [e^{A(t-\tau)}]'\xi = Ae^{A(t-\tau)}\xi = A\phi(t),$$

and $\phi(\tau) = E\xi = \xi$. This shows that (3.1) is a solution of (L). It is unique by the results of Chapter 2. ∎

Notice that the solution (3.1) of (L) such that $\phi(\tau) = \xi$ depends on t and τ only via the difference $t - \tau$. This is the typical situation for *general* autonomous systems which satisfy uniqueness conditions. Indeed, if $\phi(t)$ is a solution of

$$x' = F(x), \qquad x(0) = \xi,$$

then clearly $\phi(t - \tau)$ will be a solution of

$$x' = F(x), \qquad x(\tau) = \xi.$$

Next, consider the "forced" system of equations

$$x' = Ax + g(t) \tag{3.2}$$

where $g: J \to R^n$ is continuous. Clearly, (3.2) is a special case of (LN). In view of Theorem 3.3 (vi) and Theorem 2.14, it follows that the solution of (3.2) is given by

$$\phi(t) = e^{A(t-\tau)}\phi(\tau) + \int_\tau^t e^{A(t-\eta)}g(\eta)\,d\eta$$

$$= e^{A(t-\tau)}\xi + e^{At}\int_\tau^t e^{-A\eta}g(\eta)\,d\eta. \tag{3.3}$$

While there is no general procedure for evaluating the state transition matrix for a time-varying matrix $A(t)$, there are several such procedures for determining e^{At} when $A(t) \equiv A$. We devote the remainder of this section to this problem and to solving (L) and (3.2).

We assume that the reader is familiar with the basics of Laplace transforms. If $f(t) = [f_1(t), \ldots, f_n(t)]^T$, where $f_i: [0, \infty) \to R$, $i = 1, \ldots, n$, and if each f_i is Laplace transformable, then we define the Laplace transform of the vector f componentwise, i.e.,

$$\hat{f}(s) = [\hat{f}_1(s), \ldots, \hat{f}_n(s)]^T,$$

where

$$\hat{f}_i(s) = \mathscr{L}[f_i(t)] \triangleq \int_0^\infty f_i(t)e^{-st}\,dt.$$

We define the Laplace transform of a matrix $C(t) = [c_{ij}(t)]$ similarly. Thus, if each $c_{ij}: [0, \infty) \to R$ and if each c_{ij} is Laplace transformable,

then the Laplace transform of $C(t)$ is defined by
$$\hat{C}(s) = \mathscr{L}[c_{ij}(t)] = [\mathscr{L}c_{ij}(t)] = [\hat{c}_{ij}(s)].$$

Now consider the system
$$x' = Ax, \quad x(0) = \xi. \tag{3.4}$$

Taking the Laplace transform of both sides of (3.4), we obtain
$$s\hat{x}(s) - \xi = A\hat{x}(s) \quad \text{or} \quad (sE - A)\hat{x}(s) = \xi$$
or
$$\hat{x}(s) = (sE - A)^{-1}\xi. \tag{3.5}$$

It can be shown by analytic continuation that $(sE - A)^{-1}$ exists for all s, except at the eigenvalues of A. Taking the inverse Laplace transform of (3.5), we obtain for the solution of (3.4),
$$\phi(t) = \mathscr{L}^{-1}[(sE - A)^{-1}]\xi = \Phi(t,0)\xi = e^{At}\xi. \tag{3.6}$$

It follows from (3.4) and (3.6) that
$$\hat{\Phi}(s) = (sE - A)^{-1} \tag{3.7}$$
and
$$\Phi(t,0) \triangleq \Phi(t) = \mathscr{L}^{-1}[(sE - A)^{-1}] = e^{At}. \tag{3.8}$$

Finally, note that when the initial time $\tau \neq 0$, we can immediately compute $\Phi(t,\tau) = \Phi(t - \tau) = e^{A(t-\tau)}$.

Example 3.4. For the initial value problem
$$x'_1 = -x_1 + x_2, \quad x_1(0) = 1,$$
$$x'_2 = -2x_2, \quad x_2(0) = 2,$$
we have
$$(sE - A) = \begin{bmatrix} s+1 & -1 \\ 0 & s+2 \end{bmatrix}$$
and
$$\hat{\Phi}(s) = (sE - A)^{-1} = \begin{bmatrix} \dfrac{1}{s+1} & \dfrac{1}{(s+1)(s+2)} \\ 0 & \dfrac{1}{s+2} \end{bmatrix} = \begin{bmatrix} \dfrac{1}{s+1} & \dfrac{1}{s+1} - \dfrac{1}{s+2} \\ 0 & \dfrac{1}{s+2} \end{bmatrix}$$
and
$$\Phi(t) = e^{At} = \mathscr{L}^{-1}[\hat{\Phi}(s)] = \begin{bmatrix} e^{-t} & e^{-t} - e^{-2t} \\ 0 & e^{-2t} \end{bmatrix}.$$

3.3 Linear Systems with Constant Coefficients

Therefore

$$\phi(t) = \begin{bmatrix} \phi_1(t) \\ \phi_2(t) \end{bmatrix} = \begin{bmatrix} e^{-t} & e^{-t} - e^{-2t} \\ 0 & e^{-2t} \end{bmatrix} \begin{bmatrix} 1 \\ 2 \end{bmatrix} = \begin{bmatrix} 3e^{-t} - 2e^{-2t} \\ 2e^{-2t} \end{bmatrix}.$$

Now if we replace the original initial conditions by $x_1(1) = 1$, $x_2(1) = 2$, then we obtain for the solution

$$\phi(t) = \begin{bmatrix} \phi_1(t) \\ \phi_2(t) \end{bmatrix} = \begin{bmatrix} e^{-(t-1)} & e^{-(t-1)} - e^{-2(t-1)} \\ 0 & e^{-2(t-1)} \end{bmatrix} \begin{bmatrix} 1 \\ 2 \end{bmatrix}$$

$$= \begin{bmatrix} 3e^{-(t-1)} - 2e^{-2(t-1)} \\ 2e^{-2(t-1)} \end{bmatrix}.$$

Next, let us consider a "forced" system of the form

$$x' = Ax + g(t), \quad x(0) = \xi, \tag{3.9}$$

and let us assume that the Laplace transform of g exists. Taking the Laplace transform of both sides of (3.9), yields

$$s\hat{x}(s) - \xi = A\hat{x}(s) + \hat{g}(s) \quad \text{or} \quad (sE - A)\hat{x}(s) = \xi + \hat{g}(s)$$

or

$$\hat{x}(s) = (sE - A)^{-1}\xi + (sE - A)^{-1}\hat{g}(s) = \hat{\Phi}(s)\xi + \hat{\Phi}(s)\hat{g}(s)$$
$$\triangleq \hat{\phi}_h(s) + \hat{\phi}_p(s). \tag{3.10}$$

Taking the inverse Laplace transform of both sides of (3.10) and using (3.3), we obtain

$$\phi(t) = \phi_h(t) + \phi_p(t) = \mathscr{L}^{-1}[(sE - A)^{-1}]\xi + \mathscr{L}^{-1}[(sE - A)^{-1}\hat{g}(s)]$$
$$= \Phi(t, 0)\xi + \int_0^t \Phi(t - \eta)g(\eta)\,d\eta.$$

Therefore,

$$\phi_p(t) = \int_0^t \Phi(t - \eta)g(\eta)\,d\eta,$$

as expected (i.e., convolution of Φ and g in the time domain corresponds to multiplication of $\hat{\Phi}$ and $\hat{g}$ in the s domain).

Example 3.5. Consider the "forced" system of equations

$$x_1' = -x_1 + x_2, \quad x_1(0) = -1,$$
$$x_2' = -2x_2 + u(t), \quad x_2(0) = 0,$$

where

$$u(t) = 1 \quad \text{for} \quad t > 0,$$
$$= 0 \quad \text{elsewhere.}$$

In view of Example 3.4, we have

$$\phi_h(t) = \begin{bmatrix} e^{-t} & e^{-t} - e^{-2t} \\ 0 & e^{-2t} \end{bmatrix} \begin{bmatrix} -1 \\ 0 \end{bmatrix} = \begin{bmatrix} -e^{-t} \\ 0 \end{bmatrix}.$$

Also, in view of Example 3.4, we have

$$\hat{\phi}_p(s) = \begin{bmatrix} \dfrac{1}{s+1} & \dfrac{1}{s+1} - \dfrac{1}{s+2} \\ 0 & \dfrac{1}{s+2} \end{bmatrix} \begin{bmatrix} 0 \\ \dfrac{1}{s} \end{bmatrix} = \begin{bmatrix} \dfrac{1}{2}\left(\dfrac{1}{s}\right) + \dfrac{1}{2}\left(\dfrac{1}{s+2}\right) - \dfrac{1}{s+1} \\ \dfrac{1}{2}\left(\dfrac{1}{s}\right) - \dfrac{1}{2}\left(\dfrac{1}{s+2}\right) \end{bmatrix}$$

and

$$\phi_p(t) = \begin{bmatrix} \tfrac{1}{2} + \tfrac{1}{2}e^{-2t} - e^{-t} \\ \tfrac{1}{2} - \tfrac{1}{2}e^{-2t} \end{bmatrix}.$$

Therefore,

$$\phi(t) = \phi_h(t) + \phi_p(t) = \begin{bmatrix} \tfrac{1}{2} - 2e^{-t} + \tfrac{1}{2}e^{-2t} \\ \tfrac{1}{2} - \tfrac{1}{2}e^{-2t} \end{bmatrix}.$$

We now present a second method of evaluating e^{At} and of solving initial value problems (L) and (3.2). This method involves the transformation of A into a Jordan canonical form.

Let us consider again the initial value problem (3.4). Let P be a real $n \times n$ nonsingular matrix, as in Section 1.C on the Jordan form, and consider the transformation $x = Py$ or equivalently, $y = P^{-1}x$. Differentiating both sides with respect to t, we obtain

$$y' = P^{-1}x' = P^{-1}APy = Jy, \qquad y(\tau) = P^{-1}\xi. \tag{3.11}$$

The solution of (3.11) is given as

$$\psi(t) = e^{J(t-\tau)}P^{-1}\xi. \tag{3.12}$$

Using (3.12) and $x = Py$, we obtain for the solution of (3.4),

$$\phi(t) = Pe^{J(t-\tau)}P^{-1}\xi. \tag{3.13}$$

Now let us first consider the case when A has n distinct eigenvalues $\lambda_1, \ldots, \lambda_n$. If we choose $P = [p_1, p_2, \ldots, p_n]$ in such a way that p_i is an eigenvector corresponding to the eigenvalue λ_i, $i = 1, \ldots, n$, then the matrix $J = P^{-1}AP$ assumes the form

$$J = \begin{bmatrix} \lambda_1 & & 0 \\ & \ddots & \\ 0 & & \lambda_n \end{bmatrix}.$$

3.3 Linear Systems with Constant Coefficients

Using the power series representation

$$e^{Jt} = E + \sum_{k=1}^{\infty} \frac{t^k J^k}{k!},$$

we immediately obtain the expression

$$e^{Jt} = \begin{bmatrix} e^{\lambda_1 t} & & 0 \\ & \ddots & \\ 0 & & e^{\lambda_n t} \end{bmatrix}. \tag{3.14}$$

We therefore have the following expression for the solution of (3.4):

$$\phi(t) = P \begin{bmatrix} e^{\lambda_1(t-\tau)} & & 0 \\ & \ddots & \\ 0 & & e^{\lambda_n(t-\tau)} \end{bmatrix} P^{-1} \xi. \tag{3.15}$$

In other words, in order to solve the initial value problem (3.4) by the present method (with all eigenvalues of A assumed to be distinct), one has to determine the eigenvalues of A, compute n eigenvectors corresponding to the eigenvalues, form the matrix P, and evaluate (3.15).

In the general case when A has repeated eigenvalues, it is no longer possible to diagonalize A (see Section 1). However, we can generate n linearly independent vectors $v_1, \ldots, v_n$ and an $n \times n$ matrix $P = [v_1, v_2, \ldots, v_n]$ which transforms A into the **Jordan canonical form**. Thus $J = P^{-1}AP$. Here J is block diagonal of the form

$$J = \begin{bmatrix} J_0 & & & 0 \\ & J_1 & & \\ & & \ddots & \\ 0 & & & J_s \end{bmatrix},$$

J_0 is a diagonal matrix with diagonal elements $\lambda_1, \ldots, \lambda_k$ (not necessarily distinct), and each J_i is an $n_i \times n_i$ matrix of the form

$$J_i = \begin{bmatrix} \lambda_{k+i} & 1 & 0 & \cdots & 0 \\ & \lambda_{k+i} & 1 & \cdots & 0 \\ \vdots & & \ddots & & \vdots \\ 0 & & & & 1 \\ 0 & 0 & 0 & \cdots & \lambda_{k+i} \end{bmatrix}$$

where λ_{k+i} need not be different from λ_{k+j} if $i \neq j$, and where $k + n_1 + \cdots + n_s = n$.

Since for any block diagonal matrix

$$C = \begin{bmatrix} C_1 & & 0 \\ & \ddots & \\ 0 & & C_l \end{bmatrix}$$

we have

$$C^k = \begin{bmatrix} C_1^k & & 0 \\ & \ddots & \\ 0 & & C_l^k \end{bmatrix}, \quad k = 0, 1, 2, \ldots,$$

it follows from the power series representation of e^{Jt} that

$$e^{Jt} = \begin{bmatrix} e^{J_0 t} & & & 0 \\ & e^{J_1 t} & & \\ & & \ddots & \\ 0 & & & e^{J_s t} \end{bmatrix}, \quad -\infty < t < \infty.$$

As before, we have

$$e^{J_0 t} = \begin{bmatrix} e^{\lambda_1 t} & & 0 \\ & \ddots & \\ 0 & & e^{\lambda_k t} \end{bmatrix}.$$

Notice that for any J_i, $i = 1, \ldots, s$, we have

$$J_i = \lambda_{k+i} E_i + N_i, \tag{3.16}$$

where E_i is the $n_i \times n_i$ identity matrix and N_i is the $n_i \times n_i$ nilpotent matrix given by

$$N_i = \begin{bmatrix} 0 & 1 & \cdots & 0 \\ \vdots & & \ddots & \vdots \\ & & & 1 \\ 0 & \cdot & \cdots & 0 \end{bmatrix}.$$

Since $\lambda_{k+i} E_i$ and N_i commute, we have

$$e^{J_i t} = e^{\lambda_{k+i} t} e^{N_i t}.$$

Repeated multiplication of N_i by itself shows that $N_i^k = 0$ for all $k \geq n_i$. Therefore, the series defining e^{tN_i} terminates, and we obtain

$$e^{tJ_i} = e^{\lambda_{k+i} t} \begin{bmatrix} 1 & t & \cdots & t^{n_i-1}/(n_i-1)! \\ 0 & 1 & \cdots & t^{n_i-2}/(n_i-2)! \\ \vdots & \vdots & & \vdots \\ 0 & \cdot & \cdots & 1 \end{bmatrix}, \quad i = 1, \ldots, s.$$

3.3 Linear Systems with Constant Coefficients

From (3.13) it now follows that the solution of (3.4) is given by

$$\phi(t) = P \begin{bmatrix} e^{J_0(t-\tau)} & & & 0 \\ & e^{J_1(t-\tau)} & & \\ & & \ddots & \\ 0 & & & e^{J_s(t-\tau)} \end{bmatrix} P^{-1}\xi. \quad (3.17)$$

Example 3.6. For the initial value problem of Example 3.4, we have

$$A = \begin{bmatrix} -1 & 1 \\ 0 & -2 \end{bmatrix}$$

with eigenvalues $\lambda_1 = -1$ and $\lambda_2 = -2$. A set of corresponding eigenvectors is

$$p_1 = \begin{bmatrix} 1 \\ 0 \end{bmatrix}, \quad p_2 = \begin{bmatrix} -1 \\ 1 \end{bmatrix}$$

and therefore $y = Px$ will diagonalize the equations. Here

$$P = \begin{bmatrix} 1 & -1 \\ 0 & 1 \end{bmatrix}, \quad P^{-1} = \begin{bmatrix} 1 & 1 \\ 0 & 1 \end{bmatrix}, \quad \text{and} \quad J = \begin{bmatrix} -1 & 0 \\ 0 & -2 \end{bmatrix}.$$

Using $x(0) = [x_1(0)\ x_2(0)]^T = [1, 2]^T$ as in Example 3.4, we have

$$\begin{bmatrix} \phi_1(t) \\ \phi_2(t) \end{bmatrix} = Pe^{Jt}P^{-1}\xi = \begin{bmatrix} 1 & -1 \\ 0 & 1 \end{bmatrix}\begin{bmatrix} e^{-t} & 0 \\ 0 & e^{-2t} \end{bmatrix}\begin{bmatrix} 1 & 1 \\ 0 & 1 \end{bmatrix}\begin{bmatrix} 1 \\ 2 \end{bmatrix}$$

$$= \begin{bmatrix} 3e^{-t} - 2e^{-2t} \\ 2e^{-2t} \end{bmatrix}.$$

This checks with the answer obtained in Example 3.4.

Example 3.7. Consider the initial value problem

$$x' = Ax,$$

where

$$A = \begin{bmatrix} -1 & 0 & -1 & 1 & 1 & 3 & 0 \\ 0 & 1 & 0 & 0 & 0 & 0 & 0 \\ 2 & 1 & 2 & -1 & -1 & -6 & 0 \\ -2 & 0 & -1 & 2 & 1 & 3 & 0 \\ 0 & 0 & 0 & 0 & 1 & 0 & 0 \\ 0 & 0 & 0 & 0 & 0 & 1 & 0 \\ -1 & -1 & 0 & 1 & 2 & 4 & 1 \end{bmatrix}.$$

Evaluating the characteristic polynomial of A we have $p(\lambda) = (1 - \lambda)^7$. Following the procedure outlined in Section 1 to generate the Jordan canonical form, we obtain

$$J = P^{-1}AP,$$

where

$$J = \begin{bmatrix} 1 & 1 & 0 & 0 & 0 & 0 & 0 \\ 0 & 1 & 1 & 0 & 0 & 0 & 0 \\ 0 & 0 & 1 & 0 & 0 & 0 & 0 \\ 0 & 0 & 0 & 1 & 1 & 0 & 0 \\ 0 & 0 & 0 & 0 & 1 & 0 & 0 \\ 0 & 0 & 0 & 0 & 0 & 1 & 0 \\ 0 & 0 & 0 & 0 & 0 & 0 & 1 \end{bmatrix} = \begin{bmatrix} J_1 & & & 0 \\ & J_2 & & \\ & & J_3 & \\ 0 & & & J_4 \end{bmatrix},$$

$$P = \begin{bmatrix} -1 & 0 & 0 & -2 & 1 & 0 & 1 \\ 0 & 0 & 1 & 0 & 0 & 0 & 3 \\ 1 & 1 & 0 & 2 & 0 & -1 & 1 \\ -1 & 0 & 0 & -2 & 0 & -2 & 0 \\ 0 & 0 & 0 & 0 & 0 & 1 & 0 \\ 0 & 0 & 0 & 0 & 0 & 0 & 1 \\ 0 & -1 & 0 & -1 & 0 & 0 & 0 \end{bmatrix},$$

and

$$P^{-1} = \begin{bmatrix} 0 & 0 & 2 & 1 & 4 & -2 & 2 \\ 0 & 0 & 1 & 1 & 3 & -1 & 0 \\ 0 & 1 & 0 & 0 & 0 & -3 & 0 \\ 0 & 0 & -1 & -1 & -3 & 1 & -1 \\ 1 & 0 & 0 & -1 & -2 & -1 & 0 \\ 0 & 0 & 0 & 0 & 1 & 0 & 0 \\ 0 & 0 & 0 & 0 & 0 & 1 & 0 \end{bmatrix}.$$

Then

$$e^{tJ_1} = e^t \begin{bmatrix} 1 & t & t^2/2 \\ 0 & 1 & t \\ 0 & 0 & 1 \end{bmatrix}, \qquad e^{tJ_2} = e^t \begin{bmatrix} 1 & t \\ 0 & 1 \end{bmatrix},$$

$$e^{tJ_3} = e^t, \qquad\qquad e^{tJ_4} = e^t,$$

3.3 Linear Systems with Constant Coefficients

and

$$e^{Jt} = \begin{bmatrix} e^{J_1 t} & & & 0 \\ & e^{J_2 t} & & \\ & & e^{J_3 t} & \\ 0 & & & e^{J_4 t} \end{bmatrix}.$$

Other methods of computing e^{At} motivated by such results as Sylvester's theorem or the Cayley–Hamilton theorem have been developed. We shall give a third method of computing e^{At} which illustrates the use of algebraic techniques. Let $\{\lambda_1, \lambda_2, \ldots, \lambda_n\}$ be an enumeration of the eigenvalues of A where the $\lambda_1, \ldots, \lambda_n$ need not be distinct. Define $A_i = A - \lambda_i E$ for $i = 1, \ldots, n$. Now we guess that e^{At} can be written in the form

$$e^{At} = \sum_{i=1}^{n} P_{i-1} w_i(t), \tag{3.18}$$

where $P_0 = E$, $P_{i+1} = A_{i+1} P_i$, and the w_i are scalar functions to be determined. Differentiating (3.18), we see that we need to choose the $w_i(t)$ so that

$$\sum_{i=1}^{n} w_i'(t) P_{i-1} = A \sum_{i=1}^{n} w_i(t) P_{i-1}.$$

By the definition of the P_i's and by the Cayley–Hamilton theorem, we have

$$A P_{i-1} = \lambda_i P_{i-1} + P_i, \qquad i = 2, 3, \ldots, n,$$

and

$$A P_{n-1} = \lambda_n P_{n-1}, \qquad P_n = 0.$$

Thus, we need to choose the $w_i(t)$ so that

$$\sum_{i=1}^{n} w_i'(t) P_{i-1} = \sum_{i=1}^{n} w_i(t)(\lambda_i P_{i-1} + P_i),$$

or

$$w_1' = \lambda_1 w_1, \qquad w_i' = \lambda_i w_i + w_{i-1} \qquad \text{for} \quad i = 2, 3, \ldots, n.$$

Also notice that if $w_1(0) = 1$ and $w_i(0) = 0$ for $i \geq 2$, then

$$\sum_{i=1}^{n} w_i(0) P_{i-1} = P_0 = E.$$

This determines the w_i precisely. Indeed,

$$w_1(t) = e^{\lambda_1 t}, \qquad w_i(t) = \int_0^t e^{\lambda_i(t-s)} w_{i-1}(s)\, ds$$

for $i = 2, 3, \ldots, n$.

3.4 LINEAR SYSTEMS WITH PERIODIC COEFFICIENTS

In the present section, we consider linear homogeneous systems

$$x' = A(t)x, \qquad -\infty < t < \infty, \tag{P}$$

where the elements of A are continuous functions on R and where

$$A(t) = A(t + T) \tag{4.1}$$

for some $T > 0$. System (P) is called a **periodic system** and T is called a **period** of A.

The proof of the main result of this section involves the concept of the logarithm of a matrix, which we introduce by means of the next result.

Theorem 4.1. Let B be a nonsingular $n \times n$ matrix. Then there exists an $n \times n$ matrix A, called the **logarithm** of B, such that

$$e^A = B. \tag{4.2}$$

Proof. Let $\tilde{B}$ be similar to B. Then there is a nonsingular matrix P such that $P^{-1}BP = \tilde{B}$. If $e^{\tilde{A}} = \tilde{B}$, then

$$B = P\tilde{B}P^{-1} = Pe^{\tilde{A}}P^{-1} = e^{P\tilde{A}P^{-1}}.$$

Hence, $P\tilde{A}P^{-1}$ is also a logarithm of B. Therefore, it is sufficient to prove (4.2) when B is in a suitable canonical form.

Let $\lambda_1, \ldots, \lambda_k$ be the distinct eigenvalues of B with multiplicities $n_1, \ldots, n_k$, respectively. From above we may assume that B is in the form

$$B = \begin{bmatrix} B_0 & & & \\ & B_1 & & \\ & & \ddots & \\ & & & B_k \end{bmatrix}.$$

Clearly $\log B_0$ is a diagonal matrix with diagonal elements equal to $\log \lambda_i$. If E_{n_j} denotes the $n_j \times n_j$ identity matrix, then $(B_j - \lambda_j E_{n_j})^{n_j} = 0, j = 1, \ldots, k$, and we may therefore write

$$B_j = \lambda_j \left(E_{n_j} + \frac{1}{\lambda_j} N_j \right), \qquad N_j^{n_j} = 0.$$

Note that $\lambda_j \neq 0$, since B is nonsingular. Next, using the power series expansion

$$\log(1 + x) = \sum_{p=1}^{\infty} \frac{(-1)^{p+1}}{p} x^p, \qquad |x| < 1,$$

3.4 Linear Systems with Periodic Coefficients

we formally write

$$A_j = \log B_j = E_{n_j} \log \lambda_j + \log\left(E_{n_j} + \frac{1}{\lambda_j} N_j\right)$$

$$= E_{n_j} \log \lambda_j + \sum_{p=1}^{\infty} \frac{(-1)^{p+1}}{p} \left(\frac{N_j}{\lambda_j}\right)^p.$$

Since $N_j^{n_j} = 0$, we actually have

$$A_j = E_{n_j} \log \lambda_j + \sum_{p=1}^{n_j-1} \frac{(-1)^{p+1}}{p} \left(\frac{N_j}{\lambda_j}\right)^p, \quad j = 1, \ldots, k. \tag{4.3}$$

Note that $\log \lambda_j$ is defined, since $\lambda_j \neq 0$. Recall that $e^{\log(1+x)} = 1 + x$. If we perform the same operations with matrices, we obtain the same terms, and there is no convergence problem, since the series (4.3) for $A_j = \log B_j$ terminates. Therefore we obtain

$$e^{A_j} = \exp(E_{n_j} \log \lambda_j) \exp\left[\sum_{p=1}^{n_j-1} \frac{(-1)^{p+1}}{p} \left(\frac{N_j}{\lambda_j}\right)^p\right]$$

$$= \lambda_j \left(E_{n_j} + \frac{N_j}{\lambda_j}\right) = B_j, \quad j = 1, \ldots, k.$$

Now let

$$A = \begin{bmatrix} A_0 & & \\ & \ddots & \\ & & A_k \end{bmatrix},$$

where A_j is defined in (4.3). We now have the desired result

$$e^A = \begin{bmatrix} e^{A_0} & & 0 \\ & \ddots & \\ 0 & & e^{A_k} \end{bmatrix} = \begin{bmatrix} B_0 & & 0 \\ & \ddots & \\ 0 & & B_k \end{bmatrix} = B. \quad \blacksquare$$

Clearly, A is not unique since for example $e^{A + 2\pi k i E} = e^A e^{2\pi k i} = e^A$ for all integers k.

Theorem 4.2. *Let* (4.1) *be true and let* $A \in C(-\infty, \infty)$. *If* $\Phi(t)$ *is a fundamental matrix for* (P), *then so is* $\Phi(t + T)$, $-\infty < t < \infty$. *Moreover, corresponding to every* Φ, *there exists a nonsingular matrix* P *which is also periodic with period* T *and a constant matrix* R, *such that*

$$\Phi(t) = P(t) e^{tR}.$$

Proof. Let $\Psi(t) = \Phi(t + T)$, $-\infty < t < \infty$. Since

$$\Phi'(t) = A(t) \Phi(t), \quad -\infty < t < \infty,$$

it follows that

$$\Psi'(t) = \Phi'(t + T) = A(t + T)\Phi(t + T) = A(t)\Phi(t + T), \quad -\infty < t < \infty.$$

Hence, Ψ is a solution of (P), indeed a fundamental matrix, since $\Phi(t + T)$ is nonsingular for all $t \in (-\infty, \infty)$. Therefore, there exists a constant nonsingular matrix C such that

$$\Phi(t + T) = \Phi(t)C$$

(by Theorem 2.8) and also a constant matrix R (by Theorem 4.1) such that

$$e^{T\dot{R}} = C.$$

Therefore

$$\Phi(t + T) = \Phi(t)e^{TR}. \tag{4.4}$$

Now define a matrix P by

$$P(t) = \Phi(t)e^{-tR}. \tag{4.5}$$

Using (4.4) and (4.5), we now obtain

$$P(t + T) = \Phi(t + T)e^{-(t+T)R} = \Phi(t)e^{TR}e^{-(t+T)R} = \Phi(t)e^{-tR} = P(t).$$

Therefore $P(t)$ is nonsingular for all $t \in (-\infty, \infty)$, and periodic. This concludes the proof. ∎

Now suppose that $\Phi(t)$ is known only over the interval $[t_0, t_0 + T]$. Since $\Phi(t + T) = \Phi(t)C$, we have by setting $t = t_0$, $C = \Phi(t_0)^{-1}\Phi(t_0 + T)$ and R is given by $T^{-1}\log C$. $P(t) = \Phi(t)e^{-tR}$ is now determined over $(0, T)$. However, $P(t)$ is periodic over $(-\infty, \infty)$. Therefore, $\Phi(t)$ is given over $(-\infty, \infty)$ by $\Phi(t) = P(t)e^{tR}$. In other words, Theorem 4.2 allows us to conclude that the determination of a fundamental matrix Φ for (P) over any interval of length T, leads at once to the determination of Φ over $(-\infty, \infty)$.

Next, let Φ_1 be any other fundamental matrix for (P) with $A(t + T) = A(t)$. Then $\Phi = \Phi_1 S$ for some constant nonsingular matrix S. Since $\Phi(t + T) = \Phi(t)e^{TR}$, we have $\Phi_1(t + T)S = \Phi_1(t)Se^{TR}$, or

$$\Phi_1(t + T) = \Phi_1(t)(Se^{TR}S^{-1}) = \Phi_1(t)e^{T(SRS^{-1})}. \tag{4.6}$$

Therefore, every fundamental matrix Φ_1 determines a matrix $Se^{TR}S^{-1}$ which is similar to the matrix e^{TR}.

Conversely, let S be any constant nonsingular matrix. Then there exists a fundamental matrix of (P) such that (4.6) holds. Thus, although Φ does not determine R uniquely, the set of all fundamental matrices of (P), and hence of A, determines uniquely all quantities associated with

3.4 Linear Systems with Periodic Coefficients

e^{TR} which are invariant under a similarity transformation. Specifically, the set of all fundamental matrices of A determine a unique set of eigenvalues of the matrix e^{TR}, $\lambda_1, \ldots, \lambda_n$, which are called the **multipliers associated with** A (or sometimes, the **Floquet multipliers** associated with A). None of these vanish since $\Pi \lambda_i = \det e^{TR} \neq 0$. Also, the eigenvalues of R are called the **characteristic exponents**.

Next, we let Q be a constant nonsingular matrix such that $J = Q^{-1}RQ$, where J is the Jordan canonical form of R, i.e.,

$$J = \begin{bmatrix} J_0 & 0 & \cdots & 0 \\ 0 & J_1 & & 0 \\ \vdots & & \ddots & \vdots \\ 0 & 0 & \cdots & J_s \end{bmatrix}.$$

Let $\Phi_1 = \Phi Q$ and $P_1 = PQ$. From Theorem 4.2 we have

$$\Phi_1(t) = P_1(t)e^{tJ} \quad \text{and} \quad P_1(t + T) = P_1(t). \tag{4.7}$$

Let the eigenvalues of R be $\rho_1, \ldots, \rho_n$. Then

$$e^{tJ} = \begin{bmatrix} e^{tJ_0} & 0 & \cdots & 0 \\ 0 & e^{tJ_1} & \cdots & 0 \\ \vdots & \vdots & & \vdots \\ 0 & 0 & \cdots & e^{tJ_s} \end{bmatrix},$$

where

$$e^{tJ_0} = \begin{bmatrix} e^{t\rho_1} & 0 & \cdots & 0 \\ 0 & e^{t\rho_2} & \cdots & 0 \\ \vdots & & & \vdots \\ 0 & & \cdots & e^{t\rho_q} \end{bmatrix}$$

and

$$e^{tJ_i} = e^{t\rho_{q+i}} \begin{bmatrix} 1 & t & \frac{t^2}{2} & \cdots & \frac{t^{r_i-1}}{(r_i-1)!} \\ 0 & 1 & t & \cdots & \frac{t^{r_i-2}}{(r_i-2)!} \\ \vdots & \vdots & \vdots & & \vdots \\ 0 & 0 & 0 & \cdots & 1 \end{bmatrix}, \quad i = 1, \ldots, s, q + \sum_{i=1}^{s} r_i = n.$$

Now $\lambda_i = e^{T\rho_i}$. Thus, even though the ρ_i are not uniquely determined, their real parts are. In view of (4.7), the columns $\phi_1, \ldots, \phi_n$ of Φ_1 are linearly independent solutions of (P). Let $p_1, \ldots, p_n$ denote the periodic column vectors of P_1. Then

$$\phi_1(t) = e^{t\rho_1} p_1(t),$$
$$\phi_2(t) = e^{t\rho_2} p_2(t),$$
$$\vdots$$
$$\phi_q(t) = e^{t\rho_q} p_q(t),$$
$$\phi_{q+1}(t) = e^{t\rho_{q+1}} p_{q+1}(t),$$
$$\phi_{q+2}(t) = e^{t\rho_{q+1}}(t p_{q+1}(t) + p_{q+2}(t)), \qquad (4.8)$$
$$\vdots$$
$$\phi_{q+r_1}(t) = \left(\frac{t^{r_1-1}}{(r_1-1)!} p_{q+1}(t) + \cdots + t p_{q+r_1-1}(t) + p_{q+r_1}(t)\right) e^{t\rho_{q+1}},$$
$$\vdots$$
$$\phi_{n-r_s+1}(t) = e^{t\rho_{q+s}} p_{n-r_s+1}(t),$$
$$\vdots$$
$$\phi_n(t) = e^{t\rho_{q+s}}\left(\frac{t^{r_s-1}}{(r_s-1)!} p_{n-r_s+1}(t) + \cdots + t p_{n-1}(t) + p_n(t)\right).$$

From (4.8) it is now clear that when Re $\rho_i \triangleq \alpha_i < 0$, or equivalently, when $|\lambda_i| < 1$, then there exists a $k > 0$ such that

$$|\phi_i(t)| \le k e^{\alpha_i t} \to 0 \qquad \text{as} \quad t \to +\infty.$$

In other words, if the eigenvalues ρ_i, $i = 1, \ldots, n$, of R have negative real parts, then the norm of any solution of (P) tends to zero as $t \to +\infty$ at an exponential rate.

From (4.5) it is easy to see that $AP - P' = PR$. Thus, for the transformation

$$x = P(t)y, \qquad (4.9)$$

we compute

$$x' = A(t)x = A(t)P(t)y = P'(t)y + P(t)y' = (P(t)y)'$$

or

$$y' = P^{-1}(t)(A(t)P(t) - P'(t))y = P^{-1}(t)(P(t)R)y = Ry.$$

This computation shows that the transformation (4.9) reduces the linear, homogeneous, periodic system (P) to

$$y' = Ry,$$

a linear homogeneous system with constant coefficients. Also note that even if $A(t)$ is real (so that C is real), the matrices $P(t)$ and R may be complex.

3.5 Linear nth Order Ordinary Differential Equations

However, if C is real, then C^2 does have a real logarithm (refer to the problems at the end of the chapter).

Now if (4.1) is true for T, then it is true for $2T$ and

$$\Phi(t + 2T) = \Phi(t)\Phi(T)^2 = \Phi(t)C^2.$$

By the previous results, there is a real, $2T$-periodic matrix $S(t)$ and a real constant matrix Q such that

$$\Phi(t) = S(t)e^{-Qt}.$$

Moreover, the real transformation $x = S(t)y$ reduces (P) to the real system

$$y' = Qy.$$

3.5 LINEAR nth ORDER ORDINARY DIFFERENTIAL EQUATIONS

In this section, we consider initial value problems described by linear nth order ordinary differential equations given by

$$y^{(n)} + a_{n-1}(t)y^{(n-1)} + \cdots + a_1(t)y^{(1)} + a_0(t)y = b(t), \quad (5.1)$$

$$y^{(n)} + a_{n-1}(t)y^{(n-1)} + \cdots + a_1(t)y^{(1)} + a_0(t)y = 0, \quad (5.2)$$

and

$$y^{(n)} + a_{n-1}y^{(n-1)} + \cdots + a_1 y^{(1)} + a_0 y = 0. \quad (5.3)$$

In (5.1) and (5.2) $a_k \in C(J)$ and $b \in C(J)$ and in (5.3), $a_k \in R$, $k = 0, 1, \ldots, n - 1$. If we define the linear differential operator L_n by

$$L_n = \frac{d^n}{dt^n} + a_{n-1}(t)\frac{d^{n-1}}{dt^{n-1}} + \cdots + a_1(t)\frac{d}{dt} + a_0(t), \quad (5.4)$$

then we can rewrite (5.1) and (5.2) more compactly as

$$L_n y = b(t) \quad (5.5)$$

and

$$L_n y = 0, \quad (5.6)$$

respectively. We can rewrite (5.3) similarly by defining a differential operator L in the obvious way.

Following the procedure given in Chapter 1, we can reduce the study of Eq. (5.2) to the study of the system of n first order ordinary differential equations

$$x' = A(t)x, \quad \text{(LH)}$$

where

$$A(t) = \begin{bmatrix} 0 & 1 & 0 & \cdots & 0 \\ 0 & 0 & 1 & \cdots & 0 \\ \vdots & \vdots & \vdots & & \vdots \\ 0 & 0 & 0 & \cdots & 1 \\ -a_0(t) & -a_1(t) & -a_2(t) & \cdots & -a_{n-1}(t) \end{bmatrix}. \qquad (5.7)$$

The matrix given in (5.7) is frequently called a **companion matrix** (or A is said to be in **companion form**).

Since $A(t)$ is continuous on J, we know that there exists a unique solution ϕ, for all $t \in J$, to the initial value problem

$$x' = A(t)x, \qquad x(\tau) = \xi, \qquad \tau \in J,$$

where $\xi = (\xi_1, \ldots, \xi_n)^T \in R^n$. The first component of this solution is a solution of $L_n y = 0$ satisfying $y(\tau) = \xi_1$, $y'(\tau) = \xi_2, \ldots, y^{(n-1)}(\tau) = \xi_n$.

Now let $\phi_1, \ldots, \phi_n$ be n solutions of (5.6). Then we can easily show that the matrix

$$\Phi(t) = \begin{bmatrix} \phi_1 & \phi_2 & \cdots & \phi_n \\ \phi_1' & \phi_2' & \cdots & \phi_n' \\ \vdots & \vdots & & \vdots \\ \phi_1^{(n-1)} & \phi_2^{(n-1)} & \cdots & \phi_n^{(n-1)} \end{bmatrix}$$

is a solution of the matrix equation

$$X' = A(t)X, \qquad (5.8)$$

where $A(t)$ is defined by (5.7). We call the determinant of Φ the **Wronskian** for (5.6) with respect to the solutions $\phi_1, \ldots, \phi_n$ and we denote it by

$$W(\phi_1, \ldots, \phi_n) = \begin{vmatrix} \phi_1 & \phi_2 & \cdots & \phi_n \\ \phi_1' & \phi_2' & \cdots & \phi_n' \\ \vdots & \vdots & & \vdots \\ \phi_1^{(n-1)} & \phi_2^{(n-1)} & \cdots & \phi_n^{(n-1)} \end{vmatrix}.$$

Note that $W(\phi_1, \ldots, \phi_n)(t)$ depends on $t \in J$. Since Φ is a solution of the matrix equation (5.8), then by Theorem 2.4 we have for any $\tau \in J$, and for any $t \in J$,

$$W(\phi_1, \ldots, \phi_n)(t) = \det \Phi(t) = \det \Phi(\tau) \exp\left[\int_\tau^t \operatorname{tr} A(s)\, ds\right]$$

$$= W(\phi_1, \ldots, \phi_n)(\tau) \exp\left\{\int_\tau^t -a_{n-1}(s)\, ds\right\}. \qquad (5.9)$$

3.5 Linear nth Order Ordinary Differential Equations

Before we state and prove our first result, we consider a specific example.

Example 5.1. Consider the second order differential equation

$$t^2 y'' + t y' - y = 0, \quad 0 < t < \infty,$$

which can equivalently be written as

$$y'' + (1/t)y' - (1/t^2)y = 0, \quad 0 < t < \infty. \quad (5.10)$$

The functions $\phi_1(t) = t$ and $\phi_2(t) = 1/t$ are clearly solutions of (5.10). We now form the matrix

$$\Phi(t) = \begin{bmatrix} \phi_1 & \phi_2 \\ \phi_1' & \phi_2' \end{bmatrix} = \begin{bmatrix} t & 1/t \\ 1 & -1/t^2 \end{bmatrix}.$$

This yields

$$W(\phi_1, \phi_2)(t) = \det \Phi(t) = -2/t, \quad t > 0.$$

In the notation of (5.7) we have $a_1(t) = 1/t$, $a_0(t) = -1/t^2$ and thus $a_1(s) = 1/s$. In view of (5.9), we have for any $\tau > 0$,

$$W(\phi_1, \phi_2)(t) = \det \Phi(t) = W(\phi_1, \phi_2)(\tau) \exp\left\{\int_\tau^t -a_1(s)\,ds\right\}$$

$$= -(2/\tau)e^{\ln(\tau/t)} = -2/t, \quad t > 0,$$

as expected.

Theorem 5.2. A set of n solutions of (5.6), $\phi_1, \ldots, \phi_n$, is linearly independent on J if and only if $W(\phi_1, \ldots, \phi_n)(t) \neq 0$ for all $t \in J$. Moreover, every solution of (5.6) is a linear combination of any set of n linearly independent solutions.

Proof. The first assertion is a restatement, for the nth order equation, of Theorem 2.5. The second assertion is a restatement of (2.4) in Theorem 2.12. ∎

Theorem 5.2 enables us to make the following definition.

Definition 5.3. A set of n linearly independent solutions of (5.6) on J, $\phi_1, \ldots, \phi_n$, is called a **fundamental set of solutions** for (5.6).

Next, we turn our attention to nonhomogeneous linear nth order ordinary differential equations (as we saw in Eq. (5.1)) of the form

$$y^{(n)} + a_{n-1}(t)y^{(n-1)} + \cdots + a_1(t)y^{(1)} + a_0(t)y = b(t).$$

As shown in Chapter 1, the study of (5.1) reduces to the study of the system of n first order ordinary differential equations

$$x' = A(t)x + g(t), \qquad (5.11)$$

where $A(t)$ is given by (5.7) and $g(t) = [0, \ldots, 0, b(t)]^T$. Recall that for given $\tau \in J$ and given $x(\tau) = \xi \in R^n$, Eq. (5.11) has a unique solution given by $\phi = \phi_h + \phi_p$ where $\phi_h(t) = \Phi(t, \tau)\xi$ is a solution of (LH), $\Phi(t, \tau)$ denotes the state transition matrix of $A(t)$, and ϕ_p is a particular solution of (5.11), given by

$$\phi_p(t) = \int_\tau^t \Phi(t, s) g(s)\, ds = \Phi(t) \int_\tau^t \Phi^{-1}(s) g(s)\, ds.$$

We now specialize this result from the n-dimensional system (5.11) to the nth order equation (5.1).

Theorem 5.4. If $\phi_1, \ldots, \phi_n$ is a fundamental set for the equation $L_n y = 0$, then the unique solution ψ of the equation $L_n y = b(t)$ satisfying $\psi(\tau) = \xi_1, \ldots, \psi^{(n-1)}(\tau) = \xi_n$ is given by

$$\psi(t) = \psi_h(t) + \psi_p(t)$$

$$= \psi_h(t) + \sum_{k=1}^n \phi_k(t) \int_\tau^t \frac{W_k(\phi_1, \ldots, \phi_n)(s)}{W(\phi_1, \ldots, \phi_n)(s)} b(s)\, ds.$$

Here ψ_h is the solution of $L_n y = 0$ such that $\psi(\tau) = \xi_1$, $\psi'(\tau) = \xi_2, \ldots$, $\psi^{(n-1)}(\tau) = \xi_n$ and $W_k(\phi_1, \ldots, \phi_n)(t)$ is obtained from $W(\phi_1, \ldots, \phi_n)(t)$ by replacing the kth column in $W(\phi_1, \ldots, \phi_n)(t)$ by $(0, \ldots, 0, 1)^T$.

Proof. From the foregoing discussion, the solution of (5.11) with $y(\tau) = 0$ is

$$\phi_p(t) = \Phi(t) \int_\tau^t \Phi^{-1}(s) g(s)\, ds = \int_\tau^t \Phi(t) \Phi^{-1}(s) g(s)\, ds$$

$$\triangleq \int_\tau^t \gamma(t, s) g(s)\, ds, \qquad g(t) = [0, \ldots, 0, b(t)]^T.$$

The first component of $\phi_p(t)$, which is the solution of $L_n y = b(t)$ with $\xi_1 = 0, \ldots, \xi_n = 0$, is $\int_\tau^t \gamma_{1n}(t, s) b(s)\, ds$, where

$$\gamma_{1n}(t, s) = \sum_{k=1}^n \Phi_{1k}(t) \Phi_{kn}^{-1}(s) = \sum_{k=1}^n \phi_k(t) \frac{\tilde{\phi}_{kn}(s)}{W(\phi_1, \ldots, \phi_n)(s)}.$$

Here $\tilde{\phi}_{kn}$ is the cofactor of the knth element of Φ^T, i.e., $\tilde{\phi}_{kn}$ is the cofactor of the element $\phi_k^{(n-1)}$ in Φ. Therefore,

$$W(\phi_1, \ldots, \phi_n)(s) \gamma_{1n}(t, s) = \sum_{k=1}^n \phi_k(t) W_k(\phi_1, \ldots, \phi_n)(s),$$

3.5 Linear nth Order Ordinary Differential Equations

where $W_k(\phi_1, \ldots, \phi_n)(s)$ is defined as in the statement of the theorem. Therefore, the solution ψ_p of $L_n x = b(t)$ satisfying $\psi(\tau) = 0, \ldots, \psi^{(n-1)}(\tau) = 0$, is given by

$$\psi_p(t) = \sum_{k=1}^{n} \phi_k(t) \int_{\tau}^{t} \left\{ \frac{W_k(\phi_1, \ldots, \phi_n)(s)}{W(\phi_1, \ldots, \phi_n)(s)} \right\} b(s)\, ds.$$

The conclusion of the theorem is now obvious. ∎

Example 5.5. Consider the second order ordinary differential equation

$$y'' + (1/t)y' - (y/t^2) = b(t), \quad 0 < t < \infty,$$

where b is a real continuous function for all $t > 0$. From Example 5.1 we have $\phi_1(t) = t$, $\phi_2(t) = 1/t$, and $W(\phi_1, \phi_2)(t) = -2/t$, $t > 0$. Also,

$$W_1(\phi_1, \phi_2)(t) = \begin{vmatrix} 0 & 1/t \\ 1 & -1/t^2 \end{vmatrix} = -\frac{1}{t}, \qquad W_2(\phi_1, \phi_2)(t) = \begin{vmatrix} t & 0 \\ 1 & 1 \end{vmatrix} = t.$$

From Theorem 5.4 we have

$$\psi(t) = \psi_h(t) + \psi_p(t) = \psi_h(t) + \phi_1(t) \int_{\tau}^{t} \frac{W_1(\phi_1, \phi_2)(s)}{W(\phi_1, \phi_2)(s)} b(s)\, ds$$

$$+ \phi_2(t) \int_{\tau}^{t} \frac{W_2(\phi_1, \phi_2)(s)}{W(\phi_1, \phi_2)(s)} b(s)\, ds$$

$$= \psi_h(t) + \frac{t}{2} \int_{\tau}^{t} b(s)\, ds - \frac{1}{2t} \int_{\tau}^{t} s^2 b(s)\, ds.$$

Next, we consider nth order ordinary differential equations with constant coefficients, as was seen in Eq. (5.3):

$$L_n y \triangleq y^{(n)} + a_{n-1} y^{(n-1)} + \cdots + a_1 y^{(1)} + a_0 y = 0.$$

Here we have $J = (-\infty, \infty)$. We call

$$p(\lambda) = \lambda^n + a_{n-1} \lambda^{n-1} + \cdots + a_1 \lambda + a_0 \tag{5.12}$$

the **characteristic polynomial** of the differential equation (5.3), and we call

$$p(\lambda) = 0 \tag{5.13}$$

the **characteristic equation** of (5.3). The roots of $p(\lambda)$ are called the **characteristic roots** of (5.3).

We see that the study of Eq. (5.3) reduces to the study of the system of first order ordinary differential equations with constant coefficients given by $x' = Ax$, where

$$A = \begin{bmatrix} 0 & 1 & 0 & 0 & \cdots & 0 \\ 0 & 0 & 1 & 0 & \cdots & 0 \\ \vdots & \vdots & \vdots & \vdots & & \vdots \\ -a_0 & -a_1 & -a_2 & -a_3 & \cdots & -a_{n-1} \end{bmatrix}. \tag{5.14}$$

The following result connects (5.3) and $x' = Ax$ with A given by (5.14).

Theorem 5.6. The characteristic polynomial of A in (5.14) is precisely the characteristic polynomial $p(\lambda)$ given by (5.12).

Proof. The proof is by induction. For $k = 1$ we have $A = -a_0$ and therefore $\det(\lambda E_1 - A) = \lambda + a_0$ and so (5.12) is true for $k = 1$. Assume now that the result is true for $k = n - 1$. Then

$$\det(\lambda E_n - A) = \begin{vmatrix} \lambda & -1 & 0 & \cdots & 0 & 0 \\ 0 & \lambda & -1 & \cdots & 0 & 0 \\ \vdots & \vdots & \vdots & & \vdots & \vdots \\ 0 & 0 & 0 & \cdots & \lambda & -1 \\ a_0 & a_1 & a_2 & \cdots & a_{n-2} & \lambda + a_{n-1} \end{vmatrix}$$

$$= \lambda \det(\lambda E_{n-1} - A_1) + (-1)^{n+1} a_0 \begin{vmatrix} -1 & 0 & 0 & \cdots & 0 & 0 \\ \lambda & -1 & 0 & \cdots & 0 & 0 \\ 0 & \lambda & -1 & \cdots & 0 & 0 \\ \vdots & \vdots & \vdots & & \vdots & \vdots \\ 0 & 0 & 0 & \cdots & \lambda & -1 \end{vmatrix}$$

where

$$A_1 = \begin{bmatrix} 0 & 1 & 0 & 0 & \cdots & 0 \\ 0 & 0 & 1 & 0 & \cdots & 0 \\ \vdots & \vdots & \vdots & \vdots & & \vdots \\ -a_1 & -a_2 & -a_3 & & \cdots & -a_{n-1} \end{bmatrix}.$$

Therefore

$$\begin{aligned}\det(\lambda E_n - A) &= \lambda \det(\lambda E_{n-1} - A_1) + a_0 \\ &= \lambda(\lambda^{n-1} + a_{n-1}\lambda^{n-2} + \cdots + a_1) + a_0 \\ &= \lambda^n + a_{n-1}\lambda^{n-1} + \cdots + a_1\lambda + a_0. \quad \blacksquare \end{aligned}$$

Our next result enumerates a fundamental set for (5.3).

3.5 Linear nth Order Ordinary Differential Equations

Theorem 5.7. Let $\lambda_1, \ldots, \lambda_s$ be the distinct roots of the characteristic equation
$$p(\lambda) = \lambda^n + a_{n-1}\lambda^{n-1} + \cdots + a_1\lambda + a_0 = 0$$
and suppose that λ_i has multiplicity m_i, $i = 1, \ldots, s$, with $\sum_{i=1}^{s} m_i = n$. Then the following set of functions is a fundamental set for (5.3):
$$t^k e^{\lambda_i t}, \qquad k = 0, 1, \ldots, m_i - 1, \quad i = 1, \ldots, s. \tag{5.15}$$

Example 5.8. If
$$p(\lambda) = (\lambda - 2)(\lambda + 3)^2(\lambda + i)(\lambda - i)(\lambda - 4)^4,$$
then $n = 9$ and e^{2t}, e^{-3t}, te^{-3t}, e^{-it}, e^{+it}, e^{4t}, te^{4t}, $t^2 e^{4t}$, and $t^3 e^{4t}$ is a fundamental set for the differential equation corresponding to the characteristic equation.

Proof of Theorem 5.7. First we note that for the function $e^{\lambda t}$ we have $L_n(e^{\lambda t}) = p(\lambda)e^{\lambda t}$. Next, we observe that for the function $t^k e^{\lambda t}$ we have
$$L_n(t^k e^{\lambda t}) = L_n\left(\frac{\partial^k}{\partial \lambda^k} e^{\lambda t}\right) = \frac{\partial^k}{\partial \lambda^k} L_n(e^{\lambda t}) = \frac{\partial^k}{\partial \lambda^k}\left[p(\lambda)e^{\lambda t}\right]$$
$$= p^{(k)}(\lambda)e^{\lambda t} + kp^{(k-1)}(\lambda)te^{\lambda t} + \cdots + kp'(\lambda)t^{k-1}e^{\lambda t} + p(\lambda)t^k e^{\lambda t}.$$
Now let $\lambda = \lambda_i$. Then we have
$$L_n(t^k e^{\lambda_i t}) = 0 \qquad \text{for} \quad k = 0, 1, \ldots, m-1, \quad i = 1, \ldots, s. \tag{5.16}$$
Here we have used the fact that p is a polynomial and λ_i is a root of p of multiplicity m_i so that $p^{(k)}(\lambda_i) = 0$ for $0 \leq k \leq m_i - 1$. We have therefore shown that the functions (5.15) are indeed solutions of (5.3). We now must show that they are linearly independent.

Suppose the functions in (5.15) are not linearly independent. Then there exist constants c_{ik}, not all zero, such that
$$\sum_{i=1}^{s} \sum_{k=0}^{m_i - 1} c_{ik} t^k e^{\lambda_i t} = 0 \qquad \text{for all} \quad t \in (-\infty, \infty).$$
Thus
$$\sum_{i=1}^{\sigma} P_i(t) e^{\lambda_i t} = 0,$$
where the $P_i(t)$ are polynomials, and $\sigma \leq s$ is chosen so that $P_\sigma \not\equiv 0$ while $P_{\sigma+i}(t) \equiv 0$, $i \geq 1$. Now divide the preceding expression by $e^{\lambda_1 t}$ and obtain
$$P_1(t) + \sum_{i=2}^{\sigma} P_i(t) e^{(\lambda_i - \lambda_1)t} = 0.$$

Now differentiate this expression enough times so that the polynomial $P_1(t)$ becomes zero. This yields

$$\sum_{i=2}^{\sigma} Q_i(t)e^{\lambda_i t} = 0,$$

where $Q_i(t)$ has the same degree as $P_i(t)$ for $i \geq 2$. Continuing in this manner, we ultimately obtain a polynomial $F_\sigma(t)$ such that

$$F_\sigma(t)e^{\lambda_\sigma t} = 0,$$

where the degree of F_σ is equal to the degree of $P_\sigma(t)$. But this means that $F_\sigma(t) \equiv 0$. This is impossible, since a nonzero polynomial can vanish only at isolated points. Therefore, the indicated solutions must be linearly independent. ∎

Consider again the time varying linear operator L_n defined in (5.4). Corresponding to L_n we define a second linear operator L_n^+ of order n, which we call the **adjoint** of L_n, as follows. The domain of L_n^+ is the set of all continuous functions defined on J such that $[\bar{a}_j(t)y(t)]$ has j continuous derivatives on J. For each function y, define

$$L_n^+ y = (-1)^n y^{(n)} + (-1)^{n-1}(\bar{a}_{n-1}y)^{(n-1)} + \cdots + (-1)(\bar{a}_1 y)' + \bar{a}_0 y.$$

The equation

$$L_n^+ y = 0, \qquad t \in J,$$

is called the **adjoint equation** to $L_n y = 0$.

When (5.6) is written in companion form (LH) with $A(t)$ given by (5.7), then the adjoint system is $z' = -A^*(t)z$, where

$$A^*(t) = \begin{bmatrix} 0 & 0 & \cdots & 0 & -\bar{a}_0(t) \\ 1 & 0 & \cdots & 0 & -\bar{a}_1(t) \\ 0 & 1 & \cdots & 0 & -\bar{a}_2(t) \\ \vdots & \vdots & & \vdots & \vdots \\ 0 & 0 & \cdots & 1 & -\bar{a}_{n-1}(t) \end{bmatrix}.$$

This adjoint system can be written in component form as

$$\begin{aligned} z_1' &= \bar{a}_0(t)z_n, \\ z_j' &= -z_{j-1} + \bar{a}_{j-1}(t)z_n \qquad (2 \leq j \leq n). \end{aligned} \qquad (5.17)$$

If $\psi = [\psi_1, \psi_2, \ldots, \psi_n]^T$ is a solution of (5.17) and if $a_j \psi_n$ has j derivatives, then

$$\psi_n' - (\bar{a}_{n-1}\psi_n) = -\psi_{n-1}$$

and

$$\psi_n'' - (\bar{a}_{n-1}\psi_n)' = -\psi_{n-1}' = \psi_{n-2} - (\bar{a}_{n-2}\psi_n)$$

or
$$\psi_n'' - (\bar{a}_{n-1}\psi_n)' + (\bar{a}_{n-2}\psi_n) = \psi_{n-2}.$$

Continuing in this manner, we see that ψ_n solves $L_n^+ \psi = 0$.

The operators L_n and L_n^+ satisfy the following interesting identity called the **Lagrange Identity**.

Theorem 5.9. If a_k are real valued and $a_k \in C^k(J)$ for $k = 0, 1, \ldots, n-1$ and if u and $v \in C^n(J)$, then
$$\bar{v}L_n u - u\overline{L_n^+ v} = P(u, v)',$$
where $P(x, z)$ represents the **bilinear concomitant**
$$P(u, v) = \sum_{k=1}^{n} \sum_{j=0}^{k-1} (-1)^j u^{(k-j-1)} (a_k \bar{v})^{(j)}.$$

Proof. For $k = 0, 1, 2, \ldots, n-1$ and for any pair of smooth functions f and g, the Leibnitz formula yields
$$\left(\sum_{j=0}^{k} (-1)^j f^{(k-j)} g^{(j)} \right)' = f^{(k+1)} g + (-1)^k f g^{(k+1)}.$$

This and the definitions of L_n and L_n^+ give
$$\bar{v}L_n u - u\overline{L_n^+ v} = \sum_{k=1}^{n} (\bar{v} a_k) u^{(k)} + (-1)^{k+1} (a_k \bar{v})^{(k)} u$$
$$= \sum_{k=0}^{n-1} \left[\sum_{j=0}^{k} (-1)^j u^{(k-j-1)} (a_k \bar{v})^{(j)} \right]' = [P(u,v)]'. \quad \blacksquare$$

An immediate consequence of Theorem 5.9 is **Green's formula**.

Theorem 5.10. If a_k are real C^k functions on J for $0 \leq k \leq n-1$ and if u and $v \in C^n(J)$, then for any t and τ in J,
$$\int_\tau^t (\bar{v}L_n u - u\overline{L_n^+ v}) \, ds = P(u, v)|_\tau^t.$$

Proof. Integrate the Lagrange identity from τ to t. $\blacksquare$

3.6 OSCILLATION THEORY

In this section, we apply some of the theory developed in the foregoing sections to the study of certain oscillation properties of second order linear systems of the form
$$y'' + a_1(t) y' + a_2(t) y = 0, \tag{6.1}$$

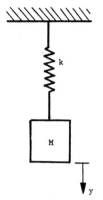

FIGURE 3.1 Linear mass–spring system.

where a_1 and a_2 are real valued functions in $C(J)$. Our study is motivated by the linear mass–spring system depicted in Fig. 3.1 and described by

$$my'' + ky = 0, \tag{6.2}$$

where m and k are positive constants. The general solution of (6.2) is $y = A\cos(\omega t + B)$ where $\omega^2 = k/m$ and where A and B are arbitrary constants. Note that for the solutions $y_1 = A_1\cos(\omega t + B_1)$ and $y_2 = A_2\cos(\omega t + B_2)$ with $A_1 \neq 0$, $A_2 \neq 0$, and $B_1 \neq B_2$ the consecutive zeros of the solutions are interlaced, i.e., they alternate along the real line. Also note that the number of zeros in any finite interval will increase with k and decrease with m, i.e., the frequency of oscillation of nontrivial solutions of (6.2) is higher for stiffer springs and higher for smaller masses. Our objective will be to generalize these results to general second order equations.

Note that if (6.1) is multiplied by

$$k(t) = \exp\left(\int_c^t a_1(s)\,ds\right),$$

then (6.1) reduces to

$$(k(t)y')' + g(t)y = 0, \tag{6.3}$$

where $g(t) = k(t)a_2(t)$. This form of the second order equation has the advantage that (for C^2 smooth k) the operator L and its adjoint L^+,

$$Lu = ku'' + k'u' + gu, \qquad L^+v = (kv)'' - (k'v)' + gv,$$

are the same. Also note that for (6.3), the identity (5.9) reduces to

$$W(\phi_1,\phi_2)(t) = W(\phi_1,\phi_2)(\tau)\exp\int_\tau^t (k'(u)/k(u))\,du$$

3.6 Oscillation Theory

or
$$k(t)W(\phi_1,\phi_2)(t) = k(\tau)W(\phi_1,\phi_2)(\tau) \tag{6.4}$$

for all $t \in J$, any fixed τ in J and all pairs $\{\phi_1,\phi_2\}$ of solutions of (6.3).

Note also that if ϕ solves (6.3) and if $\phi(t) \not\equiv 0$ on J, then any zero t_1 of ϕ must be a simple zero. Assume that this is not the case, i.e., assume $y(t_1) = 0$ and also $y'(t_1) = 0$. But the unique solution of (6.3) for these initial data at time t_1 is $\phi(t) \equiv 0$. Since $\phi(t) \not\equiv 0$, then $y(t_1)$ and $y'(t_1)$ cannot simultaneously vanish.

We can now state and prove our main oscillation results.

Theorem 6.1. Let ϕ_1 be a nonconstant solution of (6.3) on $J = (a,b)$ with consecutive zeros at points t_1 and t_2 of J and $t_1 < t_2$. If ϕ_2 is a second solution of (6.3) on J, then either $\phi_2 = c\phi_1$ for some constant c or else ϕ_2 has one and only one zero in the interval (t_1,t_2).

Proof. There is a constant c with $\phi_2 = c\phi_1$ if and only if the Wronskian $W(\phi_1,\phi_2) \equiv 0$. If $W(\phi_1,\phi_2) \not\equiv 0$, then by (6.4) we have

$$-k(t_1)\phi_1'(t_1)\phi_2(t_2) = k(t_1)W(\phi_1,\phi_2)(t_1)$$
$$= k(t_2)W(\phi_1,\phi_2)(t_2) = -k(t_2)\phi_1'(t_2)\phi_2(t_2).$$

Since $\phi_1(t)$ is of one sign on (t_1,t_2), then $\phi_1'(t_1)\phi_1'(t_2) < 0$. Also, since $k(t) > 0$, we see that $\phi_2(t_1)\phi_2(t_2) < 0$. Hence, ϕ_2 has at least one zero in (t_1,t_2). If ϕ_2 had two or more zeros there, then by reversing the roles of ϕ_1 and ϕ_2 we would see that ϕ_1 had a zero in (t_1,t_2). This is impossible. Hence, the zero of ϕ_2 is unique. ∎

Theorem 6.2. Let $k \in C^1(a,b)$ and $g_i \in C(a,b)$ for $i = 1, 2$ with $g_1(t) < g_2(t)$ and $k(t) > 0$ on $J = (a,b)$. Let ϕ_i be a solution of

$$(k(t)y')' + g_i(t)y = 0, \qquad t \in J,$$

for $i = 1, 2$. If t_1 and t_2 are consecutive zeros of ϕ_1 on J, then ϕ_2 must vanish at some point t_3 between t_1 and t_2.

Proof. For purposes of contradiction, suppose that $\phi_2(t)$ is never zero on (t_1,t_2). Without loss of generality we may assume that both ϕ_1 and ϕ_2 are positive on this interval. Multiplying the equation for ϕ_i by ϕ_j and subtracting the two resulting equations, we obtain

$$(k\phi_1')'\phi_2 - (k\phi_2')'\phi_1 = (g_2 - g_1)\phi_1\phi_2.$$

Since the term on the left is $[k(\phi_1'\phi_2 - \phi_2'\phi_1)]'$ and since $\phi_1(t_1) = \phi_1(t_2) = 0$, then on integrating we have

$$\int_{t_1}^{t_2} [g_2(s) - g_1(s)]\phi_1(s)\phi_2(s)\,ds = k(s)[\phi_1'(s)\phi_2(s) - \phi_2'(s)\phi_1(s)]\Big|_{t_1}^{t_2}$$

and
$$k(t_2)\phi_1'(t_2)\phi_2(t_2) - k(t_1)\phi_1'(t_1)\phi_2(t_1) > 0.$$
But $\phi_2(t_2) \geq 0$ and $\phi_1'(t_2) \leq 0$ while $\phi_2(t_1) \geq 0$ and $\phi_1'(t_1) \geq 0$. Thus,
$$0 \leq k(t_1)\phi_1'(t_1)\phi_2(t_1) < k(t_2)\phi_1'(t_2)\phi_2(t_2) \leq 0,$$
a contradiction. ∎

For example, if $k(t) \equiv k_0 > 0$ and $0 < g_0 \leq g(t) \leq g_1$, then consecutive zeros t_1 and t_2 of
$$k_0 y'' + g(t)y = 0$$
must satisfy $\pi(k_0/g_1)^{1/2} \leq t_2 - t_1 \leq \pi(k_0/g_0)^{1/2}$. This is seen by comparison with the constant coefficient equations
$$k_0 y'' + g_i y = 0$$
whose solutions are easy to compute.

When $k(t)$ is also allowed to vary, a somewhat different analysis is needed.

Theorem 6.3 (*Sturm–Picone*). Let $g_i \in C(J)$, $k_i \in C^1(J)$ with $g_1 < g_2$ and $k_1 > k_2 > 0$ on $\bar{J}$. Let ϕ_i be a solution on J of
$$(k_i y')' + g_i y = 0$$
and let t_1 and t_2 be two consecutive zeros of ϕ_1. Then ϕ_2 has at least one zero in the interval (t_1, t_2).

Proof. The proof is by contradiction. So assume that ϕ_2 has no zero in the interval (t_1, t_2) and without loss of generality, assume that $\phi_2(t)$ is positive on this interval. Compute
$$\begin{aligned}\{(\phi_1/\phi_2)(k_1\phi_1'\phi_2 - k_2\phi_1\phi_2')\}' \\
= \phi_1(k_1\phi_1')' - (\phi_1^2/\phi_2)(k_2\phi_2')' + (\phi_1/\phi_2)(k_1 - k_2)\phi_1'\phi_2' \\
- (\phi_1\phi_2' - \phi_1'\phi_2)(k_1\phi_1'\phi_2 - k_2\phi_1\phi_2')/\phi_2^2 \\
= (g_2 - g_1)\phi_1^2 + (k_1 - k_2)(\phi_1')^2 + k_2(\phi_1'\phi_2 - \phi_1\phi_2')^2/\phi_2^2.\end{aligned}$$

The last term is defined and continuous at the endpoints t_1 and t_2 of the interval if $\phi_2(t) \neq 0$ at $t = t_1$ and $t = t_2$. If $\phi_2(t)$ is zero at t_1, then $\phi_2'(t_1) \neq 0$ and by l'Hospital's rule we have
$$\frac{\phi_1'(t)\phi_2(t) - \phi_1(t)\phi_2'(t)}{\phi_2(t)} \to \phi_1'(t_1) - \phi_1'(t_1) = 0$$

3.6 Oscillation Theory

as $t \to t_1^+$. Similarly, if $\phi_2(t_2) = 0$, then

$$\lim(\phi_1(t)/\phi_2(t)) = \lim(\phi_1'(t)/\phi_2'(t)) \neq 0 \qquad (t \to t_2^+).$$

So

$$\frac{\phi_1(t)}{\phi_2(t)}(k_1(t)\phi_1'(t)\phi_2(t) - k_2(t)\phi_1(t)\phi_2'(t)) \to 0.$$

In any case, the third term in the last equation above is integrable. Integrating from t_1 to t_2, we have

$$(\phi_1/\phi_2)(k_1\phi_1'\phi_2 - k_2\phi_2'\phi_1)|_{t_1}^{t_2}$$
$$= \int_{t_1}^{t_2} \{(g_2 - g_1)\phi_1^2 + (k_1 - k_2)(\phi_1')^2 + k_2(\phi_1'\phi_2 - \phi_1\phi_2')^2/\phi_2^2\} \, ds. \quad (6.5)$$

Since $\phi_1(t_1) = \phi_1(t_2) = 0$, then the terms on the left are zero while the integral on the right is positive. This is a contradiction. ∎

Note that the conclusion of the above theorem remains true if $k_1 \geq k_2$, $g_2 \geq g_1$, and at least one of the $g_2 - g_1$ and $k_1 - k_2$ is not identically zero on any subinterval. The same proof as in Theorem 6.3 works in this case.

Corollary 6.4. Let $g \in C(J)$ and $k \in C^1(J)$ with g increasing and k positive and decreasing in $t \in J$. Let ϕ be a solution of (6.3) on J with consecutive zeros at points t_i, where

$$a < t_1 < t_2 < \cdots < t_n < b.$$

Then

$$t_2 - t_1 > t_3 - t_2 > \cdots > t_n - t_{n-1}.$$

Proof. Let $g_1(t) = g(t + t_j)$, $g_2(t) = g(t + t_{j+1})$ with k_1 and k_2 defined similarly. Apply now Theorem 6.3 with $\phi_1(t) = \phi(t + t_j)$ and $\phi_2(t) = \phi(t + t_{j+1})$ to complete the proof. ∎

We note that in the above result the terms "increasing" and "decreasing" can be weakened. If g is nondecreasing and k is nonincreasing, then the inequalities in the conclusion of the above corollary are no longer strict. But if at least one of the g and k is strictly monotone, then the original conclusion still holds.

Example 6.5. Consider a nontrivial solution on $0 < t < \infty$ of

$$y'' + (1 \pm (B/t)^2)y = 0, \qquad B > 0.$$

If the plus sign is used, then $(t_{j+1} - t_j)$ is decreasing as $j \to \infty$ with limit π. If the minus sign is used, then $(t_{j+1} - t_j)$ increases to π as $j \to \infty$.

PROBLEMS

1. Using the Jordan canonical form compute the solution of
$$x' = \begin{bmatrix} 1 & -4 & -1 \\ -4 & -2 & 4 \\ -1 & 4 & 1 \end{bmatrix} x, \qquad x(\tau) = \xi.$$
What can you say in general about how one computes e^{At} when A is an $n \times n$ self adjoint matrix?

2. (a) Compute e^{At} when
$$A = \begin{bmatrix} 3 & 1 \\ -1 & 1 \end{bmatrix}$$
    (i) by first computing the Jordan canonical form of A;
    (ii) by first computing P_{i-1} and $w_i(t)$ as in (3.18).
  (b) Repeat the above problem when
$$A = \begin{bmatrix} 3 & 2 & 1 \\ -1 & 3 & 2 \\ 1 & -3 & -2 \end{bmatrix}.$$

3. Let A be a constant $n \times n$ matrix. Define $\sigma = \max\{\text{Re } \lambda : \lambda \text{ is an eigenvalue of } A\}$. Show that for any $\varepsilon > 0$ there is a K such that
$$|e^{At}| \le Ke^{(\sigma + \varepsilon)t}$$
for all $t \ge 0$. Show by example that in general it is not possible to find a K which works when $\varepsilon = 0$.

4. Show that if A and B are two constant $n \times n$ matrices which commute, then $e^{(A+B)t} = e^{At}e^{Bt}$.

5. Suppose for a given continuous function $f(t)$ the equation
$$x' = \begin{bmatrix} -5 & 2 \\ -4 & 1 \end{bmatrix} x + f(t)$$
has at least one solution $\phi_p(t)$ which satisfies
$$\sup\{|\phi(t)| : \tau \le t < \infty\} < \infty.$$
Show that all solutions satisfy this boundedness condition. State and prove a generalization of this result to the n-dimensional system (3.2).

Problems

6. Let A be a continuous $n \times n$ matrix such that the system (LH) has a uniformly bounded fundamental matrix $\Phi(t)$ over $0 \leq t < \infty$.

   (i) Show that all fundamental matrices are bounded on $[0, \infty)$.

   (ii) Show that if
   $$\lim_{t \to \infty} \inf \operatorname{Re}\left[\int_0^t \operatorname{tr} A(s)\, ds\right] > -\infty,$$
   then $\Phi^{-1}(t)$ is also uniformly bounded on $[0, \infty)$. *Hint*: Use Theorem 2.4.

   (iii) Show that if the adjoint (2.6) has a fundamental matrix $\Psi(t)$ which is uniformly bounded, then for any $\xi \neq 0$ and any $\tau \in R$ the solution of (LH) satisfying $x(\tau) = \xi$ cannot tend to zero as $t \to \infty$.

7. Show that if $a(t)$ is a bounded function, if $a \in C[0, \infty)$ and if $\phi(t)$ is a nontrivial solution of
   $$y'' + a(t)y = 0 \tag{7.1}$$
   satisfying $\phi(t) \to 0$ as $t \to \infty$, then (7.1) has a solution which is not bounded over $[0, \infty)$.

8. Let g be a bounded continuous function on $(-\infty, \infty)$ and let B and $-C$ be square matrices of dimensions k and $n - k$ all of whose eigenvalues have negative real parts. Let
   $$A = \begin{bmatrix} B & 0 \\ 0 & C \end{bmatrix} \quad \text{and} \quad g(t) = \begin{bmatrix} g_1(t) \\ g_2(t) \end{bmatrix},$$
   so that (3.2) is equivalent to
   $$x_1' = Bx_1 + g_1(t), \qquad x_2' = Cx_2 + g_2(t).$$
   Show that the functions
   $$\phi_1(t) = \int_{-\infty}^{t} e^{B(t-s)} g_1(s)\, ds, \qquad \phi_2(t) = -\int_{t}^{\infty} e^{C(t-s)} g_2(s)\, ds$$
   are defined for all $t \in R$ and determine a solution of (3.2).

9. Show that if g is a bounded continuous function on R^1 and if A has no eigenvalues with zero real part, then (3.2) has at least one bounded solution. *Hint*: Use Problem 8.

10. Let $A(t)$ and $B(t)$ be $C[0, \infty)$ with $\int_0^\infty |B(t)|\, dt < \infty$. Let (LH) and its adjoint (2.6) both have bounded fundamental matrices over $[0, \infty)$. Show that all solutions of
    $$y' = A(t)y + B(t)y \tag{7.2}$$

are bounded over $[0, \infty)$. *Hint*: First show that any solution of (7.2) also satisfies
$$y(t) = x(t) + \int_\tau^t \Phi(t, s)B(s)y(s)\,ds,$$
where Φ is the state transition matrix for (LH). Then use the Gronwall inequality, cf. Chapter 2, problems.

11. In Problem 10 show that given any solution $x(t)$ of (LH) there is a unique solution $y(t)$ of (7.2) such that
$$\lim_{t\to\infty}\,[x(t) - y(t)] = 0. \tag{7.3}$$
Hint: Try $y(t) = x(t) - \int_t^\infty \Phi(t, s)B(s)y(s)\,ds$ on $\alpha \le t < \infty$ and α large.

12. In Problem 11 show that given any solution $y(t)$ of (7.2) there is a unique solution $x(t)$ of (LH) such that (7.3) holds.

13. Let $\omega > 0$, $b \in C[0, \infty)$ and $\int_0^\infty |b(t)|\,dt < \infty$. Show that $y'' + (\omega^2 + b(t))y = 0$ has a solution ϕ such that
$$[\phi(t) - \sin \omega t]^2 + [\phi'(t) - \omega \cos \omega t]^2 \to 0$$
as $t \to \infty$.

14. Let $A = P^{-1}JP$ be an $n \times n$ constant matrix whose Jordan form J is diagonal. Let $B(t) \in C[0, \infty)$ with $\int_0^\infty |B(t)|\,dt < \infty$. Let λ and v be an eigenvalue and corresponding eigenvector for A, i.e., $Av = \lambda v$, $|v| \ne 0$. Show that
$$x' = Ax + B(t)x$$
has a solution $\phi(t)$ such that $e^{-\lambda t}\phi(t) \to v$ as $t \to \infty$. *Hint*: For α large, use successive approximations on the integral equation
$$\phi(t) = e^{\lambda t}v + \int_\alpha^t X_1(t - s)B(s)\phi(s)\,ds - \int_t^\infty X_2(t - s)B(s)\phi(s)\,ds$$
for $\alpha \le t < \infty$. The matrices X_i are chosen as $X_i(t) = Pe^{J_i t}P^{-1}$, where $J = J_1 + J_2$, J_1 contains all eigenvalues of J with real parts less than Re λ, and J_2 contains all other eigenvalues of J.

15. Let $g \in C[0, \infty)$ with $\int_1^\infty t|g(t)|\,dt < \infty$. Show that $y'' + g(t)y = 0$ has solutions $\phi_1(t)$ and $\phi_2(t)$ such that
$$\phi_1(t) \to 1, \qquad \phi_1'(t) \to 0, \qquad \phi_2(t)/t \to 1, \qquad \phi_2'(t) \to 1$$
as $t \to \infty$. *Hint*: Use successive approximations to prove that the following integral equations have bounded solutions over $\alpha \le t < \infty$,
$$y_1(t) = 1 + \int_t^\infty (t - s)g(s)y_1(s)\,ds,$$
$$y_2(t) = t + \int_\alpha^t sg(s)y_2(s)\,ds + \int_t^\infty tg(s)y_2(s)\,ds$$
when α is chosen sufficiently large.

Problems

16. Let $g \in C[0, \infty)$ with $\int_1^\infty t^2|g(t)|\,dt < \infty$. Show that $y''' + g(t)y = 0$ has solutions ϕ_i which satisfy

$$\phi_1(t) \to 1, \qquad \phi_1'(t) \to 0, \qquad \phi_1''(t) \to 0,$$
$$\phi_2(t)/t \to 1, \qquad \phi_2'(t) \to 1, \qquad \phi_2''(t) \to 0,$$
$$\phi_3(t)/t^2 \to 1, \qquad \phi_3'(t)/(2t) \to 1, \qquad \phi_3''(t) \to 2$$

as $t \to \infty$.

17. Let $a_0(t)$ and $a_1(t)$ be continuous and T-periodic functions and let ϕ_1 and ϕ_2 be solutions of $y'' + a_1(t)y' + a_0(t)y = 0$ such that

$$\Phi(0) \triangleq \begin{bmatrix} \phi_1(0) & \phi_2(0) \\ \phi_1'(0) & \phi_2'(0) \end{bmatrix} = \begin{bmatrix} 1 & 0 \\ 0 & 1 \end{bmatrix} \triangleq E_2.$$

Show that the Floquet multipliers λ satisfy $\lambda^2 + \alpha\lambda + \beta = 0$, where

$$\alpha = -[\phi_1(T) + \phi_2'(T)], \qquad \beta = \exp\left[-\int_0^T a_1(t)\,dt\right].$$

18. In Problem 17 let $a_1 \equiv 0$. Show that if $-2 < \alpha < 2$, then all solutions $y(t)$ are bounded over $-\infty < t < \infty$. If $\alpha > 2$ or $\alpha < -2$, then $y(t)^2 + y'(t)^2$ must be unbounded over R. If $\alpha = -2$, show there is at least one solution $y(t)$ of period T while for $\alpha = 2$ there is at least one periodic solution of period $2T$.

19. Let $A(t), B(t) \in C(R^1)$, $A(t)$ T periodic, and $\int_0^\infty |B(t)|\,dt < \infty$. Let (LH) have n distinct Floquet multipliers and $e^{\rho t}p(t)$ be a solution of (LH) with $p(t)$ periodic. Show that there is a solution $x(t)$ of (7.2) such that

$$x(t)e^{-\rho t} - p(t) \to 0, \qquad t \to \infty.$$

20. If a_1 and a_2 are constants, find a fundamental set of solutions of

$$t^2 y'' + a_1 t y' + a_2 y = 0, \qquad 0 < t < \infty.$$

Hint: Use the change of variables $x = \log t$.

21. If a_i are real constants, find a fundamental set of solutions of

$$t^n y^{(n)} + a_{n-1} t^{n-1} y^{(n-1)} + \cdots + a_1 t y' + a_0 y = 0$$

on $0 < t < \infty$.

22. Assume that the eigenvalues λ_i, $i = 1, 2, \ldots, n$, of the companion matrix A given in (5.7) are real and distinct. Let V denote the *Vandermonde matrix*

$$V = \begin{bmatrix} 1 & 1 & \cdots & 1 \\ \lambda_1 & \lambda_2 & \cdots & \lambda_n \\ \vdots & \vdots & & \vdots \\ \lambda_1^{n-1} & \lambda_2^{n-1} & \cdots & \lambda_n^{n-1} \end{bmatrix}.$$

(a) Show that $\det V = \Pi_{i>j}(\lambda_i - \lambda_j)$. Hence $\det V \neq 0$.
(b) Show that $V^{-1}AV$ is a diagonal matrix.

23. Write $y''' - 2y'' - y' + 2y = 0$ in companion form as in (5.7). Compute all eigenvalues and eigenvectors for the resulting 3×3 matrix A.

24. Let $A = \lambda E + N$ consist of a single Jordan block [see (3.16)]. Show that for any $\alpha > 0$ A is similiar to a matrix $B = \lambda E + \alpha N$. *Hint*: Let $P = [\alpha^{i-1}\delta_{ij}]$ and compute $P^{-1}AP$.

25. Let A be a real $n \times n$ matrix. Show that there exists a real nonsingular matrix P such that $P^{-1}AP = B$ has the real Jordan canonical form (1.3), where J_k is given as before for real eigenvalues λ_j while for complex eigenvalues $\lambda = \alpha \pm i\beta$ the corresponding J_k has the form

$$J_k = \begin{bmatrix} \Lambda & E_2 & \cdots & 0_2 & 0_2 \\ 0_2 & \Lambda & \cdots & 0_2 & 0_2 \\ \vdots & \vdots & & \vdots & \vdots \\ 0_2 & 0_2 & \cdots & 0_2 & \Lambda \end{bmatrix}.$$

Here 0_2 is the 2×2 zero matrix, E_2 the 2×2 identity matrix, and

$$\Lambda = \begin{bmatrix} \alpha & -\beta \\ \beta & \alpha \end{bmatrix}.$$

26. Use the Jordan form to prove that all eigenvalues of A^2 have the form λ^2, where λ is an eigenvalue of A.

27. If $A = C^2$ in Problem 24, where C is a real nonsingular $n \times n$ matrix, show that there is a real matrix L such that $e^L = A$. *Hint*: Use Problems 25 and 26 and the fact that if $\lambda = \alpha + i\beta = re^{i\theta}$, then

$$\Lambda = \exp\begin{bmatrix} \log r & -\theta \\ \theta & \log r \end{bmatrix}.$$

28. In (6.1) make the change of variables $x = F(t)$, where

$$F'(t) = \exp\left(-\int_0^t a_1(u)\,du\right),$$

and let $t = f(x)$ be the inverse transformation. Show that this change of variables transforms (6.1) to

$$\frac{d^2y}{dx^2} + g(x)y = 0, \quad \text{where} \quad g(x) = \frac{a_2(f(x))}{F'(f(x))^2}.$$

29. In (6.1) make the change of variables

$$w = y\exp\left(\tfrac{1}{2}\int_0^t a_1(u)\,du\right).$$

Show that (6.1) becomes

$$\frac{d^2 w}{dt^2} + [a_2(t) - a_1(t)^2/4 - a_1'(t)/2]w = 0.$$

30. Let ϕ_i solve $(k(t)y')' + g_i(t)y = 0$ for $i = 1, 2$ with $g_1(t) < g_2(t)$ for all $t \in (a, b)$, $k(t) > 0$ on (a, b), and

$$\phi_1(t_1) = \phi_2(t_1) = 0, \qquad \phi_1'(t_1) = \phi_2'(t_1) > 0$$

at some point $t_1 \in (a, b)$. Let ϕ_1 increase from t_1 to a maximum at $t_2 > t_1$. Show that ϕ_2 must attain a maximum somewhere in the interval (t_1, t_2).

31. If a nontrivial solution ϕ of $y'' + (A + B\cos 2t)y = 0$ has $2n$ zeros in $(-\pi/2, \pi/2)$ and if $A, B > 0$, show that $A + B \geq (2n - 1)^2$.

32. If $k_0 y'' + g(t)y = 0$ has a nontrivial solution ϕ with at least $(n + 1)$ zeros in the interval (a, b), then show that

$$\sup\{g(t) : a < t < b\} > \left(\frac{n\pi}{b - a}\right)^2 k_0.$$

If $\inf\{g(t) : a < t < b\} > [n\pi/(b - a)]^2 k_0$, show that there is a nontrivial solution with at least n zeros.

33. In (6.3) let $x = F(t)$ (or $t = f(x)$) be given by

$$x = \frac{1}{K}\int_a^t [g(u)/k(u)]^{1/2}\,du, \qquad K = \int_a^b [g(u)/k(u)]^{1/2}\,du \qquad (g(t) \neq 0)$$

and let $Y(x) = [g(f(x))k(f(x))]^{1/4} y(f(x))$. Show that this transformation reduces (6.3) to

$$\frac{d^2 Y}{dx^2} + (K^2 - G(x))Y = 0,$$

where $G(x)$ is given by

$$G(x) = [g(f(x))k(f(x))]^{-1/4} \frac{d^2}{dx^2} [g(f(x))k(f(x))]^{1/4}.$$

34. (*Sturm*) Let ϕ_i solve $(k_i(t)y')' + g_i(t)y = 0$ where $k_i \in C^1[a, b]$, $g_i \in C[a, b]$, $k_1 > k_2 > 0$, and $g_2 > g_1$. Prove the following statements:

(a) Assume $\phi_1(a)\phi_2(a) \neq 0$ and $k_1(a)\phi_1'(a)/\phi_1(a) \geq k_2(a)\phi_2'(a)/\phi_2(a)$. If ϕ_1 has n zeros in $[a, b]$, then ϕ_2 must have at least n zeros there and the kth zero of ϕ_1 is larger than the kth zero of ϕ_2.

(b) If $\phi_1(b)\phi_2(b) \neq 0$ and if $k_1(b)\phi_1'(b)/\phi_1(b) \leq k_2(b)\phi_2'(b)/\phi_2(b)$, then the conclusions in (a) remain true.

35. In (6.3) assume the interval (a, b) is the real line R and assume that $g(t) < 0$ on R. Show that any solution ϕ of (6.3) with at least two distinct zeros must be identically zero.

36. For r a positive constant, consider the problem
$$y'' + ry(1 - y) = 0, \qquad 0 \le t \le \pi, \tag{7.4}$$
with $y(0) = y(\pi) = 0$. Prove the following:

    (i) If $r < 1$, then $\phi(t) \equiv 0$ is the only nonnegative solution of (7.4).

    (ii) If $r \ge 1$, then there is at most one solution of (7.4) which is positive on $(0, \pi)$.

    (iii) If $r > 1$, then any positive solution ϕ on $(0, \pi)$ must have a maximum $\bar{\phi}$ which satisfies $1 - r^{-1} < \bar{\phi} < 1$.

4 | BOUNDARY VALUE PROBLEMS*

In the present chapter we study certain self-adjoint boundary value problems on finite intervals. Specifically, we study the second order case in some detail. Some generalizations and refinements of the oscillation theory from the last section of Chapter 3 will be used for this purpose. We will also briefly consider nth order problems. The Green's function is constructed and we show how the nth order problem can be reduced to an equivalent integral equation problem.

A small amount of complex variable theory will be required in the discussion after Theorem 1.1 and in the proofs of Corollary 4.3 and Theorems 4.5 and 5.1. Also, in the last part of Section 4 of this chapter, the concepts of Lebesgue integral and of L^2 spaces as well as the completeness of L^2 spaces will be needed. If background is lacking, this material can be skipped—it will not be required in the subsequent chapters of this book.

4.1 INTRODUCTION

Partial differential equations occur in a variety of applications. Some simple but typical problems are the *wave equation*

$$\frac{\partial^2 u}{\partial t^2} = \frac{\partial}{\partial x}\left(k(x)\frac{\partial u}{\partial x}\right) + g(x)u,$$

and the *diffusion equation*

$$\frac{\partial v}{\partial t} = \frac{\partial}{\partial x}\left(k(x)\frac{\partial v}{\partial x}\right) + g(x)v.$$

Here $t \in [0, \infty)$, $x \in [a, b]$, and g and k are real valued functions. In solving these equations by separation of variables, one guesses a solution of the form

$$u(t, x) = e^{i\mu t}\phi(x)$$

for the wave equation, and one guesses a solution of the form

$$v(t, x) = e^{-\mu^2 t}\phi(x)$$

for the diffusion equation. In both cases, the function ϕ is seen to be a solution of the differential equation

$$\frac{d}{dx}\left(k(x)\frac{d\phi}{dx}\right) + (g(x) + \mu^2) = 0.$$

This equation must be solved for μ and ϕ subject to boundary conditions which are specified along with the original partial differential equations. Typical boundary conditions for the wave equation are

$$u(t, a) = u(t, b) = 0$$

which leads to $\phi(a) = \phi(b) = 0$; and typical boundary conditions for the diffusion equation are

$$\alpha \frac{\partial v}{\partial x}(t, a) = \beta v(t, a), \qquad \gamma \frac{\partial v}{\partial x}(t, b) = \delta v(t, b)$$

which leads to $\alpha\phi'(a) - \beta\phi(a) = 0$ and $\gamma\phi'(b) - \delta\phi(b) = 0$, where α, β, γ, and δ are constants. The *periodic boundary conditions*

$$\phi(a) = \phi(b) \qquad \text{and} \qquad \phi'(a) = \phi'(b)$$

will also be of interest.

With these examples as motivation, we now consider the real, second order, linear differential equation

$$Ly = -\lambda \rho(t)y, \qquad a \le t \le b, \qquad (1.1)$$

where

$$Ly = (k(t)y')' + g(t)y \qquad (1.2)$$

and the prime denotes differentiation with respect to t. We assume throughout for (1.1) that g and $\rho \in C[a, b]$, $k \in C^1[a, b]$, g is real valued, and both k and

4.1 Introduction

ρ are everywhere positive. For boundary conditions we take

$$L_1 y \triangleq \alpha y(a) - \beta y'(a) = 0, \qquad L_2 y \triangleq \gamma y(b) - \delta y'(b) = 0, \qquad \text{(BC)}$$

where all constants are real, $\alpha^2 + \beta^2 \neq 0$, and $\gamma^2 + \delta^2 \neq 0$. Boundary conditions of this form are called **separated boundary conditions**. Occasionally we shall use the more general boundary conditions

$$\begin{aligned} M_1 y &\triangleq d_{11} y(a) + d_{12} y'(a) - c_{11} y(b) - c_{12} y'(b) = 0, \\ M_2 y &\triangleq d_{21} y(a) + d_{22} y'(a) - c_{21} y(b) - c_{22} y'(b) = 0. \end{aligned} \qquad \text{(BC}_\text{g})$$

Now define the two real (2×2) matrices

$$D = \begin{bmatrix} d_{11} & d_{12} \\ d_{21} & d_{22} \end{bmatrix}, \qquad C = \begin{bmatrix} c_{11} & c_{12} \\ c_{21} & c_{22} \end{bmatrix}.$$

It is assumed that $M_1 y = 0$ and $M_2 y = 0$ are linearly independent conditions. Thus, either $\det D \neq 0$ or $\det C \neq 0$ or else, without loss of generality, we can assume that $d_{21} = d_{22} = c_{11} = c_{12} = 0$ so that (BC$_\text{g}$) reduces to (BC). It is also assumed that

$$k(b) \det D = k(a) \det C. \qquad (1.3)$$

This condition will ensure that the problem is *self-adjoint* (see Lemma 1.3). Notice that if $D = C = E_2$, then (BC$_\text{g}$) reduces to periodic boundary conditions and (1.3) reduces to $k(a) = k(b)$. Notice also that (BC) is a special case of (BC$_\text{g}$).

Example 1.1. Consider the problem

$$y'' + \lambda y = 0, \qquad y(0) = y(\pi) = 0.$$

This problem has no nontrivial solution when $\lambda \neq m^2$ for $m = 1, 2, 3, \ldots$. When $\lambda = m^2$ it is easy to check that there is a one-parameter family of solutions $y(t) = A \sin mt$.

Theorem 1.2. Let ϕ_1 and ϕ_2 be a fundamental set of solutions of $Ly = 0$ and let

$$\Phi = \begin{bmatrix} \phi_1 & \phi_2 \\ \phi_1' & \phi_2' \end{bmatrix}.$$

Then the problem

$$Ly = 0, \qquad M_1 y = M_2 y = 0$$

has a nontrivial solution if and only if

$$\Delta(\phi_1, \phi_2) \triangleq \det(D\Phi(a) - C\Phi(b)) = 0.$$

If $\Delta(\phi_1, \phi_2) \neq 0$, then for any $f \in C[a,b]$ and for any p and q the problem

$$Ly = f, \quad M_1 y = p, \quad \text{and} \quad M_2 y = q$$

has a unique solution.

Proof. There is a nontrivial solution ϕ of $Ly = 0$ if and only if there are constants c_1 and c_2, not both zero, such that $\phi = c_1 \phi_1 + c_2 \phi_2$ and

$$M_i \phi = c_1 M_i \phi_1 + c_2 M_i \phi_2 = 0, \quad i = 1, 2.$$

A nontrivial pair c_1 and c_2 will exist if and only if

$$\det \begin{bmatrix} M_1 \phi_1 & M_1 \phi_2 \\ M_2 \phi_1 & M_2 \phi_2 \end{bmatrix} = \Delta(\phi_1, \phi_2) = 0.$$

If $\Delta(\phi_1, \phi_2) \neq 0$ and if f, p and q are given, define

$$\phi_p(t) = \int_a^t [\phi_1(t)\phi_2(s) - \phi_2(t)\phi_1(s)][f(s)/w(\phi_1, \phi_2)(s)] \, ds.$$

Then by Theorem 3.5.4, ϕ_p is a solution of $Ly = f$. We must now pick c_1 and c_2 such that

$$\phi = (c_1 \phi_1 + c_2 \phi_2) + \phi_p$$

satisfies condition (BC$_g$), i.e.,

$$c_1 M_1 \phi_1 + c_2 M_1 \phi_2 + M_1 \phi_p = p, \quad c_1 M_2 \phi_1 + c_2 M_2 \phi_2 + M_2 \phi_p = q.$$

Since $\Delta(\phi_1, \phi_2) \neq 0$, this equation is uniquely solvable, and we have

$$\begin{bmatrix} c_1 \\ c_2 \end{bmatrix} = \begin{bmatrix} M_1 \phi_1 & M_1 \phi_2 \\ M_2 \phi_1 & M_2 \phi_2 \end{bmatrix}^{-1} \begin{bmatrix} p - M_1 \phi_p \\ q - M_2 \phi_p \end{bmatrix}. \quad \blacksquare$$

Equation (1.1) together with the boundary conditions (BC) will be called **problem (P)** and Eq. (1.1) with boundary conditions (BC$_g$) will be called **problem (P$_g$)**. Given any real or complex λ, let ϕ_1 and ϕ_2 be those solutions of (1.1) such that

$$\phi_1(a, \lambda) = 1, \quad \phi_1'(a, \lambda) = 0, \quad \phi_2(a, \lambda) = 0, \quad \phi_2'(a, \lambda) = 1.$$

Clearly ϕ_1 and ϕ_2 make up a fundamental set of solutions of (1.1). Let

$$\Delta(\lambda) = \det \begin{bmatrix} M_1 \phi_1(\cdot, \lambda) & M_1 \phi_2(\cdot, \lambda) \\ M_2 \phi_1(\cdot, \lambda) & M_2 \phi_2(\cdot, \lambda) \end{bmatrix}. \tag{1.4}$$

Then, according to Theorem 1.2, problem (P$_g$) has a nontrivial solution if and only if $\Delta(\lambda) = 0$. Since $\Delta(\lambda)$ is a holomorphic function of λ (by Theorem 2.9.1; see also Problem 2.21) and is not identically zero, then $\Delta(\lambda) = 0$ has

4.1 Introduction

solutions only at a countable, isolated (and possibly empty) set of points $\{\lambda_m\}$. The points λ_m are called **eigenvalues** of (P_g) and any corresponding nontrivial solution ϕ_m of (P_g) is called an **eigenfunction** of (P_g). An eigenvalue λ_m is called **simple** if there is only a one-parameter family $\{c\phi_m : 0 < |c| < \infty\}$ of eigenfunctions. Otherwise, λ_m is called a **multiple eigenvalue**.

Given two possibly complex functions y and z in $C[a,b]$, we define the function $\langle \cdot, \cdot \rangle : C[a,b] \times C[a,b] \to C$ by

$$\langle y, z \rangle = \int_a^b y(t)\overline{z(t)}\, dt$$

for all $y, z \in C[a,b]$. Note that this expression defines an **inner product** on $C[a,b]$, since for all $y, z, w \in C[a,b]$ and for all complex numbers α, we have

(i) $\langle y + z, w \rangle = \langle y, w \rangle + \langle z, w \rangle$,
(ii) $\langle \alpha y, z \rangle = \alpha \langle y, z \rangle$,
(iii) $\langle z, y \rangle = \overline{\langle y, z \rangle}$, and
(iv) $\langle y, y \rangle > 0$ when $y \neq 0$.

Note also that if ρ is a real, positive function defined on $[a,b]$, then the function $\langle \cdot, \cdot \rangle_\rho$ defined by

$$\langle y, z \rangle_\rho = \int_a^b y(t)\overline{z(t)}\rho(t)\, dt$$

determines an inner product on $C[a,b]$ provided that the indicated integral exists.

Next, we define the sets $\mathscr{D}$ and $\mathscr{D}_g$ by

$$\mathscr{D} = \{ y \in C^2[a,b] : L_1 y = L_2 y = 0 \}$$

and

$$\mathscr{D}_g = \{ y \in C^2[a,b] : M_1 y = M_2 y = 0 \}.$$

We call problem (P_g) a **self-adjoint problem** if and only if

$$\langle Ly, z \rangle = \langle y, Lz \rangle$$

for all $y, z \in \mathscr{D}_g$. As a special case, problem (P) is a self-adjoint problem if and only if

$$\langle Ly, z \rangle = \langle y, Lz \rangle$$

for all $y, z \in \mathscr{D}$.

We now show that under the foregoing assumptions, problem (P_g) is indeed a self-adjoint problem.

Lemma 1.3. Problem (P_g) is self-adjoint.

Proof. The definition of L and integration by parts can be used to compute the Lagrange identity given by

$$I \triangleq \int_a^b [(Ly)\bar{z} - y(L\bar{z})]\, ds = \int_a^b [(ky')'\bar{z} - (k\bar{z}')'y]\, ds$$
$$= [ky'\bar{z} - k\bar{z}'y]_a^b = k(b)\det \Phi(b) - k(a)\det \Phi(a),$$

where Φ denotes the matrix

$$\Phi = \begin{bmatrix} \bar{z} & y \\ \bar{z}' & y' \end{bmatrix}.$$

If $\det C \neq 0$, then since $\bar{z}$ and y satisfy the boundary conditions, we have $C\Phi(b) = D\Phi(a)$. Thus,

$$I = \frac{k(b)}{\det C}[\det C\, \Phi(b)] - k(a)\det \Phi(a)$$
$$= \frac{k(b)}{\det C}[\det D\, \Phi(a)] - k(a)\det \Phi(a)$$
$$= \left[\frac{k(b)\det D}{\det C} - k(a)\right]\det \Phi(a) = 0$$

by (1.3). If $\det C = \det D = 0$, then without loss of generality, problem (P_g) reduces to problem (P). Thus, $\alpha^2 + \beta^2 \neq 0$ while

$$\begin{bmatrix} \alpha\bar{z}(a) - \beta\bar{z}'(a) \\ \alpha y(a) - \beta y'(a) \end{bmatrix} = \Phi(a)^T \begin{bmatrix} \alpha \\ -\beta \end{bmatrix} = \begin{bmatrix} 0 \\ 0 \end{bmatrix}.$$

Thus, $\det \Phi(a) = 0$. Similarly, $\det \Phi(b) = 0$ so that $I = 0$. ∎

We are now in a position to prove the following result.

Theorem 1.4. *For problem (P_g) the following is true:*

(i) *All eigenvalues are real.*
(ii) *Eigenfunctions ϕ_m and ϕ_n corresponding to distinct eigenvalues λ_m and λ_n, respectively, are* **orthogonal**, *i.e., $\langle \phi_m, \phi_n \rangle_\rho = 0$.*

Also, for problem (P), all eigenvalues are simple.

Proof. Since $L\phi_m = -\lambda_m \phi_m \rho$ and $L\phi_n = -\lambda_n \phi_n \rho$, it follows that

$$\lambda_m \langle \phi_m, \phi_n \rangle_\rho = \langle \lambda_m \rho \phi_m, \phi_n \rangle = -\langle L\phi_m, \phi_n \rangle = -\langle \phi_m, L\phi_n \rangle$$
$$= \langle \phi_m, \lambda_n \rho \phi_n \rangle = \bar{\lambda}_n \langle \phi_m, \phi_n \rangle_\rho.$$

4.2 Separated Boundary Conditions

To prove (i), assume that $m = n$. Since $\langle \phi_m, \phi_m \rangle_\rho \neq 0$, we see that $\lambda_m = \overline{\lambda}_m$. Therefore λ_m is real.

To prove (ii), assume that $m \neq n$. Since $\lambda_n = \overline{\lambda}_n$, we see that

$$(\lambda_m - \lambda_n)\langle \phi_m, \phi_n \rangle_\rho = 0.$$

But $\lambda_m \neq \lambda_n$, and so $\langle \phi_m, \phi_n \rangle_\rho = 0$.

For problem (P) an eigenfunction must satisfy the boundary condition $\alpha y(a) - \beta y'(a) = 0$. If $\alpha = 0$, then $y'(a) = 0$. If $\alpha \neq 0$, then $y'(a) = +(\beta/\alpha)y(a)$. In either case $y'(a)$ is determined once $y(a)$ is known. Hence, each eigenvalue of problem (P) must be simple. ∎

Example 1.5. In problem (P_g) it is possible that the eigenvalues are not simple. For example, for the problem

$$y'' + \lambda y = 0, \qquad y(0) = y(2\pi), \qquad y'(0) = y'(2\pi),$$

we have $\lambda_m = m^2$ for $m = 0, 1, 2, \ldots$ with eigenfunctions $\phi_0(t) \equiv A$ and $\phi_m(t) = A \cos mt$ or $B \sin mt$.

Note that by Theorem 1.4, the eigenfunctions of problems (P_g) and of (P) can be taken to be real valued. Henceforth we shall assume that these eigenfunctions are real valued.

4.2 SEPARATED BOUNDARY CONDITIONS

In this section we study the existence and behavior of eigenvalues for problem (P). Our first task is to generalize the oscillation results of Section 3.6.

Given the equation

$$(k(t)y')' + g(t)y = 0 \tag{2.1}$$

and letting $x = k(t)y'$, we obtain the first order system

$$\begin{aligned} y' &= x/k(t), \\ x' &= -g(t)y. \end{aligned} \tag{2.2}$$

(Reference should be made to the preceding section for the assumptions on the functions k and g.) We can transform (2.2) using polar coordinates to obtain

$$\begin{aligned} x' &= r' \cos \theta - (r \sin \theta)\theta' = -g(t) r \sin \theta, \\ y' &= r' \sin \theta + (r \cos \theta)\theta' = r \cos \theta / k(t), \end{aligned}$$

or

$$r' = [1/k(t) - g(t)]r\cos\theta\sin\theta,$$
$$\theta' = g(t)\sin^2\theta + \cos^2\theta/k(t). \tag{2.3}$$

If $y \neq 0$, then y and y' cannot simultaneously vanish and $r^2 = x^2 + y^2 > 0$. Thus we can take $r(t)$ as always positive [or else as $r(t) \equiv 0$]. Therefore, Eq. (2.1) is equivalent to Eq. (2.2) or to Eq. (2.3).

We now state and prove our first result.

Theorem 2.1. Let $k_i \in C^1[a,b]$ and $g_i \in C[a,b]$ for $i = 1, 2$ with $0 < k_2 \leq k_1$ and $g_1 \leq g_2$. Let ϕ_i be a solution of $(k_i y')' + g_i y = 0$ and let r_i and θ_i satisfy the corresponding problem in polar coordinates, i.e.,

$$r'_i = [1/k_i(t) - g_i(t)]r_i \cos\theta_i \sin\theta_i,$$
$$\theta'_i = g_i(t)\sin^2\theta_i + \cos^2\theta_i/k_i(t).$$

If $\theta_1(a) \leq \theta_2(a)$, then $\theta_1(t) \leq \theta_2(t)$ for all $t \in J = [a,b]$. If in addition $g_2 > g_1$ on J, then $\theta_1(t) < \theta_2(t)$ for all $t \in (a,b]$.

Proof. Define $v = \theta_2 - \theta_1$ so that

$$v' = [g_2(t) - g_1(t)]\sin^2\theta_2 + [1/k_2(t) - 1/k_1(t)]\cos^2\theta_2$$
$$+ \{g_1(t)[\sin^2\theta_2 - \sin^2\theta_1] + [1/k_1(t)][\cos^2\theta_2 - \cos^2\theta_1]\}.$$

Define

$$h(t) = [g_2(t) - g_1(t)]\sin^2\theta_2 + [1/k_2(t) - 1/k_1(t)]\cos^2\theta_2$$

and note that in view of the hypotheses, $h(t) \geq 0$. Also note that

$$\{g_1(t)[\sin^2\theta_2 - \sin^2\theta_1] + [1/k_1(t)][\cos^2\theta_2 - \cos^2\theta_1]\}$$
$$= [g_1(t) - 1/k_1(t)][\sin^2\theta_2 - \sin^2\theta_1].$$

This term can be written in the form $f(t, \theta_1, \theta_2)v$ where

$$f(t, u_1, u_2) = \begin{cases} [g_1(t) - 1/k_1(t)](\sin u_1 + \sin u_2)(\sin u_2 - \sin u_1)/(u_2 - u_1) \\ \qquad\qquad\qquad\qquad\qquad\qquad\qquad \text{if } u_2 \neq u_1, \\ [g_1(t) - 1/k_1(t)]2\sin u_1 \cos u_2 \qquad \text{if } u_2 = u_1. \end{cases}$$

This means that

$$v' = f(t, \theta_1(t), \theta_2(t))v + h(t), \qquad v(a) \geq 0.$$

If we define $F(t) = f(t, \theta_1(t), \theta_2(t))$, then

$$v(t) = \exp\left[\int_a^t F(s)\,ds\right]v(a) + \int_a^t \exp\left[\int_s^t F(u)\,du\right]h(s)\,ds \geq 0$$

4.2 Separated Boundary Conditions

for $t \in J$. If $g_2 > g_1$, then $h > 0$ except possibly at isolated points and so $v(t) > 0$ for $t > a$. ∎

In problem (P) we consider the first boundary condition

$$L_1 y = \alpha y(a) - \beta y'(a) = 0.$$

There is no loss of generality in assuming that $0 \le |\alpha| \le 1$, $0 \le \beta/k(a) \le 1$, and $\alpha^2 + \beta^2/k(a)^2 = 1$. This means that there is a unique constant A in the range $0 \le A < \pi$ such that the expression $L_1 y = 0$ can be written as

$$\cos A \, y(a) - \sin A \, k(a) y'(a) = 0. \tag{2.4}$$

Similarly, there is a B in the range $0 < B \le \pi$ such that $L_2 y = 0$ can be written as

$$\cos B \, y(b) - k(b) \sin B \, y'(b) = 0. \tag{2.5}$$

Condition (2.4) will determine a solution of (1.1) up to a multiplicative constant. If a nontrivial solution also satisfies (2.5), it will be an eigenfunction and the corresponding value of λ will be an eigenvalue.

Theorem 2.2. Problem (P) has an infinite number of eigenvalues $\{\lambda_m\}$ which satisfy $\lambda_0 < \lambda_1 < \lambda_2 < \cdots$, and $\lambda_m \to \infty$ as $m \to \infty$. Each eigenvalue λ_m is simple. The corresponding eigenfunction ϕ_m has exactly m zeros in the interval $a < t < b$. The zeros of ϕ_m separate those of ϕ_{m+1} (i.e., the zeros of ϕ_m lie between the zeros of ϕ_{m+1}).

Proof. Let $\phi(t, \lambda)$ be the unique solution of (1.1) which satisfies

$$\phi(a, \lambda) = \sin A, \qquad k(a) \phi'(a, \lambda) = \cos A.$$

Then ϕ satisfies (2.4). Let $r(t, \lambda)$ and $\theta(t, \lambda)$ be the corresponding polar form of the solution $\phi(t, \lambda)$. The initial conditions are then transformed to $\theta(a, \lambda) = A$, $r(a, \lambda) = 1$. Eigenvalues are those values λ for which $\phi(t, \lambda)$ satisfies (2.5), that is, those values λ for which $\theta(b, \lambda) = B + m\pi$ for some integer m.

By Theorem 2.1 it follows that for any $t \in [a, b]$, $\theta(t, \lambda)$ is monotone increasing in λ. Note that $\theta(t, \lambda) = 0$ modulo π if and only if $\phi(t, \lambda)$ is zero. From (2.3) it is clear that $\theta' = 1/k > 0$ at a zero of ϕ and hence $\theta(t, \lambda)$ is strictly increasing in a neighborhood of a zero.

We claim that for any fixed constant c, $a < c \le b$, we have

$$\lim_{\lambda \to \infty} \theta(c, \lambda) = \infty \quad \text{and} \quad \lim_{\lambda \to -\infty} \theta(c, \lambda) = 0.$$

To prove the first of these statements, note that $\theta(a, \lambda) = A \ge 0$ and that $\theta' > 0$ if $\theta = 0$. Hence $\theta(t, \lambda) \ge 0$ for all t and λ. Fix $c_0 \in (a, c)$. We shall show that $\theta(c, \lambda) - \theta(c_0, \lambda) \to \infty$ as $\lambda \to \infty$. This will suffice.

Pick constants K, R, and G such that $k(t) \leq K$, $\rho(t) \geq R > 0$, and $g(t) \geq -G$. Consider the equation

$$Ky'' + (\lambda R - G)y = 0 \qquad (\lambda > 0) \tag{2.6}$$

with $y(c_0) = \phi(c_0, \lambda)$, $Ky'(c_0) = k(c_0)\phi'(c_0, \lambda)$. If $\psi(t, \lambda)$ is the solution of (2.6) in its corresponding polar form, then by Theorem 2.1 and the choice of K, R, and G, it follows that $\theta(t, \lambda) \geq \psi(t, \lambda)$ for $c_0 < t \leq c$. Since $\theta(c_0, \lambda) = \psi(c_0, \lambda)$, this gives

$$\theta(c, \lambda) - \theta(c_0, \lambda) \geq \psi(c, \lambda) - \psi(c_0, \lambda).$$

The successive zeros of (2.6) are easily computed. They occur at intervals $T(\lambda) = \pi[K(\lambda R - G)^{-1}]^{1/2}$. Since $T(\lambda) \to 0$ as $\lambda \to \infty$, then for any integer $j > 1$, ψ will have j zeros between c_0 and c for λ large enough, for example, for $(c - c_0) \geq T(\lambda)j$. Then $\psi(c, \lambda) - \psi(c_0, \lambda) \geq j\pi$. Since j is arbitary, it follows that $\theta(c, \lambda) - \theta(c_0, \lambda) \to \infty$ as $\lambda \to \infty$.

To prove that $\theta(c, \lambda) \to 0$ as $\lambda \to -\infty$, first fix $\varepsilon > 0$. We may, without loss of generality, choose ε so small that $\pi - \varepsilon > A \geq 0$. Choose K, R, and G so that $0 < K \leq k(t)$, $0 < R \leq \rho(t)$, and $G \geq |g(t)|$. If $\lambda < 0$, $A > \varepsilon$ and $\varepsilon \leq \theta \leq \pi - \varepsilon$, then

$$\theta'(t, \lambda) \leq G + \lambda R \sin^2 \varepsilon + 1/K \leq -(A - \varepsilon)/(c - a) < 0$$

as soon as $\lambda < \{(A - \varepsilon)/(a - c) - G - 1/K\}(R \sin^2 \varepsilon)^{-1} < 0$. Since $\theta(a, \lambda) = A > \varepsilon$, then for $-\lambda$ sufficiently large,

$$\theta(t, \lambda) \leq A - [(A - \varepsilon)/(c - a)](t - a),$$

for as long as $\theta(t, \lambda) \geq \varepsilon$. Let $t = c$ to see that $\theta(t, \lambda)$ must go below ε by the time $t = c$. If θ starts less than ε or becomes less than ε, then $\theta'(t, \lambda) < 0$ at $\theta = \varepsilon$ guarantees that it will remain there.

With these preliminaries completed, we now proceed to the main argument. Since $0 < B \leq \pi$, since $\theta(b, \lambda) \to 0$ as $\lambda \to -\infty$ and since $\theta(t, \lambda)$ is monotone increasing to $+\infty$ with λ, then there is a unique $\lambda = \lambda_0$ at which $\theta(b, \lambda) = B$. Notice that $0 < A = \theta(a, \lambda_0)$ and $B = \theta(b, \lambda_0) \leq \pi$ while $\theta(t, \lambda_0)$ is increasing in a neighborhood of $\theta = 0$ and $\theta = \pi$. Hence θ must satisfy

$$0 < \theta(t, \lambda_0) < \pi \qquad \text{when} \quad a < t < b.$$

Thus, $\phi_0(t) = \phi(t, \lambda_0)$ is not zero on $a < t < b$.

Let λ increase from λ_0 to the unique λ_1 where $\theta(b, \lambda_1) = B + \pi$. Since $A = \theta(a, \lambda_1) < \pi < \theta(b, \lambda_1) = B + \pi$ and since $\theta'(t_1, \lambda_1) > 0$ at any point

4.3 Asymptotic Behavior of Eigenvalues

t_1 where $\theta(t_1, \lambda_1) = \pi$, then $\phi_1(t) = \phi(t, \lambda_1)$ will have exactly one zero in $a < t < b$. Continue in this manner to obtain λ_m where $\theta(b, \lambda_m) = B + m\pi$ and $\phi_m(t) = \phi(t, \lambda_m)$. That the zeros of ϕ_m and ϕ_{m+1} interlace follows immediately from Theorem 3.6.1. ∎

4.3 ASYMPTOTIC BEHAVIOR OF EIGENVALUES

In the present section we shall require the notation $O(\cdot)$ and $o(\cdot)$ encountered in the calculus. Recall that for a function $g: R \to R$, and for $\beta \geq 0$, the notation $g(x) = O(|x|^\beta)$ as $|x| \to \infty$ means that

$$\lim_{|x| \to \infty} \sup \frac{|g(x)|}{|x|^\beta} < \infty.$$

Also, recall that $g(x) = o(|x|^\beta)$ as $|x| \to \infty$ means that

$$\lim_{|x| \to \infty} \frac{|g(x)|}{|x|^\beta} = 0.$$

If in the above, the continuous variable x is replaced by an integer valued variable $m \geq 0$, then $g(m) = O(m^\beta)$ as $m \to \infty$ and $g(m) = o(m^\beta)$ as $m \to \infty$ are defined in the obvious way.

In this section we study in detail the behavior as $m \to \infty$ of the eigenvalues λ_m and eigenfunctions ϕ_m of problem (P). We assume here that $k, \rho \in C^3[a, b]$, $g \in C^1[a, b]$ and that the constants β and δ in (BC) are not zero.

Let K be the constant defined by

$$K = \pi^{-1} \int_a^b [\rho(v)/k(v)]^{1/2} \, dv.$$

Then under the **Liouville transformation**

$$s = K^{-1} \int_a^t [\rho(v)/k(v)]^{1/2} \, dv, \qquad Y = K^{1/2}(\rho k)^{1/4} y, \qquad \mu^2 = K^2 \lambda, \quad (3.1)$$

Eq. (1.1) assumes the form

$$(d^2 Y/ds^2) + [\mu^2 - q(s)] Y = 0. \tag{3.2}$$

Here $Q_1 = (\rho k)^{1/4}$, $Q_2 = g/\rho$ (expressed as functions of s), and $q = (Q_1''/Q_1) + Q_2$. The boundary conditions (BC) have the same general form. Hence we shall use

$$\alpha Y(0) - \beta Y'(0) = 0, \qquad \gamma Y(\pi) - \delta Y'(\pi) = 0. \tag{3.3}$$

We are now in a position to prove the following result.

Theorem 3.1. Let $q \in C^1[0, \pi]$ and let $\beta\delta \neq 0$. Then there is a $j \geq 0$ such that for all large m

$$\mu_{m+j} = m + c/(m\pi) + O(m^{-2}),$$

and

$$Y_{m+j}(s) = (\cos ms)[1 + O(m^{-2})] + (\sin ms)[H(s)/m + O(m^{-2})].$$

Expressions for $H(s)$ and c are specified in the proof.

Proof. If (3.2) is written as

$$(d^2 Y/ds^2) + \mu^2 Y = q(s)Y,$$

then by Theorem 3.5.4 the solution satisfying $Y(0) = 1$, $Y'(0) = \alpha/\beta$ is

$$Y(s) = \cos \mu s + \alpha(\beta\mu)^{-1} \sin \mu s + \mu^{-1} \int_0^s \sin \mu(s-v) q(v) Y(v)\, dv. \tag{3.4}$$

The solution is continuous on $[0, \pi]$ so that $|Y(s)| \leq M$ on $[0, \pi]$ for some constant M. Hence

$$|Y(s)| \leq [1 + \alpha^2(\beta\mu)^{-2}]^{1/2} + \mu^{-1} M \int_0^\pi |q(v)|\, dv.$$

Since $|Y|$ attains its upper bound, we can replace $|Y(s)|$ by M and solve for M. This yields, for μ large,

$$M \leq [1 + \alpha^2(\beta\mu)^{-2}]^{1/2} \left(1 - \mu^{-1} \int_0^\pi |q(v)|\, dv\right)^{-1}.$$

The solution of (3.4) will automatically solve the first boundary condition in (3.3). In order to solve the second boundary condition, it is necessary and sufficient that μ solve the equation

$$\tan \pi\mu = S_1/(\mu - S_2), \tag{3.5}$$

where

$$S_1 = \alpha/\beta - \gamma/\delta + \int_0^\pi [\cos \mu s + \gamma(\delta\mu)^{-1} \sin \mu s] q(s) Y(s)\, ds$$

and

$$S_2 = -\alpha\gamma(\beta\delta\mu)^{-1} + \int_0^\pi [\sin \mu s - \gamma(\delta\mu)^{-1} \cos \mu s] q(s) Y(s)\, ds.$$

4.3 Asymptotic Behavior of Eigenvalues

Since $|Y(s)| \leq M = 1 + O(\mu^{-1})$, then both S_1 and S_2 are bounded for μ sufficiently large. Also, the bound on M and Eq. (3.4) yield

$$Y(s) = \cos \mu s + \mu^{-1} D(s, \mu),$$

where D is bounded for $0 \leq s \leq \pi$ and μ large. Use this in (3.4) to see that

$$Y(s) = \cos \mu s \left\{ 1 - \mu^{-1} \int_0^s \sin \mu v [\cos \mu v + \mu^{-1} D(v, \mu)] q(v) \, dv \right\}$$

$$+ \sin \mu s \left\{ \alpha(\beta\mu)^{-1} + \mu^{-1} \int_0^s \cos \mu v [\cos \mu v + \mu^{-1} D(v, \mu)] q(v) \, dv \right\}.$$

Note that by a trigonometric identity and integration by parts

$$\int_0^s \sin \mu v \cos \mu v \, q(v) \, dv = \tfrac{1}{2} \int_0^s \sin(2\mu v) q(v) \, dv$$

$$= -(4\mu)^{-1} \cos(2\mu v) q(v)|_0^s + (4\mu)^{-1} \int_0^s \cos(2\mu v) q'(v) \, dv = O(\mu^{-1}).$$

Similarly

$$\int_0^s \cos^2 \mu v \, q(v) \, dv = \tfrac{1}{2} \int_0^s (1 + \cos(2\mu v)) q(v) \, dv = \tfrac{1}{2} \int_0^s q(v) \, dv + O(\mu^{-1})$$

and so

$$Y(s) = (\cos \mu s)[1 + O(\mu^{-2})] + (\sin \mu s)\left(\left[\alpha\beta^{-1} + \tfrac{1}{2} \int_0^s q(v) \, dv \right] \mu^{-1} + O(\mu^{-2}) \right).$$

This can be used in S_1 and S_2 above to see that

$$S_1 = \alpha/\beta - \gamma/\delta + \tfrac{1}{2} \int_0^\pi q(v) \, dv + O(\mu^{-1}), \qquad S_2 = O(\mu^{-1}).$$

Thus Eq. (3.5) has the form

$$\tan \mu \pi = [c + O(\mu^{-1})]/[\mu + O(\mu^{-1})], \tag{3.6}$$

where $c = \alpha/\beta - \gamma/\delta + \tfrac{1}{2} \int_0^\pi q(v) \, dv$.

Equation (3.2) can be rewritten as

$$(d^2 Y/ds^2) + [\hat{\mu}^2 - (q(s) + L)]Y = 0, \qquad \hat{\mu}^2 = \mu^2 + L,$$

for any $L > 0$. Hence it is no loss of generality to assume that

$$\left(\tfrac{1}{2} \int_0^\pi q(s) \, ds \right)((\gamma/\delta) + c) > 0.$$

Since

$$dS_1/d\mu = \int_0^\pi (-\sin \mu s - \gamma(\delta\mu^2)^{-1} \sin \mu s + \gamma\delta^{-1} \cos \mu s) q(s) Y(s) \, ds,$$

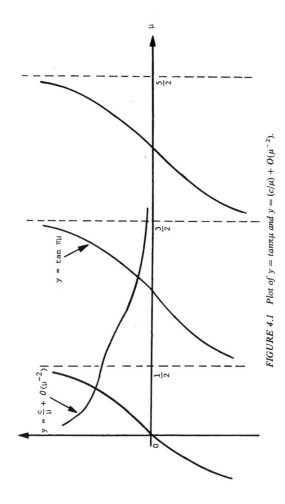

FIGURE 4.1 Plot of $y = \tan \pi \mu$ and $y = (c/\mu) + O(\mu^{-2})$.

4.3 Asymptotic Behavior of Eigenvalues

then by the same type of argument as above,

$$(\mu - S_2)^{-1} dS_1/d\mu = \gamma/(2\delta\mu) \int_0^\pi q(s)\,ds + O(\mu^{-2}).$$

Similarly,

$$dS_2/d\mu = (\mu/2) \int_0^\pi q(s)\,ds + O(1),$$

so that

$$\frac{d}{d\mu}\left(\frac{S_1}{\mu - S_2}\right) = \frac{S_1'}{\mu - S_2} - \frac{S_1(1 - S_2)}{(\mu - S_2)^2} = \left(\frac{1}{2\mu}\int_0^\pi q(s)\,ds\right)\left(\frac{\gamma}{\delta} + c\right) + O(\mu^{-2}).$$

Thus for μ sufficiently large, the curve $y = S_1(\mu - S_2)^{-1}$ is monotone. This proves that there is an integer $j \geq 0$ such that for all integers m sufficiently large, there is one and only one eigenvalue $\mu_{m+j} = m + o(1)$ (see Fig. 4.1).

Near $\mu = m$ we can expand $\tan \pi\mu$ in the series

$$\tan \pi\mu = \pi(\mu - m) + (\pi^3/3)(\mu - m)^3 + O(\mu - m)^5.$$

Since μ_{m+j} is near m, this means that

$$\tan \pi\mu_{m+j} = \pi(\mu_{m+j} - m) + O(\mu_{m+j} - m)^3 = \frac{c}{\mu_{m+j}} + O(\mu_m^{-2}) = \frac{c}{m} + O(m^{-2}).$$

Thus

$$\mu_{m+j} - m = c/(\pi m) + O(m^{-2}).$$

The expression for μ_{m+j} can be used to estimate the shape of the eigenfunction Y_{m+j} as follows:

$$\cos \mu_{m+j} s = (\cos ms)[1 + O(m^{-2})] - (\sin ms)[cs/m + O(m^{-2})],$$
$$\sin \mu_{m+j} s = (\sin ms)[1 + O(m^{-2})] + (\cos ms)[cs/m + O(m^{-2})],$$

and so

$$Y_{m+j}(s) = (\cos ms)[1 + O(m^{-2})] + (\sin ms)[H(s)/m + O(m^{-2})],$$

where

$$H(s) = \alpha/\beta + \tfrac{1}{2}\int_0^s q(v)\,dv - cs. \quad\blacksquare$$

The analysis can be modified to cover the case in which β or δ or both are zero. This will be left to the reader in the problems at the end of the chapter. When we refer to Theorem 3.1 we have in mind this extension of the theorem.

As a simple example consider
$$y'' + \lambda y = 0, \qquad y(0) = 0, \quad y(\pi) = y'(\pi).$$
Then $\phi(t) = A \sin \mu t$ where μ is a solution of $\tan \pi \mu = \mu$ and $\mu^2 = \lambda$. From a plot of $y = \tan \pi \mu$ superimposed over $y = \mu$, it is clear that $\mu_{m+j} = m + \frac{1}{2} + o(1)$. Thus we see that Theorem 3.1 must be slightly modified if β or δ is zero.

Another example which illustrates this extension of Theorem 3.1 is
$$y'' + (\lambda/(1 + t)^2)y = 0, \qquad y(0) = y(1) = 0.$$
Solutions of this differential equation are of the form $y = (1 + t)^d$. It is easy to compute that $d = (1 \pm \sqrt{1 - 4\lambda})/2$. Upon working through the boundary conditions, one finds that
$$\lambda_m = m^2 \pi^2 (\log 2)^{-2} + \tfrac{1}{4},$$
and
$$\phi_m(t) = A_m (1 + t)^{1/2} \sin[m\pi \log(1 + t)/\log 2].$$

4.4 INHOMOGENEOUS PROBLEMS

In this section we study inhomogeneous second order boundary value problems. We begin with the consideration of
$$Ly = -(\lambda \rho y + f), \qquad L_1(y) = L_2(y) = 0. \tag{4.1}$$
Here L_1, L_2, k, ρ, and g are real valued and satisfy the hypotheses of Section 1 while f can be complex valued. Let y_1 and y_2 be the solutions of $Ly = -\lambda \rho y$ satisfying
$$y_1(a, \lambda) = 1, \qquad y_1'(a, \lambda) = 0, \qquad y_2(a, \lambda) = 0, \qquad y_2'(a, \lambda) = 1. \tag{4.2}$$
Then $\beta y_1 + \alpha y_2$ will satisfy $L_1 y = 0$. The eigenvalues of problem (P) exist when $L_2(\beta y_1 + \alpha y_2) = 0$ also, i.e., when
$$\Delta(\lambda) = \beta L_2(y_1) + \alpha L_2(y_2) = 0,$$
where Δ is the function defined by (1.4).

Theorem 4.1. Problem (4.1) has a solution when λ_m is an eigenvalue if and only if f is orthogonal to the corresponding eigenvector ϕ_m, i.e., if and only if $\langle f, \phi_m \rangle = 0$.

4.4 Inhomogeneous Problems

Proof. Let y_1 and y_2 be the solutions determined by (4.2) and let c_1 and c_2 be arbitrary constants. (Since λ will be fixed, the dependence of y_i on λ will be suppressed.) Try a solution y of the form

$$y(t) = c_1 y_1(t) + c_2 y_2(t) - y_1(t) \int_a^t y_2(s) \frac{f(s)}{k(a)} ds - y_2(t) \int_t^b y_1(s) \frac{f(s)}{k(a)} ds. \quad (4.3)$$

Clearly $Ly = -(\lambda \rho y + f)$. In order to satisfy the boundary conditions $L_1(y) = L_2(y) = 0$, it is necessary and sufficient that

$$\begin{aligned} \alpha c_1 - \beta c_2 &= -\beta \langle y_1, f \rangle / k(a), \\ L_2(y_1) c_1 + L_2(y_2) c_2 &= L_2(y_1) \langle y_2, f \rangle / k(a). \end{aligned} \quad (4.4)$$

Now λ_m is an eigenvalue, so $\Delta(\lambda_m) = 0$ and there is an eigenvector $\phi_m = \beta y_1 + \alpha y_2$. Hence, $\alpha L_2(y_2) = -\beta L_2(y_1)$, i.e., the determinant of the matrix on the right side in (4.4) is zero. Then (4.4) will have a solution if and only if the matrix

$$\begin{bmatrix} \alpha & -\beta & -\beta \langle y_1, f \rangle \\ L_2(y_1) & L_2(y_2) & L_2(y_1) \langle y_2, f \rangle \end{bmatrix}$$

has rank one. This requirement can be checked by cases. For $\alpha \neq 0$ we use $\Delta(\lambda_m) = 0$ to see that the second row of the matrix is $[L_2(y_1), -L_2(y_1)(\beta/\alpha), L_2(y_1) \langle y_2, f \rangle]$. If $L_2(y_1) = 0$, then since $\alpha \neq 0$ and $\Delta(\lambda_m) = 0$ we get $L_2(y_2) = 0$. Thus

$$\begin{bmatrix} y_1(b) & y_1'(b) \\ y_2(b) & y_2'(b) \end{bmatrix} \begin{bmatrix} \gamma \\ -\delta \end{bmatrix} = \begin{bmatrix} 0 \\ 0 \end{bmatrix}, \qquad \gamma^2 + \delta^2 \neq 0.$$

By (3.6.4) the determinant of the coefficient matrix is $k(b)/k(a) \neq 0$. Thus $L_2(y_1) \neq 0$ and the rank one condition reduces to

$$\alpha \langle f, y_2 \rangle = -\beta \langle f, y_2 \rangle \quad \text{or} \quad \langle f, \phi_m \rangle = 0.$$

If $\alpha = 0$, then $L_2(y_1) = 0$ and $\beta y_1 = \phi_m$. Since the eigenvalue λ_m is simple, y_2 is not an eigenfunction, i.e., $L_2(y_2) \neq 0$. The rank one condition is $\beta \langle f, y \rangle = 0$ or $f \perp y_1$ (i.e., f is orthogonal to y_1). ∎

Corollary 4.2. *For any eigenvalue λ_m, $\Delta'(\lambda_m) \neq 0$.*

Proof. Consider the problem

$$Ly = -(\lambda \rho y + \rho \phi_m), \qquad L_1(y) = L_2(y) = 0. \quad (4.5)$$

By Theorem 4.1 there is a solution of (4.5) if and only if

$$\int_a^b \rho(t) [\phi_m(t)]^2 \, dt = 0.$$

But this is impossible.

From $\Delta(\lambda) = L_2(\phi(\cdot,\lambda))$, where $\phi = \beta y_1 + \alpha y_2$, we compute

$$\Delta'(\lambda_m) = L_2\left(\frac{\partial \phi}{\partial \lambda}(\cdot,\lambda_m)\right).$$

If $\Delta'(\lambda_m) = 0$, then $y(t) = \partial \phi(t,\lambda_m)/\partial \lambda$ will solve (4.5). But (4.5) has no solution. ∎

Corollary 4.3. Let $\langle f, \phi_m \rangle = 0$ for all eigenfunctions ϕ_m. Then (4.1) has a solution $y(t,\lambda)$ which is an entire function of λ for each fixed $t \in [a,b]$.

Proof. For $\lambda \neq \lambda_m$ try a solution of the form (4.3). Then in order to solve the boundary conditions [i.e., solve (4.4)], one must have

$$c_1(\lambda) = \beta(-L_2(y_2)\langle f, y_1 \rangle + L_2(y_1)\langle f, y_2 \rangle)/[\Delta(\lambda)k(a)]$$

and

$$c_2(\lambda) = L_2(y_1)(\beta\langle f, y_1 \rangle + \alpha\langle f, y_2 \rangle)/[\Delta(\lambda)k(a)].$$

By Theorem 2.9.1, $c_1(\lambda)$ and $c_2(\lambda)$ are holomorphic functions except possibly when $\Delta(\lambda) \neq 0$, i.e., except possibly when λ is an eigenvalue. At $\lambda = \lambda_m$ the numerator is zero since by Theorem 4.1 $\langle f, \phi_m \rangle = 0$, $\phi_m = \alpha y_2 + \beta y_1$, and $\alpha L_2(y_2) + \beta L_2(y_1) = 0$. Since the zero of $\Delta(\lambda)$ at λ_m is simple, by Corollary 4.2, then $c_1(\lambda)$ and $c_2(\lambda)$ have removable singularities at λ_m. ∎

Before proceeding further, we need to recall the following concepts from linear algebra.

Definition 4.4. A set $\{\psi_m\}$ of functions, $\psi_m: [a,b] \to R$, is called **orthogonal** (with respect to the weight ρ) if the constant defined by

$$\langle \psi_m, \psi_k \rangle_\rho = \int_a^b \psi_m(t)\overline{\psi_k(t)}\rho(t)\,dt$$

is zero when $m \neq k$ and positive if $m = k$. An orthogonal set $\{\psi_m\}$ is **orthonormal** if $\langle \psi_m, \psi_m \rangle_\rho = 1$ for all m. An orthogonal set $\{\psi_m\}$ is **complete** if no nonzero function f is orthogonal to all elements of the set.

Now let $\{\phi_m\}$ be the set of eigenfunctions for problem (P). These functions are orthogonal by Theorem 1.4. Moreover, since the function ϕ_m can be multiplied by the nonzero constants $\langle \phi_m, \phi_m \rangle_\rho^{-1/2}$, there is no loss of generality in assuming that they are orthonormal. Finally, we note that under the Liouville transformation (3.1), we have

$$\int_a^b y(t)\overline{z(t)}\rho(t)\,dt = \int_a^b [K^{1/2}y(t)(\rho k)^{1/4}][K^{1/2}z(t)(\rho k)^{1/4}][(\rho/k)^{1/2}K^{-1}\,dt]$$
$$= \int_0^\pi Y(s)Z(s)\,ds. \qquad (4.6)$$

4.4 Inhomogeneous Problems

Thus, the Liouville transformation preserves the inner product and, in particular, it preserves the orthonormality of the transformed eigenfunctions $\{Y_m\}$. For this reason it is enough to prove completeness for (3.2) and (3.3) rather than for problem (P).

Consider the problem

$$(d^2y/dt^2) + (\lambda - q(t))y = 0 \tag{4.7}$$

and

$$\alpha y(0) - \beta y'(0) = 0, \qquad \gamma y(\pi) - \delta y'(\pi) = 0, \tag{4.8}$$

where $q \in C^1[0, \pi]$. Let this problem have eigenvalues λ_m and eigenfunctions ϕ_m. We are now in a position to prove the following result.

Theorem 4.5. *The set $\{\phi_m\}$ is complete.*

Proof. Suppose f is real valued and $\langle f, \phi_m \rangle = 0$ for all eigenfunctions ϕ_m and consider

$$(d^2v/dt^2) + [\lambda - q(t)]v = -f(t)$$

with boundary conditions (4.8). By Corollary 4.3 this problem has a unique solution $v(t, \lambda)$ which is an entire function of λ for each $t \in [0, \pi]$. Thus we can expand v in a convergent power series

$$v(t, \lambda) = v_0(t) + \lambda v_1(t) + \lambda^2 v_2(t) + \cdots, \tag{4.9}$$

where the functions $v_i(t)$ satisfy

$$v_0'' - q(t)v_0 = -f(t) \quad \text{or} \quad v_i'' - q(t)v_i = -v_{i-1}, \quad i = 1, 2, 3, \ldots,$$

and the boundary conditions (4.8). Thus

$$\int_0^\pi (v_{m+1}v_k'' - v_k v_{m+1}'')\, dt = \int_0^\pi [v_{m+1}(qv_k - v_{k-1}) - v_k(qv_{m+1} - v_m)]\, dt$$

$$= \int_0^\pi (-v_{m+1}v_{k-1} + v_k v_m)\, dt.$$

On the other hand, the left side reduces to

$$[v_{m+1}v_k' - v_k v_{m+1}']_0^\pi = 0$$

by (4.8). Thus

$$\int_0^\pi v_{m+1}(t)v_{k-1}(t)\, dt = \int_0^\pi v_m(t)v_k(t)\, dt,$$

i.e., the value of the integral depends only on the sum $m + k$ of the subscripts. Call this common value $I(k + m)$. The expression

$$\int_0^\pi (Av_{m-1} + Bv_{m+1})^2\, dt = I(2m - 2)A^2 + 2I(2m)AB + I(2m + 2)B^2$$

is a positive semidefinite quadratic form in A and B. Thus $I(2m + 2) \geq 0$, $I(2m - 2) \geq 0$ and

$$I(2m)^2 - I(2m + 2)I(2m - 2) \leq 0. \tag{4.10}$$

We see from this and a simple induction argument that either $I(2m) > 0$ for all $m \geq 0$ or else $I(2m) = 0$ for all $m \geq 1$.

Suppose $I(2m) > 0$. By (4.10) we have

$$I(2)/I(0) \leq I(4)/I(2) \leq I(6)/I(4) \leq \cdots. \tag{4.11}$$

From (4.9) we see that

$$\int_0^\pi v_0(t)v(t, \lambda)\,dt = I(0) + \lambda I(1) + \lambda^2 I(2) + \cdots$$

is an entire function of λ. Hence

$$I(0) + I(2)\lambda^2 + I(4)\lambda^4 + \cdots$$

is also an entire function. We can use the lim sup test to compute the radius of convergence of this function. However, the ratio test and (4.11) imply that the radius of convergence is at most $(I(2)/I(0))^{-1/2} < \infty$. This contradiction implies that $I(2) = I(4) = \cdots = 0$. In particular, $I(2) = 0$ means $v_1(t) = 0$ on $0 \leq t \leq \pi$. Since $v_0 = qv_1 - v_1''$ and $f = qv_0 - v_0''$, then $f \equiv 0$ on $0 \leq t \leq \pi$. ∎

We also have the following result.

Corollary 4.6. *The sequence $\{\phi_m\}$ of eigenfunctions for problem (P) is complete with respect to the weight ρ provided $k, \rho \in C^3[a,b]$ and $g \in C^1[a,b]$.*

We shall define $L^2((a,b), \rho)$ as the set of all complex valued measurable functions on (a,b) such that $\langle f, f \rangle_\rho < \infty$, where

$$\langle f, f \rangle_\rho = \int_a^b |f(t)|^2 \rho(t)\,dt$$

denotes a Lebesgue integral. For such an f we define the function $\|\cdot\|$ by

$$\|f\| = \langle f, f \rangle_\rho^{1/2}.$$

Note that the function $\|\cdot\|$ is a norm, since it satisfies all the axioms of a norm (see Section 2.6). It is known that $L^2((a,b), \rho)$ is complete in this norm. Also, we define the **generalized Fourier coefficients** of f as

$$f_m = \langle f, \phi_m \rangle_\rho = \int_a^b f(t)\overline{\phi_m(t)}\rho(t)\,dt$$

4.4 Inhomogeneous Problems

and we define the **generalized Fourier series** for f as

$$f(t) \sim \sum_{m=0}^{\infty} f_m \phi_m(t).$$

We shall require the following results.

Lemma 4.7 (*Bessel inequality*). If $f \in L^2((a,b), \rho)$, then

$$\sum_{m=0}^{\infty} |f_m|^2 \le \|f\|^2.$$

Proof. Since $\langle \phi_m, \phi_k \rangle_\rho = 0$ if $m \ne k$ and since $\langle \phi_k, \phi_k \rangle_\rho = 1$, we have

$$0 \le \left\langle f - \sum_{m=0}^{M} f_m \phi_m, f - \sum_{m=0}^{M} f_m \phi_m \right\rangle_\rho = \langle f, f \rangle_\rho - \sum_{m=0}^{M} |f_m|^2$$

for any integer $M \ge 1$. ∎

Theorem 4.8. If $f \in L^2((a,b), \rho)$, then the generalized Fourier series for f converges to f in the L^2 sense, i.e.,

$$\lim_{M \to \infty} \left\| f - \sum_{m=0}^{M} f_m \phi_m \right\| = 0.$$

Proof. Define $S_M = \sum_{m=0}^{M} f_m \phi_m$. Then S_M is a Cauchy sequence since for $M > N$,

$$\|S_M - S_N\|^2 = \sum_{m=N+1}^{M} |f_m|^2 \le \sum_{m=N+1}^{\infty} |f_m|^2 \to 0$$

as $N, M \to \infty$ by Lemma 4.7. By the completeness of the space $L^2((a,b), \rho)$, there is a function $g \in L^2((a,b), \rho)$ such that

$$g = \lim_{M \to \infty} S_M.$$

But $\langle (f-g), \phi_m \rangle_\rho = 0$ for all eigenfunctions ϕ_m. Since $\{\phi_m\}$ is complete, $f - g = 0$. ∎

Let us now return to the subject at hand.

Theorem 4.9. For problem (P) let $k, \rho \in C^3[a,b]$ and let $g \in C^1[a,b]$. If $f \in C^2[a,b]$ and if f satisfies the boundary conditions (BC), then the generalized Fourier series for f converges uniformly and absolutely to f.

Proof. Since Lf is defined and continuous on $[a,b]$, then for any eigenfunction ϕ_m we have

$$\langle Lf, \phi_m \rangle = \langle f, L\phi_m \rangle = \langle f, -\lambda_m \phi_m \rangle_\rho = -\lambda_m f_m.$$

Here we have used the fact that L is self-adjoint. Since the coefficients $\alpha_m = \langle Lf, \phi_m \rangle$ are square summable (by Lemma 4.7), then this sequence is bounded, say $|\langle Lf, \phi_m \rangle| \leq M$ for all m. By Theorem 3.1, $\lambda_m = O(m^2)$ so that

$$|f_m| \leq |\langle Lf, \phi_m \rangle / \lambda_m| \leq M_1/m^2$$

for some constant M_1. Again by Theorem 3.1 the eigenfunctions $\phi_m(t)$ are uniformly bounded, say $|\phi_m(t)| \leq K$ for all m and all t. Thus

$$\left| \sum_{m=0}^{\infty} f_m \phi_m(t) \right| \leq |f_0| + \sum_{m=1}^{\infty} (M_1/m^2)K < \infty.$$

The Weierstrass test (Theorem 2.1.3) completes the proof. ∎

We now give the last result of this section.

Theorem 4.10. Let $k, \rho \in C^3[a,b]$ and let $g \in C^1[a,b]$. For any $f \in C[a,b]$ and for any complex λ not an eigenvalue, the problem

$$Ly = -\rho(\lambda y + f), \qquad L_1 y = L_2 y = 0 \tag{4.12}$$

has a unique solution y. This solution can be written as the uniformly and absolutely convergent series

$$y(t) = \sum_{m=0}^{\infty} f_m (\lambda_m - \lambda)^{-1} \phi_m(t). \tag{4.13}$$

Proof. The series (4.13) is derived by assuming that

$$f = \sum_{m=0}^{\infty} f_m \phi_m, \qquad y = \sum_{m=0}^{\infty} y_m \phi_m,$$

putting these series into Eq. (4.12), and solving for y_m. Since the $\phi_m(t)$ are uniformly bounded and $\lambda_m = O(m^2)$, the proof that the series converges uniformly and absolutely follows along similar lines as the proof of Theorem 4.9. By Theorem 1.2, problem (4.12) has a unique solution $z(t)$. Thus, for any m we have

$$0 = \langle Lz + \rho(\lambda z + f), \phi_m \rangle = \langle Lz, \phi_m \rangle + \lambda \langle z, \phi_m \rangle_\rho + \langle f, \phi_m \rangle_\rho.$$

Since L is self-adjoint, it follows that

$$\langle Lz, \phi_m \rangle = \langle z, L\phi_m \rangle = \langle z, -\lambda_m \rho \phi_m \rangle = -\lambda_m \langle z, \phi_m \rangle_\rho.$$

4.5 General Boundary Value Problems

Thus we can solve for $\langle z, \phi_m \rangle_\rho$ and find that

$$\langle z, \phi_m \rangle_\rho = f_m(\lambda_m - \lambda)^{-1} = y_m.$$

Since z and y have the same Fourier coefficients, then, by completeness, they are the same function. ∎

For the problem

$$y'' + \lambda y = 0, \qquad y(0) = y(\pi) = 0$$

it is easy to compute that $\lambda_m = m^2$ and $\phi_m(t) = \sqrt{2/\pi} \sin mt$, $m = 1, 2, 3, \ldots$. These eigenfunctions form a complete set on $(0, \pi)$. Moreover, if $f(t) \sim \sqrt{2/\pi} \sum_{m=1}^\infty f_m \sin mt$, then for $\lambda \neq m^2$,

$$y'' + \lambda y = -f(t), \qquad y(0) = y(\pi) = 0$$

has the solution

$$y(t) = \sqrt{\frac{2}{\pi}} \sum_{m=1}^\infty \frac{f_m}{\lambda - m^2} \sin mt.$$

For the problem

$$y'' + \lambda y = 0, \qquad y'(0) = y'(\pi) = 0$$

it is easy to compute $\lambda_m = m^2$ for $m = 0, 1, 2, \ldots$ while

$$\phi_0(t) = 1/\sqrt{\pi} \quad \text{and} \quad \phi_m(t) = \sqrt{2/\pi} \cos mt \qquad (m \geq 1).$$

These eigenfunctions also form a complete set on $(0, \pi)$.

4.5 GENERAL BOUNDARY VALUE PROBLEMS

Many of the results of the previous section can be generalized to a wide class of boundary value problems. The generalization is made by transforming these boundary value problems into integral equations. These equations can then be studied using integral equations or even functional analytic techniques. The price paid for this generality is that the information obtained is not so detailed as in the second order problem (P).

On a finite interval $[a, b]$, let $a_j \in C^j[a, b]$ be given for $j = 0, 1, 2, \ldots, n$ with $a_n(t) > 0$ for all t. Consider the nth order linear differential operator defined by

$$Ly = a_n y^{(n)} + a_{n-1} y^{(n-1)} + a_{n-2} y^{(n-2)} + \cdots + a_1 y' + a_0 y.$$

Let U be the boundary operator defined by

$$U_i y = \sum_{j=1}^{n} [\alpha_{ij} y^{(j-1)}(a) + \beta_{ij} y^{(j-1)}(b)]$$

for $i = 1, 2, \ldots, n$ and

$$Uy = [U_1 y, U_2 y, \ldots, U_n y].$$

Let $\rho \in C[a,b]$ be a fixed positive function. Consider the boundary value problem

$$Ly = -\lambda \rho y, \qquad Uy = 0. \tag{5.1}$$

If for some λ there is a nontrivial solution ϕ of (5.1), then we shall call λ an **eigenvalue** of (5.1) and ϕ an **eigenfunction** of (5.1). Clearly, for any scalar $\alpha \neq 0$, $\alpha \phi$ is also an eigenfunction of (5.1).

We shall take the point of view that the operator L is not completely specified until its domain and range are given. The range will be the set $C[a,b]$ while the domain will be

$$\mathcal{D} = \{ y \in C^n[a,b] : Uy = 0 \}.$$

Problem (5.1) will be called **self-adjoint** if $\langle Ly, z \rangle = \langle y, Lz \rangle$ for all $y, z \in \mathcal{D}$. Here we use the notation

$$\langle y, z \rangle = \int_a^b y(t) \overline{z(t)} \, dt,$$

and

$$\langle y, z \rangle_\rho = \int_a^b y(t) \overline{z(t)} \rho(t) \, dt, \tag{5.2}$$

as before. For example, if L^+ denotes the adjoint of L, then by the Lagrange identity (see Theorem 3.5.9) we have

$$\langle Ly, z \rangle = \langle y, L^+ z \rangle + P(y,z)|_a^b.$$

So L will be self-adjoint, for example, if $L = L^+$ and $P(y,z)|_a^b = 0$ for all $y, z \in \mathcal{D}$.

By the same proof as in Section 4.1, if L is self-adjoint, then all eigenvalues are real and all eigenfunctions corresponding to distinct eigenvalues are orthogonal with respect to the inner product (5.2).

The inhomogeneous problem

$$Ly + \lambda \rho y + f = 0, \qquad Uy = 0 \tag{5.3}$$

will be solved by constructing a **Green's function**.

4.5 General Boundary Value Problems

Theorem 5.1. Suppose there exists at least one complex number λ_0 which is not an eigenvalue of (5.1). Then there is a unique function $G(t, s, \lambda)$ which is defined for $a \le t, s \le b$ and for all complex numbers λ which are not eigenvalues, and which has the following properties:

(i) $\partial^j G/\partial t^j$ exists and is continuous on the set S, where

$$S = \{(t, \tau, \lambda): a \le t, s \le b, \lambda \text{ not an eigenvalue}\}$$

for $j = 0, 1, 2, \ldots, n$ expect on the line $t = s$ when $j = n - 1$ and n.

(ii) The $(n - 1)$st derivative has a jump discontinuity at $t = s$ such that

$$\frac{\partial^{n-1} G}{\partial t^{n-1}}(s^+, s, \lambda) - \frac{\partial^{n-1} G}{\partial t^{n-1}}(s^-, s, \lambda) = \frac{1}{a_n(s)}.$$

(iii) As a function of t, $y(t) = G(t, s, \lambda)$ satisfies $Uy = 0$ and also $Ly + \lambda \rho y = 0$ for $a \le t \le b, t \ne s$.

The solution of (5.3) at any λ not an eigenvalue is

$$y(t) = -\int_a^b G(t, s, \lambda) f(s)\, ds \triangleq (\mathscr{G}_\lambda f)(t). \tag{5.4}$$

Proof. Let $\phi_j(t, \lambda)$ be the solution of $Ly + \lambda \rho y = 0$ which satisfies the initial conditions $\phi_{j+1}^{(k)}(a, \lambda) = \delta_{kj}$ for $0 \le j, k \le n - 1$. Here δ_{ij} denotes the Kronecker delta, i.e., $\delta_{ij} = 0$ when $i \ne j$ and $\delta_{ij} = 1$ when $i = j$. Define

$$H(t, s, \lambda) = \det \begin{bmatrix} \phi_1(s, \lambda) & \cdots & \phi_n(s, \lambda) \\ \phi_1'(s, \lambda) & \cdots & \phi_n'(s, \lambda) \\ \vdots & & \vdots \\ \phi_1^{(n-2)}(s, \lambda) & \cdots & \phi_n^{(n-2)}(s, \lambda) \\ \phi_1(t, \lambda) & \cdots & \phi_n(t, \lambda) \end{bmatrix} \div \{a_n(s) W(\phi_1, \ldots, \phi_n)(s)\}$$

for $t \ge s$ and $H(t, s, \lambda) = 0$ if $t < s$. Since the Wronskian satisfies

$$W(\phi_1, \phi_2, \ldots, \phi_n)(s) = \exp \int_a^s -[a_{n-1}(s)/a_n(s)]\, ds,$$

the denominator is never zero.

Clearly $\partial^j H/\partial t^j = 0$ for $j = 0, 1, 2, \ldots, n$ and for $a \le t \le s$. Also a determinant is zero when two rows are equal. Hence $\partial H^j/\partial t^j$ exists for $a \le t \le s$ and is zero at $t = s$ for $j = 0, 1, 2, \ldots, n - 2$. When $j = n - 1$,

$$\frac{\partial^{n-1} H}{\partial t^{n-1}}(s^+, s, \lambda) - \frac{\partial^{n-1} H}{\partial t^{n-1}}(s^-, s, \lambda) = \frac{W(\phi_1, \ldots, \phi_n)(s)}{W(\phi_1, \ldots, \phi_n)(s) a_n(s)} - 0 = a_n(s)^{-1}.$$

Thus H satisfies (i)–(iii) of the theorem expect $UH(\cdot, s, \lambda) = 0$. It will need modification in order to satisfy this last property.

Define $G(t, s, \lambda) = H(t, s, \lambda) + \sum_{j=1}^{n} c_j \phi_j(t, \lambda)$, where $c_i = c_i(s, \lambda)$ are to be chosen later. We need

$$U_i G = U_i H + \sum_{j=1}^{n} c_j U_i \phi_j = 0$$

or

$$\sum_{j=1}^{n} c_j U_i \phi_j = -U_i H(\cdot, s, \lambda). \tag{5.5}$$

Since the determinant $\Delta(\lambda)$ of the matrix $[U_i \phi_j(\cdot, \lambda)]$ is an entire function of λ and since it is not zero at the value λ_0, then $\Delta(\lambda)^{-1}$ is a memomorphic function. Hence (5.5) has a unique and continuous solution set $c_j(s, \lambda)$ for $a \le s \le b$ and all λ with $\Delta(\lambda) \ne 0$.

Finally, we note that by Theorem 3.5.4

$$\int_a^t \sum_{k=1}^{n} \frac{\phi_k(t, \lambda) W_k(\phi_1, \ldots, \phi_n)(s)}{W(\phi_1, \ldots, \phi_n)(s)} \left(\frac{-f(s)}{a_n(s)} \right) ds$$

is a solution of $Ly + \lambda \rho y = -f$. Since

$$\sum_{k=1}^{n} \frac{\phi_k(t, \lambda) W_k(\phi_1, \ldots, \phi_n)(s)}{[W(\phi_1, \ldots, \phi_n)(s)] a_n(s)} = H(t, s, \lambda)$$

when $s \le t$, we see that y defined as

$$y(t) = -\int_a^b G(t, s, \lambda) f(s)\, ds = -\int_a^t H(t, s, \lambda) f(s)\, ds$$
$$- \sum_{j=1}^{n} \left(\int_a^b c_j(s, \lambda) f(s)\, ds \right) \phi_j(t, \lambda)$$

consists of a solution of the inhomogeneous problem $Ly + \lambda \rho y = -f$ and a solution of the homogeneous problem $Ly + \lambda \rho y = 0$. Hence y solves the inhomogeneous problem. Moreover

$$Uy = -\int_a^b UG(\cdot, s, \lambda) f(s)\, ds = -\int_a^b 0 \cdot f(s)\, ds = 0. \quad \blacksquare$$

We note that the values λ where $\Delta(\lambda) = 0$ are the eigenvalues of L. Since Δ is an entire function of λ (cf. Section 2.9), there is at most a countable set $\{\lambda_m\}$ of eigenvalues, and these eigenvalues cannot cluster at any finite value in the complex plane. Note also that if L is self-adjoint, then the existence of λ_0 in Theorem 5.1 is trivial—any λ_0 with nonzero imaginary part will do.

4.5 General Boundary Value Problems

The function $G(t, \tau, \lambda)$ must have the form

$$G(t, s, \lambda) = \sum_{j=1}^{n} A_j(s, \lambda)\phi_j(t, \lambda), \qquad a \le t < s,$$

and

$$G(t, s, \lambda) = \sum_{j=1}^{n} B_j(s, \lambda)\phi_j(t, \lambda), \qquad s < t \le b.$$

The conditions of the theorem can be used to determine A_j and B_j. For example, at $\lambda = 0$ the problem

$$y'' = -f(t), \qquad y(0) = y(1) = 0$$

has the Green's function

$$G(t, s, 0) = \begin{cases} (s-1)t, & 0 \le t \le s, \\ s(t-1), & s \le t \le 1. \end{cases}$$

If $f(t) \equiv 1$, then the solution (5.4) has the form

$$y(t) = \int_0^t (1-t)sf(s)\,ds + \int_t^1 t(1-s)f(s)\,ds$$

$$= (1-t)\int_0^t s\,ds + t\int_t^1 (1-s)\,ds = t(1-t)/2.$$

In the self-adjoint case we shall have $\langle Ly, z \rangle = \langle y, Lz \rangle$ for all $y, z \in \mathscr{D}$. At a given λ, let $y = \mathscr{G}_\lambda f$ and $z = \mathscr{G}_{\bar\lambda} h$ where $\mathscr{G}$ is the integral operator defined in (5.4) and λ is any complex number which is not an eigenvalue. Then $Ly + \lambda \rho y = f$ and $Lz + \bar\lambda \rho z = h$ so that

$$\langle f, \mathscr{G}_{\bar\lambda} h \rangle = \langle Ly + \lambda \rho y, z \rangle = \langle y, Lz + \bar\lambda \rho z \rangle = \langle \mathscr{G}_\lambda f, h \rangle$$

for any $f, h \in C[a, b]$. This can be written as

$$\int_a^b \int_a^b \overline{G(t, s, \bar\lambda)} f(t)\overline{h(s)}\,ds\,dt = \int_a^b \int_a^b G(t, s, \lambda) f(s)\overline{h(t)}\,ds\,dt.$$

Interchanging the order of integration in the first integral, we see that

$$\int_a^b \int_a^b \overline{G(s, t, \bar\lambda)} f(s)\overline{h(t)}\,ds\,dt = \int_a^b \int_a^b G(t, s, \lambda) f(s)\overline{h(t)}\,ds\,dt.$$

Since f and h can run over all continuous functions, one can argue in a variety of ways that this implies that

$$G(t, s, \lambda) = \overline{G(s, t, \bar\lambda)}.$$

The Green's function provides an inverse for L in the sense that $L\mathscr{G}f = f$ and $gLy = y$ for all y in $\mathscr{D}$ and all f in $C[a, b]$. (We are assuming without loss of generality that $\lambda_0 = 0$.) Using $\mathscr{G}$ at $\lambda = 0$, the boundary value problem (5.3) may be restated in the equivalent form

$$y + \lambda \mathscr{G}(y\rho) = F,$$

where $F = \mathscr{G}f$. This operator equation can also be written as the integral equation

$$y(t) = F(t) - \lambda \int_a^b G(t,s,0)\rho(s)y(s)\,ds. \tag{5.6}$$

In case L is self-adjoint, (5.6) can be modified to preserve the symmetry of the terms multiplying $y(s)$ under the integral sign. Let $z(t) = \sqrt{\rho(t)}y(t)$, multiply (5.6) by $\sqrt{\rho(t)}$, and compute

$$z(t) = \sqrt{\rho(t)}F(t) - \lambda \int_a^b G(t,s,0)\sqrt{\rho(t)\rho(s)}z(s)\,ds. \tag{5.7}$$

Now

$$G(t,s,0)\sqrt{\rho(t)\rho(s)} = \overline{G(s,t,0)}\sqrt{\rho(s)\rho(t)}$$

for $a \le t, s \le b$.

The integral equation (5.6), and even more so, the symmetric case (5.7), can most efficiently be studied under rather weak assumptions on G using integral equation techniques and/or functional analytic techniques. Since no more theory concerning differential equations is involved, we shall not pursue this subject further.

PROBLEMS

1. Let $k \in C^1[a,b]$ and $g \in C[a,b]$ be real valued functions and let α, β, γ, and δ in (BC) be complex numbers. Show that problem (P) is self-adjoint if and only if $\alpha\bar{\beta} = \beta\bar{\alpha}$ and $\gamma\bar{\delta} = \bar{\gamma}\delta$. Show that this condition is true if and only if (P) is equivalent to a problem with all coefficients real.

2. For what values of a and b, with $0 \le a < b \le \pi$, is the problem

$$[(2 + \sin t)y']' + (\cos t)y = 0, \qquad y(a) = y(b), \quad y'(a) = y'(b)$$

self-adjoint?

3. Find all eigenvalues and eigenfunctions for the problem

$$[(1+t)^2 y']' + \lambda y = 0, \qquad y(0) = y(1) = 0.$$

Hint: Try $y = (1+t)^{i\alpha+c} = (\cos[\alpha \log(1+t)] + i\sin[\alpha \log(1+t)])(1+t)^c$.

Problems

4. Suppose the boundary conditions (BC$_g$) are such that

$$\int_a^b (ky')'y\, dt = -\int_a^b k(y')^2\, dt$$

for all $y \in \mathscr{D}^*$. Show that there exists a constant $G > 0$ such that $\lambda_m \geq -G$ for all eigenvalues.

5. Show that for any $f \in C[0, \pi]$

$$\lim_{A \to \infty} \int_0^\pi (\sin At)f(t)\, dt = \lim_{A \to \infty} \int_0^\pi (\cos At)f(t)\, dt = 0.$$

Hint: f can be uniformly approximated by functions from $C^1[0, \pi]$.

6. In Theorem 3.1 show that if $g \in C[0, \pi]$, then

$$\mu_m = m + o(1).$$

7. In Theorem 3.1 show that

(a) if $\delta = 0$ or $\beta = 0$ but not both, then $\mu_{m+j} = m + \tfrac{1}{2} + O(1/m)$, and

(b) if $\delta = \beta = 0$, then $\mu_{m+j} = m + 1 + O(1/m)$.

(c) Compute the asymptotic form $Y_{m+j}(s)$ for both cases.

8. (*Rayleigh quotients*) Define

$$V = \{\phi \in C^2[a,b] : L_1\phi L_2\phi = 0,\ \langle \phi, \phi \rangle_\rho = 1\}.$$

Let $k, \rho \in C^3[a,b]$ and $g \in C^1[a,b]$. Show that

$$\inf\{\langle -Ly, y \rangle : y \in V\} = \lambda_0,$$

the smallest eigenvalue. *Hint*: Use Theorem 4.9.

9. Find the asymptotic form of the eigenvalues and the eigenfunctions of the problem

$$(ty')' + (\lambda t - A^2/t)y = 0, \qquad y(1) = y(2) = 0.$$

10. Find the Green's function at $\lambda = 0$ for
  (a) $Ly = y''$, $y(0) = y'(1) = 0$,
  (b) $Ly = y''$, $y(0) = y'(0)$, $y(1) = -y'(1)$.
  (c) $Ly = y''$, $y(0) + y(1) = 0$, $y'(0) + y'(1) = 0$, and
  (d) $Ly = y'' + Ay$, $y(0) = y(\pi) = 0$, $A > 0$.

11. Show that $\lambda = 0$ is not an eigenvalue for

$$y''' + \lambda y = 0, \qquad y(0) = y'(0) = y''(1) = 0.$$

At $\lambda = 0$, compute the Green's function. Is this problem self-adjoint?

12. In problem (P), suppose that $\beta = \gamma = 0$, and suppose that $\lambda = 0$ is not an eigenvalue. Let $G(t,s)$ be the Green's function at $\lambda = 0$. Prove the following:

    (a) $u(t) \triangleq \partial G(t,a)/\partial s$ solves $Lu = 0$ on $a < t < b$, with $u(a^+) = -k(a)^{-1}$, $u(b) = 0$.

    (b) $v(t) \triangleq \partial G(t,b)/\partial s$ solves $Lv = 0$ on $a < t < b$, with $v(a) = 0$, $v(b^-) = k(b)^{-1}$.

    (c) For any $f \in C[a,b]$ and any constants A and B, the solution of $Ly = -f$, $L_1 y = A$, $L_2 y = B$ is

$$y(t) = -\int_a^b G(t,s)f(s)\,ds - Ak(a)\frac{\partial G}{\partial s}(t,a) + Bk(b)\frac{\partial G}{\partial s}(t,b).$$

13. Solve $y'' = -1$, $y(0) = 3$, $y(1) = 2$.

14. For the problem

$$(ky')' + g(t)y = -f(t), \qquad a < t < b, \qquad L_1 y = A, \qquad L_2 y = B,$$

show that if $\lambda = 0$ is not an eigenvalue and if $\beta\delta \neq 0$, then

$$y(t) = -\int_a^b G(t,s,0)f(s)\,ds - \frac{k(a)}{\beta}AG(t,a,0) + \frac{k(b)}{\delta}BG(t,b,0)$$

is the unique solution. Solve $u'' = -t$, $u'(0) = -2$, $u'(1) + u(1) = 3$ by this method.

15. Solve $y'' = -t$, $y(0) = 2$, $y'(1) + y(1) = 1$.

16. (*Singular problem*) Show that $\lambda = 0$ is not an eigenvalue of

$$ty'' + y' + \lambda y = 0, \qquad y(t) \text{ bounded as } t \to 0^+, \qquad y(1) = 0.$$

Compute the Green's function at $\lambda = 0$.

17. Prove that the set $\{1, \cos t, \sin t, \cos 2t, \sin 2t, \ldots\}$ is complete over the interval $[-\pi, \pi]$. *Hint*: Use the two examples at the end of Section 4.

18. Let $k \in C^1[a,b]$ and $g \in C[a,b]$ with both functions complex valued and with $k(t) \neq 0$ for all t. Let α, β, γ, and δ be complex numbers. Show that if (P) is self-adjoint, then

    (i) $k(t)$ and $g(t)$ are real valued, and

    (ii) $\alpha\bar{\beta} = \beta\bar{\alpha}$, $\gamma\bar{\delta} = \bar{\gamma}\delta$.

5 | STABILITY

In Chapter 2 we established sufficient conditions for the existence, uniqueness, and continuous dependence on initial data of solutions to initial value problems described by ordinary differential equations. In Chapter 3 we derived explicit closed-form expressions for the solution of linear systems with constant coefficients and we determined the general form and properties of linear systems with time-varying coefficients. Since there are no general rules for determining explicit formulas for the solutions of such equations, nor for systems of nonlinear equations, the analysis of initial value problems of this type is usually accomplished along two lines: (a) a quantitative approach is used which usually involves the numerical solution of such equations by means of simulations on a digital computer, and (b) a qualitative approach is used which is usually concerned with the behavior of families of solutions of a given differential equation and which usually does not seek specific explicit solutions. In applications, both approaches are usually employed to complement each other. Since there are many excellent texts on the numerical solution of ordinary differential equations, and since a treatment of this subject is beyond the scope of this book, we shall not pursue this topic. The principal results of the qualitative approach include stability properties of an equilibrium point (rest position) and the boundedness of solutions of ordinary differential equations. We shall consider these topics in the present chapter and the next chapter.

In Section 1, we recall some essential notation that we shall use throughout this chapter. In Section 2, we introduce the concept of an equilibrium point, while in Section 3 we define the various types of stability, instability, and boundedness concepts which will be the basis of the entire development of the present chapter. In Section 4, we discuss the stability properties of autonomous and periodic systems, and in Sections 5 and 6 we discuss the stability properties of linear systems. The main stability results of this chapter involve the existence of certain real valued functions (called Lyapunov functions) which we introduce in Section 7. In Sections 8 and 9 we present the main stability, instability, and boundedness results which constitute the direct method of Lyapunov (of stability analysis). Linear systems are discussed again in Section 10, this time in the context of the Lyapunov theory. Extensions and improvements of the Lyapunov theory are presented in Sections 11 (invariance theorems) and 12 (extent of asymptotic stability). The stability results, as given in Section 9, constitute sufficient conditions. It turns out that some of these results are also necessary conditions. This is demonstrated in Section 13, where a sample result of a so-called converse theorem is presented. Comparison theorems, as they arise in the context of stability theory, are treated in Section 14. In Section 15, the stability properties of an important class of problems that arise in applications (regulator systems) are discussed.

5.1 NOTATION

We begin by recalling some of the notation which we shall require throughout this chapter.

If $x \in R^n$, then $|x|$ will denote the norm of x, where $|\cdot|$ represents any one of the equivalent norms on R^n. Also, if A is any real (or complex) $m \times n$ matrix, then $|A|$ will denote the norm of the matrix of A induced by the norm on R^n, i.e.,

$$|A| = \sup_{|x|=1} |Ax| = \sup_{0<|x|\le 1} \frac{|Ax|}{|x|} = \sup_{x\ne 0} \frac{|Ax|}{|x|}$$

(see Section 2.6 for further details). Note in particular that

$$|Ax| \le |A||x|.$$

Recall also that $B(x_0, h)$ and $B(h)$ denote the spheres with radius $h > 0$ and centers $x = x_0$ and $x = 0$, respectively, i.e.,

$$B(x_0, h) = \{x \in R^n : |x - x_0| < h\} \quad \text{and} \quad B(h) = \{x \in R^n : |x| < h\}.$$

5.2 THE CONCEPT OF AN EQUILIBRIUM POINT

We concern ourselves with systems of equations

$$x' = f(t, x), \qquad \text{(E)}$$

where $x \in R^n$. When discussing global results, such as global asymptotic stability, we shall always assume that $f: R^+ \times R^n \to R^n$. On the other hand, when considering local results, we shall usually assume that $f: R^+ \times B(h) \to R^n$ for some $h > 0$. On some occassions we may assume that $t \in R$, rather than $t \in R^+$. Unless otherwise stated, we shall assume that for every (t_0, ξ), $t_0 \in R^+$, the initial value problem

$$x' = f(t, x), \qquad x(t_0) = \xi \qquad \text{(I)}$$

possesses a unique solution $\phi(t, t_0, \xi)$ which depends continuously on the initial data (t_0, ξ). Since it is very natural in this chapter to think of t as representing time, we shall use the symbol t_0 in (I) to represent the initial time (rather than using τ as was done earlier). Furthermore, we shall frequently use the symbol x_0 in place of ξ to represent the initial state. This nomenclature is standard in the literature on stability.

Definition 2.1. A point $x_e \in R^n$ is called an **equilibrium point** of (E) (at time $t^* \in R^+$) if

$$f(t, x_e) = 0 \qquad \text{for all} \quad t \geq t^*.$$

Other terms for equilibrium point include *stationary point, singular point, critical point*, and *rest position*. Note that if x_e is an equilibrium point of (E) at t^*, then it is an equilibrium point at all $\tau \geq t^*$. Note also that in the case of autonomous systems

$$x' = f(x) \qquad \text{(A)}$$

and in the case of T-periodic systems

$$x' = f(t, x), \qquad f(t, x) = f(t + T, x), \qquad \text{(P)}$$

a point $x_e \in R^n$ is an equilibrium at some time t^* if and only if it is an equilibrium point at all times. Also note that if x_e is an equilibrium (at t^*) of (E), then the transformation $s = t - t^*$ reduces (E) to

$$dx/ds = f(s + t^*, x),$$

and x_e is an equilibrium (at $s = 0$) of this system. For this reason, we shall henceforth assume that $t^* = 0$ in Definition 2.1 and we shall not mention t^*

further. Note also that if x_e is an equilibrium point of (E), then for any $t_0 \geq 0$

$$\phi(t, t_0, x_e) = x_e \quad \text{for all} \quad t \geq t_0,$$

i.e., x_e is a unique solution of (E) with initial data given by $\phi(t_0, t_0, x_e) = x_e$.

Example 2.2. In Chapter 1 we considered the simple pendulum described by the equations

$$\begin{align} x_1' &= x_2, \\ x_2' &= -k \sin x_1, \quad k > 0. \end{align} \tag{2.1}$$

Physically, the pendulum has two equilibrium points. One of these is located as shown in Fig. 5.1a and the second point is located as shown in Fig. 5.1b. However, the *model* of this pendulum, described by Eq. (2.1), has countably infinitely many equilibrium points which are located in R^2 at the points $(\pi n, 0), n = 0, \pm 1, \pm 2, \ldots$.

Definition 2.3. An equilibrium point x_e of (E) is called an **isolated equilibrium point** if there is an $r > 0$ such that $B(x_e, r) \subset R^n$ contains no equilibrium points of (E) other than x_e itself.

Both equilibrium points in Example 2.2 are isolated equilibrium points in R^2. Note however that none of the equilibrium points in our next example are isolated.

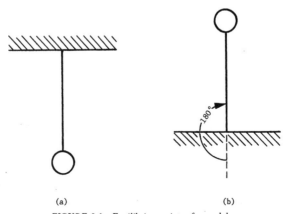

FIGURE 5.1 Equilibrium points of a pendulum.

5.2 The Concept of an Equilibrium Point

Example 2.4. In Chapter 1 we considered a simple epidemic model in a given population described by the equations

$$x_1' = -ax_1 + bx_1x_2,$$
$$x_2' = -bx_1x_2, \quad (2.2)$$

where $a > 0, b > 0$ are constants. [Only the case $x_1 \geq 0, x_2 \geq 0$ is of physical interest, though Eq. (2.2) is mathematically well defined on all of R^2.] In this case, every point on the positive x_2 axis is an equilibrium point for (2.2).

There are systems with no equilibrium points at all, as is the case, e.g., in the system

$$x_1' = 2 + \sin(x_1 + x_2) + x_1,$$
$$x_2' = 2 + \sin(x_1 + x_2) - x_1. \quad (2.3)$$

We leave it to the reader to verify the next two examples.

Example 2.5. The linear homogeneous system

$$x' = A(t)x \quad \text{(LH)}$$

has a unique equilibrium which is at the origin if $A(t_0)$ is nonsingular for all $t_0 \geq 0$.

Example 2.6. Assume that for

$$x' = f(x), \quad \text{(A)}$$

f is continuously differentiable with respect to all of its arguments, and let

$$J(x_e) = \left.\frac{\partial f(x)}{\partial x}\right|_{x=x_e},$$

where $\partial f/\partial x$ is the $n \times n$ **Jacobian matrix** defined by

$$\partial f/\partial x = [\partial f_i/\partial x_j].$$

If $f(x_e) = 0$ and $J(x_e)$ is nonsingular, then x_e is an isolated equilibrium of (E).

Unless stated otherwise, we shall assume throughout this chapter that a given equilibrium point is an isolated equilibrium. Also, we shall usually find it extremely useful to assume that in a given discussion, the equilibrium of interest is located at the origin of R^n. This assumption can be made without any loss of generality. To see this, assume that $x_e \neq 0$ is an equilibrium point of

$$x' = f(t, x), \quad \text{(E)}$$

i.e., $f(t, x_e) = 0$ for all $t \geq 0$. Let $w = x - x_e$. Then $w = 0$ is an equilibrium of the transformed system

$$w' = F(t, w), \qquad (2.4)$$

where

$$F(t, w) = f(t, w + x_e). \qquad (2.5)$$

Since (2.5) establishes a one-to-one correspondence between the solutions of (E) and (2.4), we may assume henceforth that (E) possesses the equilibrium of interest located at the origin. This equilibrium $x = 0$ will sometimes be referred to as the **trivial solution** of (E).

5.3 DEFINITIONS OF STABILITY AND BOUNDEDNESS

We now give precise definitions of several stability, instability, and boundedness concepts. Throughout this section, we consider systems of equations (E),

$$x' = f(t, x), \qquad \text{(E)}$$

and we assume that (E) possesses an isolated equilibrium at the origin. Thus, $f(t, 0) = 0$ for all $t \geq 0$.

Definition 3.1. The equilibrium $x = 0$ of (E) is **stable** if for every $\varepsilon > 0$ and any $t_0 \in R^+$ there exists a $\delta(\varepsilon, t_0) > 0$ such that

$$|\phi(t, t_0, \xi)| < \varepsilon \qquad \text{for all} \quad t \geq t_0 \qquad (3.1)$$

whenever

$$|\xi| < \delta(\varepsilon, t_0). \qquad (3.2)$$

Note that if the equilibrium point $x = 0$ satisfies (3.1) for a single t_0 when (3.2) is true, then it also satisfies this condition at every initial time $t'_0 > t_0$, where a different value of δ may be required. To see this, we note that the spherical neighborhood $B(\delta(\varepsilon, t_0))$ is mapped by the solutions $\phi(t, t_0, \xi)$ onto a neighborhood of the origin at t'_0. This neighborhood contains in its interior a spherical neighborhood centered at the origin and with a radius δ'. If we choose $\xi' \in B(\delta')$, then (3.1) implies that $|\phi(t, t', \xi')| < \varepsilon$ for all $t \geq t'_0$. Hence, in Definition 3.1 it would have been enough to take the single value $t_0 = 0$ in (3.1) and (3.2).

5.3 Definitions of Stability and Boundedness

In Fig. 5.2 we depict the behavior of the trajectories in the vicinity of a stable equilibrium for the case $x \in R^2$. By choosing the initial points in a sufficiently small spherical neighborhood, we can force the graph of the solution for $t \geq t_0$ to lie entirely inside a given cylinder.

In Definition 3.1, δ depends on ε and t_0 [i.e., $\delta = \delta(\varepsilon, t_0)$]. If δ is independent of t_0, i.e., if $\delta = \delta(\varepsilon)$, then the equilibrium $x = 0$ of (E) is said to be **uniformly stable**.

Definition 3.2. The equilibrium $x = 0$ of (E) is **asymptotically stable** if

(i) it is stable, and
(ii) for every $t_0 \geq 0$ there exists an $\eta(t_0) > 0$ such that $\lim_{t \to \infty} \phi(t, t_0, \xi) = 0$ whenever $|\xi| < \eta$.

The set of all $\xi \in R^n$ such that $\phi(t, t_0, \xi) \to 0$ as $t \to \infty$ for some $t_0 \geq 0$ is called the **domain of attraction** of the equilibrium $x = 0$ of (E). Also, if for (E) condition (ii) is true, then the equilibrium $x = 0$ is said to be **attractive**.

Definition 3.3. The equilibrium $x = 0$ of (E) is **uniformly asymptotically stable** if

(i) it is uniformly stable, and
(ii) there is a $\delta_0 > 0$ such that for every $\varepsilon > 0$ and for any $t_0 \in R^+$, there exists a $T(\varepsilon) > 0$, independent of t_0, such that

$$|\phi(t, t_0, \xi)| < \varepsilon \quad \text{for all} \quad t \geq t_0 + T(\varepsilon) \quad \text{whenever} \quad |\xi| < \delta_0.$$

In Fig. 5.3 we depict property (ii) of Definition 3.3 pictorially. By choosing the initial points in a sufficiently small spherical neighborhood at $t = t_0$, we can force the graph of the solution to lie inside a given cylinder for all $t > t_0 + T(\varepsilon)$. Condition (ii) can be paraphrased by saying that there exists a $\delta_0 > 0$ such that

$$\lim_{t \to \infty} \phi(t + t_0, t_0, \xi) = 0$$

uniformly in (t_0, ξ) for $t_0 \geq 0$ and for $|\xi| \leq \delta_0$.

Frequently, in applications, we are interested in the following special case of uniform asymptotic stability.

Definition 3.4. The equilibrium $x = 0$ of (E) is **exponentially stable** if there exists an $\alpha > 0$, and for every $\varepsilon > 0$, there exists a $\delta(\varepsilon) > 0$, such that

$$|\phi(t, t_0, \xi)| \leq \varepsilon e^{-\alpha(t-t_0)} \quad \text{for all} \quad t \geq t_0$$

whenever $|\xi| < \delta(\varepsilon)$ and $t_0 \geq 0$.

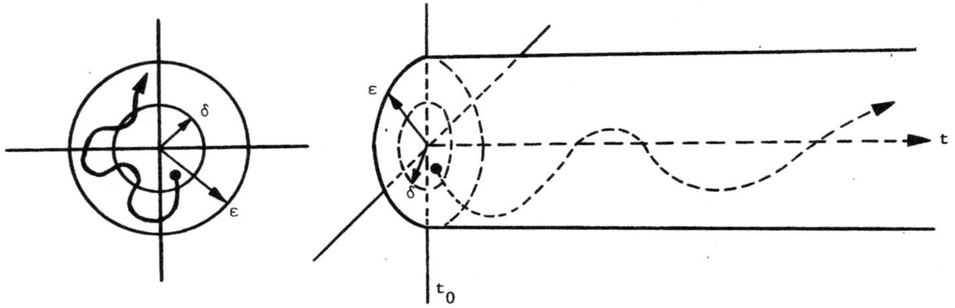

FIGURE 5.2

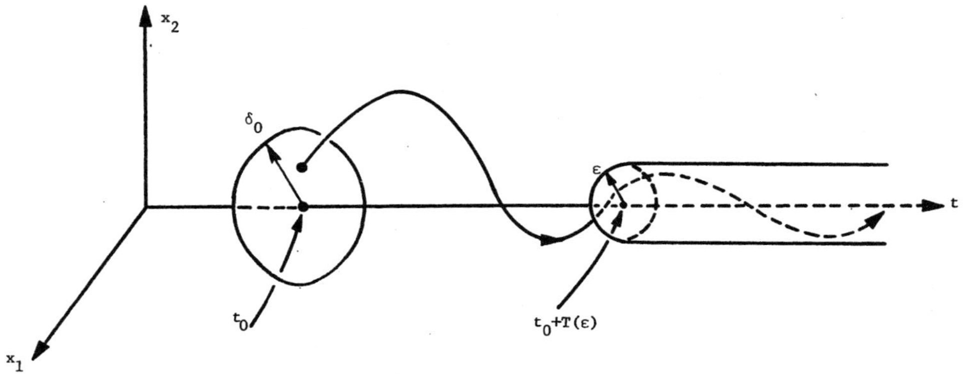

FIGURE 5.3

5.3 Definitions of Stability and Boundedness

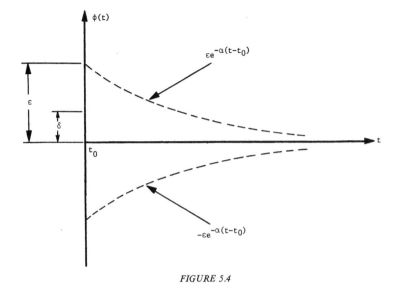

FIGURE 5.4

In Fig. 5.4, the behavior of a solution in the vicinity of an exponentially stable equilibrium $x = 0$ is shown.

Definition 3.5. The equilibrium $x = 0$ of (E) is **unstable** if it is not stable. In this case, there exists a $t_0 \geq 0$ and a sequence $\xi_m \to 0$ of initial points and a sequence $\{t_m\}$ such that $|\phi(t_0 + t_m, t_0, \xi_m)| \geq \varepsilon$ for all m, $t_m \geq 0$.

If $x = 0$ is an unstable equilibrium of (E), it still can happen that all the solutions tend to zero with increasing t. Thus, instability and attractivity are compatible concepts. Note that the equilibrium $x = 0$ is necessarily unstable if every neighborhood of the origin contains initial points corresponding to unbounded solutions (i.e., solutions whose norm $|\phi(t, t_0, \xi)|$ grows to infinity on a sequence $t_m \to \infty$ (cf. Definition 3.6). However, it can happen that a system (E) with unstable equilibrium $x = 0$ may have only bounded solutions.

The preceding concepts pertain to local properties of an equilibrium. In the following definitions, we consider some global characterizations of an equilibrium.

Definition 3.6. A solution $\phi(t, t_0, \xi)$ of (E) is **bounded** if there exists a $\beta > 0$ such that $|\phi(t, t_0, \xi)| < \beta$ for all $t \geq t_0$, where β may depend on each solution. System (E) is said to possess **Lagrange stability** if for each $t_0 \geq 0$ and ξ the solution $\phi(t, t_0, \xi)$ is bounded.

Definition 3.7. The solutions of (E) are **uniformly bounded** if for any $\alpha > 0$ and $t_0 \in R^+$, there exists a $\beta = \beta(\alpha) > 0$ (independent of t_0) such that if $|\xi| < \alpha$, then $|\phi(t, t_0, \xi)| < \beta$ for all $t \geq t_0$.

Definition 3.8. The solutions of (E) are **uniformly ultimately bounded** (with bound B) if there exists a $B > 0$ and if corresponding to any $\alpha > 0$ and $t_0 \in R^+$, there exists a $T = T(\alpha) > 0$ (independent of t_0) such that $|\xi| < \alpha$ implies that $|\phi(t, t_0, \xi)| < B$ for all $t \geq t_0 + T$.

In contrast to the boundedness properties defined in Definitions 3.6–3.8, the concepts introduced in Definitions 3.1–3.5 as well as those to follow in Definitions 3.9–3.11 are usually referred to as stability (respectively, instability) **in the sense of Lyapunov**.

Definition 3.9. The equilibrium $x = 0$ of (E) is **asymptotically stable in the large** if it is stable and if every solution of (E) tends to zero as $t \to \infty$.

In the case of Definition 3.9, the domain of attraction of the equilibrium $x = 0$ of (E) is all of R^n. Note that in this case, $x = 0$ is the *only* equilibrium of (E).

Definition 3.10. The equilibrium $x = 0$ of (E) is **uniformly asymptotically stable in the large** if

(i) it is uniformly stable, and
(ii) for any $\alpha > 0$ and any $\varepsilon > 0$, and $t_0 \in R^+$, there exists $T(\varepsilon, \alpha) > 0$, independent of t_0, such that if $|\xi| < \alpha$, then $|\phi(t, t_0, \xi)| < \varepsilon$ for all $t \geq t_0 + T(\varepsilon, \alpha)$.

Definition 3.11. The equilibrium $x = 0$ of (E) is **exponentially stable in the large** if there exists $\alpha > 0$ and for any $\beta > 0$, there exists $k(\beta) > 0$ such that

$$|\phi(t, t_0, \xi)| \leq k(\beta)|\xi|e^{-\alpha(t-t_0)} \qquad \text{for all} \quad t \geq t_0$$

whenever $|\xi| < \beta$.

We conclude this section by considering a few examples.

Example 3.12. The scalar equation

$$x' = 0 \qquad (3.3)$$

has for any initial condition $x(0) = c$ the solution $\phi(t, 0, c) = c$, i.e., all solutions are equilibria of (3.3). The trivial solution is stable; in fact, it is uniformly stable. However, it is not asymptotically stable.

5.3 Definitions of Stability and Boundedness

Example 3.13. The scalar equation
$$x' = ax, \quad a > 0, \tag{3.4}$$
has for every $x(0) = c$ the solution $\phi(t, 0, c) = ce^{at}$ and $x = 0$ is the only equilibrium of (3.4). This equilibrium is unstable.

Example 3.14. The scalar equation
$$x' = -ax, \quad a > 0, \tag{3.5}$$
has for every $x(0) = c$ the solution $\phi(t, 0, c) = ce^{-at}$ and $x = 0$ is the only equilibrium of (3.5). This equilibrium is exponentially stable in the large.

Example 3.15. The scalar equation
$$x' = \frac{-1}{t+1} x \tag{3.6}$$
has for every $x(t_0) = c$, $t_0 \geq 0$, a unique solution of the form
$$\phi(t, t_0, c) = (1 + t_0)c \frac{1}{t+1} \tag{3.7}$$
and $x = 0$ is the only equilibrium of (3.6). This equilibrium is uniformly stable and asymptotically stable in the large, but it is not uniformly asymptotically stable.

Example 3.16. As mentioned before, a system
$$x' = f(t, x) \tag{E}$$
can have all solutions approaching its critical point $x = 0$ without the critical point being asymptotically stable. An example of this type of behavior is given by the system
$$x_1' = \frac{x_1^2(x_2 - x_1) + x_2^5}{(x_1^2 + x_2^2)[1 + (x_1^2 + x_2^2)^2]},$$
$$x_2' = \frac{x_2^2(x_2 - 2x_1)}{(x_1^2 + x_2^2)[1 + (x_1^2 + x_2^2)^2]}.$$

For a detailed discussion of this system, see the book by Hahn [17, p. 84].

We shall consider the stability properties of higher order systems in much greater detail in the subsequent sections of this chapter and the next chapter, after we have developed the background required to analyze such systems.

5.4 SOME BASIC PROPERTIES OF AUTONOMOUS AND PERIODIC SYSTEMS

In this section, we show that in the case of autonomous systems

$$x' = f(x) \tag{A}$$

and periodic systems

$$x' = f(t, x), \quad f(t, x) = f(t + T, x), \tag{P}$$

stability of the equilibrium $x = 0$ is equivalent to uniform stability, and asymptotic stability of the equilibrium $x = 0$ is equivalent to uniform asymptotic stability. Since an autonomous system may be viewed as a periodic system with arbitrary period, it suffices to prove these statements only for the case of periodic systems.

Theorem 4.1. If the equilibrium $x = 0$ of (P) [or of (A)] is stable, then it is uniformly stable.

Proof. For purposes of contradiction, assume that the equilibrium $x = 0$ of (P) is not uniformly stable. Then there is an $\varepsilon > 0$ and sequences $\{t_{0m}\}$ with $t_{0m} \geq 0$, $\{\xi_m\}$, and $\{t_m\}$ such that $\xi_m \to 0$, $t_m \geq t_{0m}$, and $|\phi(t_m, t_{0m}, \xi_m)| \geq \varepsilon$. Let $t_{0m} = k_m T + \tau_m$, where k_m is a nonnegative integer and $0 \leq \tau_m < T$ and define $t_m^* = t_m - k_m T \geq \tau_m$. Then by uniqueness and periodicity of (P), we have $\phi(t + k_m T, t_{0m}, \xi_m) \equiv \phi(t, \tau_m, \xi_m)$ since both of these solve (P) and satisfy the initial condition $x(\tau_m) = \xi_m$. Thus

$$|\phi(t_m^*, \tau_m, \xi_m)| \geq \varepsilon. \tag{4.1}$$

We claim that the sequence $t_m^* \to \infty$. For if it did not, then by going to a convergent subsequence and relabeling, we could assume that $\tau_m \to \tau^*$ and $t_m^* \to t^*$. Then by continuity with respect to initial conditions, $\phi(t_m^*, \tau_m, \xi_m) \to \phi(t^*, \tau^*, 0) = 0$. This contradicts (4.1).

Since $x = 0$ is stable by assumption, then at $t_0 = T$ there is a $\delta > 0$ such that if $|\xi| < \delta$ then $|\phi(t, T, \xi)| < \varepsilon$ for $t \geq T$. Since $\xi_m \to 0$, then by continuity with respect to initial conditions, $|\phi(T, \tau_m, \xi_m)| < \delta$ for all $m \geq m(\delta)$. But then by the choice of δ and by (4.1), we have

$$\varepsilon > |\phi(t_m^*, T, \phi(T, \tau_m, \xi_m))| = |\phi(t_m^*, \tau_m, \xi_m)| \geq \varepsilon.$$

This contradiction completes the proof. ∎

Theorem 4.2. If the equilibrium $x = 0$ of (P) [or of (A)] is asymptotically stable, then it is uniformly asymptotically stable.

5.5 Linear Systems

Proof. The uniform stability is already proved. To prove attractivity, i.e., Definition 3.3 (ii), fix $\varepsilon > 0$. By hypothesis, there is an $\eta(T) > 0$ and a $t(\varepsilon, T) > 0$ such that if $|\xi| \leq \eta(T)$, then $|\phi(t, T, \xi)| < \varepsilon$ for all $t \geq T + t(\varepsilon, T)$. Uniform stability and attractivity imply $t(\varepsilon, T)$ is independent of $|\xi| \leq \eta$. By continuity with respect to initial conditions, there is a $\delta' > 0$ such that $|\phi(T, \tau, \xi)| < \eta(T)$ if $|\xi| < \delta'$ and $0 \leq \tau \leq T$. So $|\phi(t + T, \tau, \xi)| < \varepsilon$ if $|\xi| < \delta'$, $0 \leq \tau \leq T$, and $t \geq t(\varepsilon, T)$. Thus for $0 \leq \tau \leq T$, $|\xi| < \delta'$ and $t \geq (T - \tau) + t(\varepsilon, T)$, we have $|\phi(t + \tau, \tau, \xi)| < \varepsilon$. Put $\delta(\varepsilon) = \delta'$ and $t(\varepsilon) = t(\varepsilon, T) + T$. If $kT \leq \tau < (k + 1)T$, then $\phi(t, \tau, \xi) = \phi(t - kT, \tau - kT, \xi)$. Thus, if $|\xi| < \delta(\varepsilon)$ and $t \geq \tau + t(\varepsilon)$, then $t - kT \geq \tau - kT + t(\varepsilon)$ and

$$|\phi(t, \tau, \xi)| = |\phi(t - kT, \tau - kT, \xi)| < \varepsilon. \qquad \blacksquare$$

5.5 LINEAR SYSTEMS

In this section, we shall first study the stability properties of the equilibrium of linear autonomous homogeneous systems

$$x' = Ax, \quad t \geq 0, \qquad (L)$$

and linear homogeneous systems [with $A(t)$ continuous]

$$x' = A(t)x, \quad t \geq t_0, \quad t_0 \geq 0. \qquad (LH)$$

Recall that $x = 0$ is always an equilibrium of (L) and (LH) and that $x = 0$ is the only equilibrium of (LH) if $A(t)$ is nonsingular for all $t \geq 0$. Recall also that the solution of (LH) for $x(t_0) = \xi$ is of the form

$$\phi(t, t_0, \xi) = \Phi(t, t_0)\xi, \quad t \geq t_0,$$

where Φ denotes the state transition matrix of $A(t)$. Recall further that the solution of (L) for $x(t_0) = \xi$ is given by

$$\phi(t, t_0, \xi) = \Phi(t, t_0)\xi = \Phi(t - t_0)\xi = e^{A(t - t_0)}\xi.$$

We first consider some of the properties of system (LH).

Theorem 5.1. The equilibrium $x = 0$ of (LH) is **stable** if and only if the solutions of (LH) are bounded. Equivalently, the equilibrium $x = 0$ of (LH) is stable if and only if

$$\sup_{t \geq t_0} |\Phi(t, t_0)| \triangleq c(t_0) < \infty,$$

where $|\Phi(t, t_0)|$ denotes the matrix norm induced by the vector norm used on R^n.

Proof. Suppose that the equilibrium $x = 0$ of (LH) is stable. Then for any $t_0 \geq 0$ and for $\varepsilon = 1$ there is a $\delta = \delta(t_0, 1) > 0$ such that $|\phi(t, t_0, \xi)| < 1$ for all $t \geq t_0$ and all ξ with $|\xi| \leq \delta$. But then

$$|\phi(t, t_0, \xi)| = |\Phi(t, t_0)\xi| = |\Phi(t, t_0)(\xi\delta)/|\xi|| (|\xi|/\delta) < |\xi|/\delta$$

for all $\xi \neq 0$ and all $t \geq t_0$. Using the definition of matrix norm (see Section 2.6 or 5.1), we see that this is equivalent to

$$|\Phi(t, t_0)| \leq \delta^{-1}, \qquad t \geq t_0.$$

Conversely, suppose that all solutions $\phi(t, t_0, \xi) = \Phi(t, t_0)\xi$ are bounded. Let $\{e_1, \ldots, e_n\}$ be the natural basis for n space and let $|\phi(t, t_0, e_j)| < \beta_j$ for all $t \geq t_0$. For any vector $\xi = \sum_{j=1}^{n} \alpha_j e_j$ we have

$$|\phi(t, t_0, \xi)| = \left|\sum_{j=1}^{n} \alpha_j \phi(t, t_0, e_j)\right| \leq \sum_{j=1}^{n} |\alpha_j|\beta_j \leq \left(\max_j \beta_j\right) \sum_{j=1}^{n} |\alpha_j| \leq K|\xi|$$

for some constant $K > 0$ when $t \geq t_0$. Given $\varepsilon > 0$, we choose $\delta = \varepsilon/K$. If $|\xi| < \delta$, then $|\phi(t, t_0, \xi)| \leq K|\xi| < \varepsilon$ for all $t \geq t_0$. ∎

Theorem 5.2. *The equilibrium $x = 0$ of (LH) is **uniformly stable** if and only if*

$$\sup_{t_0 \geq 0} c(t_0) \triangleq \sup_{t_0 \geq 0} \left(\sup_{t \geq t_0} |\Phi(t, t_0)|\right) \triangleq c_0 < \infty.$$

The proof of this theorem is very similar to the proof of Theorem 5.1 and is left to the reader as an exercise.

For the asymptotic stability of the equilibrium $x = 0$ of (LH) we have the following result.

Theorem 5.3. *The following statements are equivalent.*

(i) *The equilibrium $x = 0$ of (LH) is asymptotically stable.*
(ii) *The equilibrium $x = 0$ of (LH) is asymptotically stable in the large.*
(iii) $\lim_{t \to \infty} |\Phi(t, t_0)| = 0.$

Proof. Suppose (i) is true. Then there is an $\eta(t_0) > 0$ such that when $|\xi| \leq \eta(t_0)$, then $\phi(t, t_0, \xi) \to 0$ as $t \to \infty$. But then we have for any $\xi \neq 0$,

$$\phi(t, t_0, \xi) = \phi(t, t_0, \eta(t_0)\xi/|\xi|)(|\xi|/\eta(t_0)) \to 0$$

as $t \to \infty$. Therefore (ii) is true.

Next, assume that (ii) is true and fix $t_0 \geq 0$. For any $\varepsilon > 0$ there must exist a $T(\varepsilon) > 0$ such that for all $t \geq t_0 + T(\varepsilon)$, we have $|\phi(t, t_0, \xi)| = |\Phi(t, t_0)\xi| < \varepsilon$. To see this, let $\{e_j\}$ be the natural basis for R^n. Thus for some

5.5 Linear Systems

fixed constant $K > 0$, if $\xi = (\alpha_1, \alpha_2, \ldots, \alpha_n)^T$ and if $|\xi| \leq 1$, then $\xi = \sum_{j=1}^{n} \alpha_j e_j$ and $\sum_{j=1}^{n} |\alpha_j| \leq K$. For each j there is a $T_j(\varepsilon)$ such that $|\Phi(t, t_0)e_j| < \varepsilon/K$ and $t \geq t_0 + T_j(\varepsilon)$. Define $T(\varepsilon) = \max\{T_j(\varepsilon): j = 1, \ldots, n\}$. For $|\xi| \leq 1$ and $t \geq t_0 + T(\varepsilon)$, we have

$$|\Phi(t, t_0)\xi| = \left|\sum_{j=1}^{n} \alpha_j \Phi(t, t_0)e_j\right| \leq \sum_{j=1}^{n} |\alpha_j|(\varepsilon/K) \leq \varepsilon.$$

By the definition of the matrix norm, this means that $|\Phi(t, t_0)| \leq \varepsilon$ for $t \geq t_0 + T(\varepsilon)$. Hence (iii) is true.

Assume now that (iii) is true. Then $|\Phi(t, t_0)|$ is bounded in t for $t \geq t_0$. By Theorem 5.1 the trivial solution is stable. To prove asymptotic stability, fix $t_0 \geq 0$ and $\varepsilon > 0$. If $|\xi| < \eta(t_0) = 1$, then $|\phi(t, t_0, \xi)| \leq |\Phi(t, t_0)||\xi| \to 0$ as $t \to \infty$. Hence (i) is true. ∎

For the exponential stability of the equilibrium $x = 0$ of (LH), we have the following result.

Theorem 5.4. The equilibrium $x = 0$ of (LH) is uniformly asymptotically stable if and only if it is exponentially stable.

Proof. Exponential stability implies uniform asymptotic stability of the equilibrium $x = 0$ for all systems (E) and hence for systems (LH) in particular.

For the converse, assume that the trivial solution of (LH) is uniformly asymptotically stable. Thus there is a $\delta > 0$ and a $T > 0$ such that if $|\xi| \leq \delta$, then

$$|\Phi(t + t_0 + T, t_0)\xi| < \delta/2$$

for all $t, t_0 \geq 0$. This means that

$$|\Phi(t + t_0 + T, t_0)| \leq \tfrac{1}{2} \quad \text{if} \quad t, t_0 \geq 0. \tag{5.1}$$

Since by Theorem 3.2.12, $\Phi(t, s) = \Phi(t, \tau)\Phi(\tau, s)$ for any t, s, and τ, then

$$|\Phi(t + t_0 + 2T, t_0)| = |\Phi(t + t_0 + 2T, t + t_0 + T)\Phi(t + t_0 + T, t_0)| \leq \tfrac{1}{4}$$

by (5.1.). By induction for $t, t_0 \geq 0$ we have

$$|\Phi(t + t_0 + nT, t_0)| \leq 2^{-n}. \tag{5.2}$$

Let $\alpha = \log 2/T$. Then (5.2) implies that for $0 \leq t < T$ we have

$$|\phi(t + t_0 + nT, t_0, \xi)| \leq 2|\xi|2^{-(n+1)} = 2|\xi|e^{-\alpha(n+1)T} \leq 2|\xi|e^{-\alpha(t+nT)}. \quad ∎$$

In the next theorem, we summarize the principal stability results of linear autonomous homogeneous systems (L).

Theorem 5.5. (i) The equilibrium $x = 0$ of (L) is **stable** if all eigenvalues of A have nonpositive real parts and every eigenvalue of A

which has a zero real part is a simple zero of the characteristic polynomial of A.

(ii) The equilibrium $x = 0$ of (L) is **asymptotically stable** if and only if all eigenvalues of A have negative real parts. In this case, there exist constants $k > 0$, $\sigma > 0$ such that

$$|\Phi(t, t_0)| \leq k \exp[-\sigma(t - t_0)] \qquad (t_0 \leq t < \infty),$$

where $\Phi(t, t_0)$ denotes the state transition matrix of (L).

Proof. We shall have to make reference to the discussion in Section 3.3 concerning the use of the Jordan canonical form J to compute $\exp(At)$. Let $P^{-1}AP = J$ and define $x = Py$. Then

$$y' = P^{-1}APy = Jy.$$

Note that $x = 0$ is stable if and only if $y = 0$ is stable, and furthermore, $x = 0$ is asymptotically stable if and only if $y = 0$ is asymptotically stable. Hence, we can assume without loss of generality that the "system matrix A" is in Jordan canonical form, i.e., A is in block diagonal form

$$A = \text{diag}(J_0, J_1, \ldots, J_s),$$

where

$$J_0 = \text{diag}(\lambda_1, \ldots, \lambda_k) \qquad \text{and} \qquad J_k = \lambda_{k+i} E_i + N_i$$

for the Jordan blocks $J_1, \ldots, J_s$.

As in (3.3.17) we see that

$$e^{J_0 t} = \text{diag}(e^{\lambda_1 t}, \ldots, e^{\lambda_k t}) \tag{5.3}$$

and

$$e^{J_i t} = e^{\lambda_{k+i} t} \begin{bmatrix} 1 & t & t^2 & \cdots & t^{n_i - 1}/(n_i - 1)! \\ 0 & 1 & t & \cdots & t^{n_i - 2}/(n_i - 2)! \\ \vdots & \vdots & \vdots & & \vdots \\ 0 & 0 & 0 & \cdots & 1 \end{bmatrix} \tag{5.4}$$

for $i = 1, \ldots, s$. Clearly $|e^{J_0 t}| = O(e^{\beta t})$ if $\text{Re}\,\lambda_i \leq \beta$ for all i. Also $|e^{J_i t}| = O(t^{n_i - 1} e^{\beta t}) = O(e^{(\beta + \varepsilon)t})$ for any $\varepsilon > 0$ when $\beta = \text{Re}\,\lambda_{k+i}$.

From the foregoing statements, it is clear that if $\text{Re}\,\lambda_i \leq 0$ for $1 \leq i \leq k$ and if $\text{Re}\,\lambda_{i+k} < 0$ for $1 \leq i \leq s$, then $|e^{At}| \leq K$ for some constant $K > 0$. Thus $|\Phi(t, t_0)| = |e^{A(t-t_0)}| \leq K$ for $t \geq t_0 \geq 0$. Hence, by Theorem 5.1, $y = 0$ (and therefore $x = 0$) is stable. The hypotheses of part (i) guarantee that the eigenvalues λ_i satisfy the stated conditions.

If all eigenvalues of A have negative real parts, then from the preceding discussion, there is a $K > 0$ and an $\alpha > 0$ such that

$$|e^{A(t-t_0)}| < K e^{-\alpha(t-t_0)}.$$

5.5 Linear Systems

Hence $y = 0$ (and therefore $x = 0$) is exponentially stable. Conversely, if there is an eigenvalue λ_i with nonnegative real part, then either one term in (5.3) does not tend to zero or else a term in (5.4) is unbounded as $t \to \infty$. In either case, $\exp(Jt)\xi$ will not tend to zero when ξ is properly chosen. Hence, $y = 0$ (and therefore $x = 0$) cannot be asymptotically stable. ∎

It can be shown that the equilibrium $x = 0$ of (L) is stable if and only if all eigenvalues of A have nonpositive real parts and those with zero real part occur in the Jordan form J only in J_0 and not in any of the Jordan blocks J_i, $1 \leq i \leq s$. The proof of this is left as an exercise to the reader.

We shall find it convenient to use the following convention.

Definition 5.6. A real $n \times n$ matrix A is called **stable** or a **Hurwitz matrix** if all of its eigenvalues have negative real parts. If at least one of the eigenvalues has a positive real part, then A is called **unstable**. A matrix A which is neither stable nor unstable is called **critical** and the eigenvalues of A with zero real parts are called **critical eigenvalues**.

Thus, the equilibrium $x = 0$ of (L) is asymptotically stable if and only if A is stable. If A is unstable, then $x = 0$ is unstable. If A is critical, then the equilibrium is stable if the eigenvalues with zero real parts correspond to a simple zero of the characteristic polynomial of A; otherwise, the equilibrium may be unstable.

Next, we consider the stability properties of linear periodic systems

$$x' = A(t)x, \qquad A(t) = A(t + T), \tag{PL}$$

where $A(t)$ is a continuous real matrix for all $t \in R$. We recall from Chapter 3 that if $\Phi(t, t_0)$ is the state transition matrix for (PL), then there exists a constant $n \times n$ matrix R and an $n \times n$ matrix $\Psi(t, t_0)$ such that

$$\Phi(t, t_0) = \Psi(t, t_0) \exp[R(t - t_0)], \tag{5.5}$$

where

$$\Psi(t + T, t_0) = \Psi(t, t_0) \qquad \text{for all} \quad t \geq 0.$$

Theorem 5.7. (i) The equilibrium $x = 0$ of (PL) is **uniformly stable** if all eigenvalues of R [in Eq. (5.5)] have nonpositive real parts and any eigenvalue of R having zero real part is a simple zero of the characteristic polynomial of R.

(ii) The equilibrium $x = 0$ of (PL) is **uniformly asymptotically stable** if and only if all eigenvalues of R have negative real parts.

Proof. According to the discussion at the end of Section 3.4, the change of variables $x = \Psi(t, t_0)y$ transforms (PL) to the system $y' = Ry$.

Moreover, $\Psi(t, t_0)^{-1}$ exists over $t_0 \leq t \leq t_0 + T$ so that the equilibrium $x = 0$ is stable (respectively, asymptotically stable) if and only if $y = 0$ is also stable (respectively, asymptotically stable). The results now follow from Theorem 5.5, applied to $y' = Ry$. ∎

The final result of this section is known as the **Routh–Hurwitz criterion**. It applies to nth order linear autonomous homogeneous ordinary differential equations of the form

$$a_0 x^{(n)} + a_1 x^{(n-1)} + \cdots + a_{n-1} x^{(1)} + a_n x = 0, \qquad a_0 \neq 0, \qquad (5.6)$$

where the coefficients $a_0, \ldots, a_n$ are all real. We recall from Chapter 1 that (5.6) is equivalent to the system of first order ordinary differential equations

$$x' = Ax, \qquad (5.7)$$

where A denotes the companion-form matrix given by

$$A = \begin{bmatrix} 0 & 1 & 0 & \cdots & 0 \\ 0 & 0 & 1 & \cdots & 0 \\ \vdots & \vdots & \vdots & & \vdots \\ -a_n/a_0 & -a_{n-1}/a_0 & \cdot & \cdots & -a_1/a_0 \end{bmatrix}.$$

To determine whether or not the equilibrium $x = 0$ of (5.7) is asymptotically stable, it suffices to determine if all eigenvalues of A have negative real parts, or what amounts to the same thing, if the roots of the polynomial

$$p(s) = a_0 s^n + a_1 s^{n-1} + \cdots + a_{n-1} s + a_n = 0 \qquad (5.8)$$

all have negative real parts. Similarly as in Definition 5.6, we shall find it convenient to use the following nomenclature.

Definition 5.8. An nth order polynomial $p(s)$ with real coefficients [such as (5.8)] is called **stable** if all zeros of $p(s)$ have negative real parts. It is called **unstable** if at least one of the zeros of $p(s)$ has a positive real part. It is called **critical** if $p(s)$ is neither stable nor unstable. A stable polynomial is also called a **Hurwitz polynomial**.

It turns out that we can determine whether or not a polynomial is Hurwitz by examining its coefficients without actually solving for the roots of the polynomial explicitly. This is demonstrated in the final theorem of this section. We first state the following necessary conditions.

Theorem 5.9. For (5.4) to be a Hurwitz polynomial, it is necessary that
$$a_1/a_0 > 0, \quad a_2/a_0 > 0, \ldots, a_n/a_0 > 0. \qquad (5.9)$$

The proof of this result is simple and is left as an exercise to the reader.

5.5 Linear Systems

Without loss of generality, we assume in the following that $a_0 > 0$. We will require the following array, called a **Routh array**.

$$
\begin{array}{llllll}
 & c_{10} = a_0 & c_{20} = a_2 & c_{30} = a_4 & c_{40} = a_6 & \cdots \\
 & c_{11} = a_1 & c_{21} = a_3 & c_{31} = a_5 & c_{41} = a_7 & \cdots \\
b_2 = a_0/a_1 & c_{12} = a_2 - b_2 a_3 & c_{22} = a_4 - b_2 a_5 & c_{32} = a_6 - b_2 a_7 & & \cdots \\
b_3 = c_{11}/c_{12} & c_{13} = c_{21} - b_3 c_{22} & c_{23} = c_{31} - b_3 c_{32} & c_{33} = c_{41} - b_3 c_{42} & & \cdots \\
\vdots & \vdots & \vdots & \vdots & & \\
b_j = \dfrac{c_{1,j-2}}{c_{1,j-1}} & c_{ij} = c_{i+1,j-2} - b_j c_{i+1,j-1}, & i = 1, 2, \ldots; & j = 2, 3, \ldots & & \\
\vdots & \vdots & & & & \\
 & c_{1n} = a_n. & & & &
\end{array}
$$

Note that if $n = 2m$, then we have $c_{m+1,0} = c_{m+1,2} = a_n$, $c_{m+1,1} = c_{m+1,3} = 0$. Also, if $n = 2m - 1$, then we have $c_{m0} = a_{n-1}$, $c_{m1} = a_n$, $c_{m2} = c_{m3} = 0$. The foregoing array terminates after $n - 1$ steps if all the numbers in c_{ij} are not zero and the last line determines c_{1n}.

In addition to inequalities (5.9), we shall require in the next result the inequalities

$$c_{11} > 0, \quad c_{12} > 0, \ldots, c_{1n} > 0. \tag{5.10}$$

Theorem 5.10. The polynomial $p(s)$ given in (5.8) is a Hurwitz polynomial if and only if the inequalities (5.9) and (5.10) hold.

The usual proof of this result involves some background from complex variables and an involved algebraic argument. The proof will not be given here. The reader should refer to the book by Hahn [17, pp. 16–22] for a proof.

An alternate form of the foregoing criterion can be given in terms of the **Hurwitz determinants** defined by

$$D_1 = a_1, \quad D_2 = \det \begin{bmatrix} a_1 & a_3 \\ a_0 & a_2 \end{bmatrix}, \quad D_3 = \det \begin{bmatrix} a_1 & a_3 & a_5 \\ a_0 & a_2 & a_4 \\ 0 & a_1 & a_3 \end{bmatrix}, \ldots,$$

$$D_k = \det \begin{bmatrix} a_1 & a_3 & a_5 & \cdots & a_{2k-1} \\ a_0 & a_2 & a_4 & \cdots & a_{2k-2} \\ 0 & a_1 & a_3 & \cdots & a_{2k-3} \\ 0 & a_0 & a_2 & \cdots & a_{2k-4} \\ \vdots & \vdots & \vdots & & \vdots \\ 0 & 0 & 0 & \cdots & a_k \end{bmatrix},$$

where we take $a_j = 0$ if $j > n$.

Corollary 5.11. The polynomial $p(s)$ given in (5.8) is a Hurwitz polynomial if and only if the inequalities (5.9) and the inequalities

$$D_j > 0 \quad \text{for} \quad j = 1, \ldots, n \tag{5.11}$$

are true.

For example, for the polynomial $p(s) = s^6 + 3s^5 + 2s^4 + 9s^3 + 5s^2 + 12s + 20$, the Routh array is given by

$$\begin{array}{cccc} 1 & 2 & 5 & 20 \\ 3 & 9 & 12 & 0 \\ -1 & 1 & 20 & \\ 12 & 72 & 0 & \\ 7 & 20 & & \\ 264/7 & 0 & & \\ 20 & & & \end{array}$$

Since $-1 < 0$, the polynomial $p(s)$ has a root with positive real part.

5.6 SECOND ORDER LINEAR SYSTEMS

In the present section, we study the stability properties of second order linear autonomous homogeneous systems given by

$$\begin{aligned} x_1' &= a_{11}x_1 + a_{12}x_2, \\ x_2' &= a_{21}x_1 + a_{22}x_2 \end{aligned} \tag{6.1}$$

or in matrix form

$$x' = Ax, \tag{6.2}$$

where

$$A = \begin{bmatrix} a_{11} & a_{12} \\ a_{21} & a_{22} \end{bmatrix}. \tag{6.3}$$

Recall that when $\det A \neq 0$, the system (6.1) will have one and only one equilibrium point, namely $x = 0$. We shall classify this equilibrium point [and hence, system (6.1)] according to the following cases which the eigen-

5.6 Second Order Linear Systems

values λ_1, λ_2 of A can assume:

(a) λ_1, λ_2 are real and $\lambda_1 < 0, \lambda_2 < 0 : x = 0$ is a **stable node**.
(b) λ_1, λ_2 are real and $\lambda_1 > 0, \lambda_2 > 0 : x = 0$ is an **unstable node**.
(c) λ_1, λ_2 are real and $\lambda_1 \lambda_2 < 0 : x = 0$ is a **saddle**.
(d) λ_1, λ_2 are complex conjugates and $\operatorname{Re} \lambda_1 = \operatorname{Re} \lambda_2 < 0 : x = 0$ is a **stable focus**.
(e) λ_1, λ_2 are complex conjugates and $\operatorname{Re} \lambda_1 = \operatorname{Re} \lambda_2 > 0 : x = 0$ is an **unstable focus**.
(f) λ_1, λ_2 are complex conjugates and $\operatorname{Re} \lambda_1 = \operatorname{Re} \lambda_2 = 0 : x = 0$ is a **center**.

The reason for the foregoing nomenclature will become clear shortly. Note that in accordance with the results of Section 5, stable nodes and stable foci are asymptotically stable equilibrium points, centers are stable equilibrium points (but not asymptotically stable ones), and saddles, unstable foci, and unstable nodes are unstable equilibrium points.

In the following, we let P denote a real constant nonsingular 2×2 matrix and we let

$$y = P^{-1} x. \tag{6.4}$$

Under this similarity transformation, system (6.2) assumes the equivalent form

$$y' = \Lambda y, \tag{6.5}$$

where

$$\Lambda = P^{-1} A P. \tag{6.6}$$

Note that if an initial condition for (6.2) is given by $x(0) = x_0$, then the corresponding initial condition for (6.5) will be given by

$$y(0) = y_0 = P^{-1} x_0. \tag{6.7}$$

We shall assume without loss of generality that when λ_1, λ_2 are real and not equal, then $\lambda_1 > \lambda_2$.

We begin our discussion by assuming that λ_1 and λ_2 are real and that A can be diagonalized, so that

$$\Lambda = \begin{bmatrix} \lambda_1 & 0 \\ 0 & \lambda_2 \end{bmatrix},$$

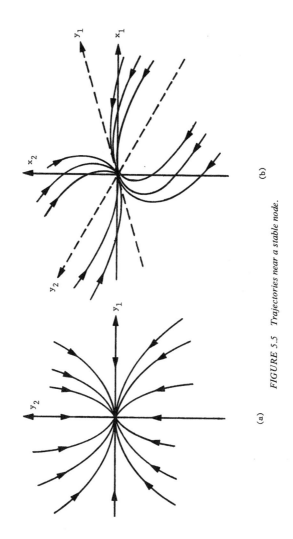

FIGURE 5.5 Trajectories near a stable node.

5.6 Second Order Linear Systems

where the λ_1, λ_2 are not necessarily distinct. Then (6.5) assumes the form

$$\begin{aligned} y'_1 &= \lambda_1 y_1, \\ y'_2 &= \lambda_2 y_2. \end{aligned} \quad (6.8)$$

For a given set of initial conditions $(y_{10}, y_{20}) = (y_1(0), y_2(0))$, the solution of (6.8) is given by

$$\begin{aligned} y_1(t) &\triangleq \phi_1(t, 0, y_{10}) = y_{10} e^{\lambda_1 t}, \\ y_2(t) &\triangleq \phi_2(t, 0, y_{20}) = y_{20} e^{\lambda_2 t}. \end{aligned} \quad (6.9)$$

We can study the qualitative properties of the equilibrium of (6.8) [resp., (6.1)] by considering a family of solutions of (6.8) which have initial points near the origin. By eliminating t in (6.9), we can express (6.9) equivalently as

$$y_2(t) = y_{20} [y_1(t)/y_{10}]^{\lambda_2/\lambda_1}. \quad (6.10)$$

Using either (6.9) or (6.10), we can sketch families of trajectories in the $y_1 y_2$ plane for a stable node (Fig. 5.5a), for an unstable node (Fig. 5.6a), and for a saddle (Fig. 5.7a). Using (6.4) in conjunction with (6.9) or (6.10), we can sketch corresponding families of trajectories in the $x_1 x_2$ plane. In all of these figures, the arrows signify increasing time t. Note that the qualitative shapes of the trajectories in the $y_1 y_2$ plane and in the $x_1 x_2$ plane are the same, i.e., the qualitative behavior of corresponding trajectories in the $y_1 y_2$ plane and in the $x_1 x_2$ plane has been preserved under the similarity transformation (6.4). However, under a given transformation, the trajectories shown in the (canonical) $y_1 y_2$ coordinate frame are generally subjected to a rotation and distortions, resulting in corresponding trajectories in the original $x_1 x_2$ coordinate frame.

Next, let us assume that *matrix A has two real repeated eigenvalues*, $\lambda_1 = \lambda_2 = \lambda$, and that Λ is in the Jordan canonical form

$$\Lambda = \begin{bmatrix} \lambda & 1 \\ 0 & \lambda \end{bmatrix}.$$

In this case, (6.5) assumes the form

$$\begin{aligned} y'_1 &= \lambda y_1 + y_2, \\ y'_2 &= \lambda y_2. \end{aligned} \quad (6.11)$$

For an initial point, we obtain for (6.11) the solution

$$\begin{aligned} y_1(t) &\triangleq \phi_1(t, 0, y_{10}, y_{20}) = y_{10} e^{\lambda t} + y_{20} t e^{\lambda t}, \\ y_2(t) &\triangleq \phi_2(t, 0, y_{20}) = y_{20} e^{\lambda t}. \end{aligned} \quad (6.12)$$

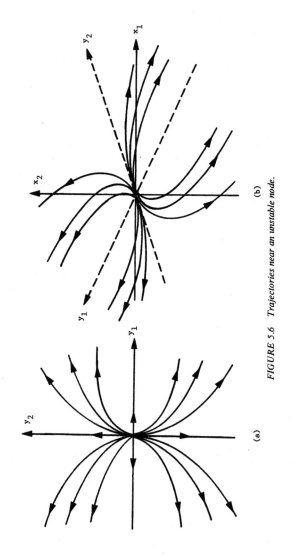

FIGURE 5.6 *Trajectories near an unstable node.*

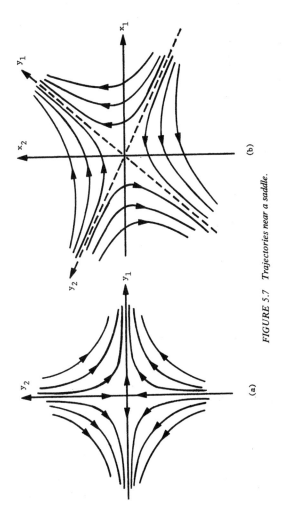

FIGURE 5.7 *Trajectories near a saddle.*

As before, we can eliminate the parameter t, and we can plot trajectories in the $y_1 y_2$ plane (resp., $x_1 x_2$ plane) for different sets of initial data near the origin. We leave these details as an exercise to the reader. In Fig. 5.8 we have typical trajectories near a stable node ($\lambda < 0$) for repeated eigenvalues.

Next, we consider the case when matrix A has two complex conjugate eigenvalues

$$\lambda_1 = \delta + i\tau, \qquad \lambda_2 = \delta - i\tau.$$

In this case, there exists a similarity transformation P such that the matrix $\Lambda = P^{-1} A P$ assumes the form

$$\Lambda = \begin{bmatrix} \delta & \tau \\ -\tau & \delta \end{bmatrix} \tag{6.13}$$

so that

$$\begin{aligned} y_1' &= \delta y_1 + \tau y_2, \\ y_2' &= -\tau y_1 + \delta y_2. \end{aligned} \tag{6.14}$$

The solution for the case $\delta > 0$, for initial data (y_{10}, y_{20}) is

$$\begin{aligned} y_1(t) &= \phi_1(t, 0, y_{10}, y_{20}) = e^{\delta t}[y_{10} \cos \tau t + y_{20} \sin \tau t], \\ y_2(t) &= \phi_2(t, 0, y_{10}, y_{20}) = e^{\delta t}[-y_{10} \sin \tau t + y_{20} \cos \tau t]. \end{aligned} \tag{6.15}$$

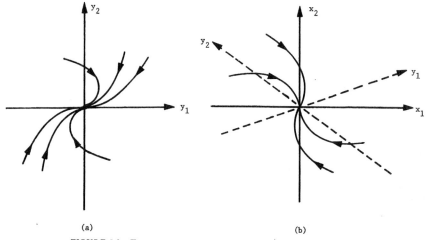

(a)   (b)

FIGURE 5.8   *Trajectories near a stable node (repeated eigenvalue case).*

5.6 Second Order Linear Systems

Letting $\rho = (y_{10}^2 + y_{20}^2)^{1/2}$, $\cos\alpha = y_{10}/\rho$, and $\sin\alpha = y_{20}/\rho$, we can rewrite (6.15) as

$$y_1(t) = \phi_1(t, 0, y_{10}, y_{20}) = e^{\delta t}\rho\cos(\tau t - \alpha),$$
$$y_2(t) = \phi_2(t, 0, y_{10}, y_{20}) = -e^{\delta t}\rho\sin(\tau t - \alpha). \tag{6.16}$$

If we let r and θ be the polar coordinates, $y_1 = r\cos\theta$ and $y_2 = r\sin\theta$, we may rewrite the solution as

$$r(t) = \rho e^{\delta t}, \qquad \theta(t) = -(\tau t - \alpha). \tag{6.17}$$

If, as before, we eliminate the parameter t, we obtain

$$r = C\exp(-\delta/\tau)\theta, \qquad C = \rho\exp[(\delta/\tau)\alpha]. \tag{6.18}$$

For different initial conditions near the origin (the origin is in this case an unstable focus), Eq. (6.18) yields a family of trajectories (in the form of spirals tending away from the origin as t increases) as shown in Fig. 5.9 (for $\tau > 0$).

When $\delta < 0$, we obtain in a similar manner, for different initial conditions near the origin, a family of trajectories as shown in Fig. 5.10 (for $\tau > 0$). In this case, the origin is a stable focus and the trajectories are in the form of spirals which tend toward the origin as t increases.

Finally, if $\delta = 0$, the origin is a center and the preceding formulas yield in this case, for different initial data near the origin, a family of concentric circles of radius ρ, as shown in Fig. 5.11 (for the case $\tau > 0$).

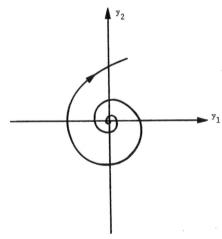

FIGURE 5.9 *Trajectory near an unstable focus.*

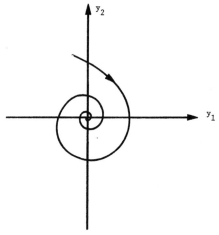

FIGURE 5.10   Trajectory near a stable focus.

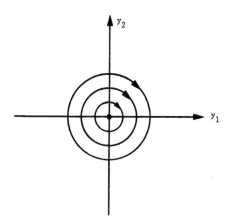

FIGURE 5.11   Trajectory near a center.

5.7 LYAPUNOV FUNCTIONS

In Section 9, we shall present stability results for the equilibrium $x = 0$ of a system

$$x' = f(t, x). \tag{E}$$

5.7 Lyapunov Functions

Such results involve the existence of real valued functions $v: D \to R$. In the case of local results (e.g., stability, instability, asymptotic stability, and exponential stability results), we shall usually only require that $D = B(h) \subset R^n$ for some $h > 0$, or $D = R^+ \times B(h)$. On the other hand, in the case of global results (e.g., asymptotic stability in the large, exponential stability in the large, and uniform boundedness of solutions), we have to assume that $D = R^n$ or $D = R^+ \times R^n$. Unless stated otherwise, we shall always assume that $v(t, 0) = 0$ for all $t \in R^+$ [resp., $v(0) = 0$].

Now let ϕ be an arbitrary solution of (E) and consider the function $t \mapsto v(t, \phi(t))$. If v is continuously differentiable with respect to all of its arguments, then we obtain (by the chain rule) the derivative of v with respect to t along the solutions of (E), $v'_{(E)}$, as

$$v'_{(E)}(t, \phi(t)) = \frac{\partial v}{\partial t}(t, \phi(t)) + \nabla v(t, \phi(t))^T f(t, \phi(t)).$$

Here ∇v denotes the gradient vector of v with respect to x. For a solution $\phi(t, t_0, \xi)$ of (E), we have

$$v(t, \phi(t)) = v(t_0, \xi) + \int_{t_0}^{t} v'_{(E)}(\tau, \phi(\tau, t_0, \xi)) \, d\tau.$$

These observations lead us to the following definition.

Definition 7.1. Let $v: R^+ \times R^n \to R$ [resp., $v: R^+ \times B(h) \to R$] be continuously differentiable with respect to all of its arguments and let ∇v denote the gradient of v with respect to x. Then $v'_{(E)}: R^+ \times R^n \to R$ [resp., $v'_{(E)}: R^+ \times B(h) \to R$] is defined by

$$\begin{aligned} v'_{(E)}(t, x) &= \frac{\partial v}{\partial t}(t, x) + \sum_{i=1}^{n} \frac{\partial v}{\partial x_i}(t, x) f_i(t, x) \\ &= \frac{\partial v}{\partial t}(t, x) + \nabla v(t, x)^T f(t, x). \end{aligned} \quad (7.1)$$

We call $v'_{(E)}$ the **derivative of v** (with respect to t) **along the solutions of** (E) [or along the trajectories of (E)].

It is important to note that in (7.1), the derivative of v with respect to t, along the solutions of (E), is evaluated *without having to solve Eq.* (E). The significance of this will become clear in the next few sections. We also note that when $v: R^n \to R$ [resp., $v: B(h) \to R$], then (7.1) reduces to $v'_{(E)}(t, x) = \nabla v(x)^T f(t, x)$. Also, in the case of autonomous systems,

$$x' = f(x), \quad (A)$$

if $v: R^n \to R$ [resp., $v: B(h) \to R$], we have

$$v'_{(A)}(x) = \nabla v(x)^T f(x). \quad (7.2)$$

Occasionally we shall only require that v be continuous on its domain of definition and that it satisfy locally a Lipschitz condition with respect to x. In such cases we define the **upper right-hand derivative of v with respect to t along the solutions of** (E) by

$$v'_{(E)}(t,x) = \lim_{\theta \to 0^+} \sup(1/\theta)\{v(t+\theta, \phi(t+\theta,t,x)) - v(t,x)\}$$
$$= \lim_{\theta \to 0^+} \sup(1/\theta)\{v(t+\theta, x+\theta \cdot f(t,x)) - v(t,x)\}. \quad (7.3)$$

When v is continuously differentiable, then (7.3) reduces to (7.1). Whether v is continuous or continuously differentiable will either be clear from the context or it will be specified.

We now give several important properties which v functions may possess.

Definition 7.2. A continuous function $w: R^n \to R$ [resp., $w: B(h) \to R$] is said to be **positive definite** if

(i) $w(0) = 0$, and
(ii) $w(x) > 0$ for all $x \neq 0$ [resp., $0 < |x| \leq r$ for some $r > 0$].

Definition 7.3. A continuous function $w: R^n \to R$ is said to be **radially unbounded** if

(i) $w(0) = 0$,
(ii) $w(x) > 0$ for all $x \in R^n - \{0\}$, and
(iii) $w(x) \to \infty$ as $|x| \to \infty$.

Definition 7.4. A function w is said to be **negative definite** if $-w$ is a positive definite function.

Definition 7.5. A continuous function $w: R^n \to R$ [resp., $w: B(h) \to R$] is said to be **indefinite** if

(i) $w(0) = 0$, and
(ii) in every neighborhood of the origin $x = 0$, w assumes negative and positive values.

Definition 7.6. A continuous function $w: R^n \to R$ [resp., $w: B(h) \to R$] is said to be **positive semidefinite** if

(i) $w(0) = 0$, and
(ii) $w(x) \geq 0$ for all $x \in B(r)$ for some $r > 0$.

Definition 7.7. A function w is said to be **negative semidefinite** if $-w$ is positive semidefinite.

Next, we consider the case $v: R^+ \times R^n \to R$, resp., $v: R^+ \times B(h) \to R$.

5.7 Lyapunov Functions

Definition 7.8. A continuous function $v: R^+ \times R^n$ [resp., $v: R^+ \times B(h)] \to R$ is said to be **positive definite** if there exists a positive definite function $w: R^n \to R$ [resp., $w: B(h) \to R$] such that

(i) $v(t, 0) = 0$ for all $t \geq 0$, and
(ii) $v(t, x) \geq w(x)$ for all $t \geq 0$ and for all $x \in B(r)$ for some $r > 0$.

Definition 7.9. A continuous function $v: R^+ \times R^n \to R$ is **radially unbounded** if there exists a radially unbounded function $w: R^n \to R$ such that

(i) $v(t, 0) = 0$ for all $t \geq 0$, and
(ii) $v(t, x) \geq w(x)$ for all $t \geq 0$ and for all $x \in R^n$.

Definition 7.10. A continuous function $v: R^+ \times R^n \to R$ [resp., $v: R^+ \times B(h) \to R$] is said to be **decrescent** if there exists a positive definite function $w: R^n \to R$ [resp., $w: B(h) \to R$] such that

$$|v(t, x)| \leq w(x) \quad \text{for all} \quad t \geq 0 \quad \text{and for all} \quad x \in B(r)$$

for some $r > 0$.

The definitions of **positive semidefinite**, **negative semidefinite**, and **negative definite**, when $v: R^+ \times R^n \to R$ or $v: R^+ \times B(h) \to R$, involve obvious modifications of Definitions 7.4, 7.6, and 7.7.

Some of the preceding characterizations of v functions (and w functions) can be rephrased in equivalent and very useful ways. In doing so, we employ certain comparison functions which we introduce next.

Definition 7.11. A continuous function $\psi: [0, r_1] \to R^+$ [resp., $\psi: [0, \infty) \to R^+$] is said to belong to **class** K, i.e., $\psi \in K$, if $\psi(0) = 0$ and if ψ is strictly increasing on $[0, r_1]$ [resp., on $[0, \infty)$]. If $\psi: R^+ \to R^+$, if $\psi \in K$, and if $\lim_{r \to \infty} \psi(r) = \infty$, then ψ is said to belong to **class** KR.

We are now in a position to state and prove the following results.

Theorem 7.12. A continuous function $v: R^+ \times R^n \to R$ [resp., $v: R^+ \times B(h) \to R$] is **positive definite** if and only if

(i) $v(t, 0) = 0$ for all $t \geq 0$, and
(ii) for any $r > 0$ [resp., some $r > 0$] there exists a $\psi \in K$ such that $v(t, x) \geq \psi(|x|)$ for all $t \geq 0$ and for all $x \in B(r)$.

Proof. If $v(t, x)$ is positive definite, then there is a function $w(x)$ satisfying the conditions of Definition 7.2 such that $v(t, x) \geq w(x)$ for $t \geq 0$ and $|x| \leq r$. Define $\psi_0(s) = \inf\{w(x) : s \leq |x| \leq r\}$ for $0 < s \leq r$. Clearly

ψ_0 is a positive and nondecreasing function such that $\psi_0(|x|) \leq w(x)$ on $0 < |x| \leq r$. Since ψ_0 is continuous, it is Riemann integrable. Define the function ψ by $\psi(0) = 0$ and

$$\psi(u) = u^{-1} \int_0^u (s/r)\psi_0(s)\,ds, \qquad 0 \leq u \leq r.$$

Clearly $0 < \psi(u) \leq \psi_0(u) \leq w(x) \leq v(t,x)$ if $t \geq 0$ and $|x| = u$. Moreover, ψ is continuous and increasing by construction.

Conversely, assume that (i) and (ii) are true and define $w(x) = \psi(|x|)$. ∎

We remark that both of the equivalent definitions of positive definite just given will be used. One of these forms is often easier to use in specific examples (to establish whether or not a given function is positive definite), while the second form will be very useful in proving stability results.

The proofs of the next two results are similar to the foregoing proof and are left as an exercise to the reader.

Theorem 7.13. A continuous function $v: R^+ \times R^n \to R$ is **radially unbounded** if and only if

(i) $v(t, 0) = 0$ for all $t \geq 0$, and
(ii) there exists a $\psi \in KR$ such that $v(t,x) \geq \psi(|x|)$ for all $t \geq 0$ and for all $x \in R^n$.

Theorem 7.14. A continuous function $v: R^+ \times R^n \to R$ [resp., $v: R^+ \times B(h) \to R$] is **decrescent** if and only if there exists a $\psi \in K$ such that

$$|v(t,x)| \leq \psi(|x|) \qquad \text{for all} \quad t \geq 0 \quad \text{and for all} \quad x \in B(r)$$
$$\text{for some} \quad r > 0.$$

We now consider several specific cases to illustrate the preceding concepts.

Example 7.15. (a) The function $w: R^3 \to R$ given by $w(x) = x^T x = x_1^2 + x_2^2 + x_3^2$ is positive definite and radially unbounded.

(b) The function $w: R^3 \to R$ given by $w(x) = x_1^2 + (x_2 + x_3)^2$ is positive semidefinite. It is not positive definite since it vanishes for all $x \in R^3$ such that $x_1 = 0$ and $x_2 = -x_3$.

(c) The function $w: R^2 \to R$ given by $w(x) = x_1^2 + x_2^2 - (x_1^2 + x_2^2)^3$ is positive definite (in the interior of the unit circle given by $x_1^2 + x_2^2 < 1$); however, it is not radially unbounded. In fact, if $x^T x > 1$, then $w(x) < 0$.

(d) The function $w: R^3 \to R$ given by $w(x) = x_1^2 + x_2^2$ is positive semidefinite. It is not positive definite.

5.7 Lyapunov Functions

(e) The function $w: R^2 \to R$ given by $w(x) = x_1^4/(1 + x_1^4) + x_2^4$ is positive definite but not radially unbounded.

Note that when $w: R^n \to R$ [resp., $B(h) \to R$] is positive or negative definite, then it is also decrescent, for in this case we can always find $\psi_1, \psi_2 \in K$ such that

$$\psi_1(|x|) \leq |w(x)| \leq \psi_2(|x|) \quad \text{for all} \quad x \in B(r) \quad \text{for some} \quad r > 0.$$

On the other hand, in the case when $v: R^+ \times R^n \to R$ [resp., $v: R^+ \times B(h) \to R$], care must be taken in establishing whether or not v is decrescent.

Example 7.16. (a) For the function $v: R^+ \times R^2 \to R$ given by $v(t, x) = (1 + \cos^2 t)x_1^2 + 2x_2^2$, we have

$$\psi_1(|x|) \triangleq x^T x \leq v(x) \leq 2x^T x \triangleq \psi_2(|x|), \qquad \psi_1, \psi_2 \in KR,$$

for all $t \geq 0$, $x \in R^2$. Therefore, v is positive definite, decrescent, and radially unbounded.

(b) For $v: R^+ \times R^2 \to R$ given by $v(t, x) = (x_1^2 + x_2^2)\cos^2 t$, we have

$$0 \leq v(t, x) \leq x^T x \triangleq \psi(|x|), \qquad \psi \in K,$$

for all $x \in R^2$ and for all $t \geq 0$. Thus, v is positive semidefinite and decrescent.

(c) For $v: R^+ \times R^2 \to R$ given by $v(t, x) = (1 + t)(x_1^2 + x_2^2)$, we have

$$\psi(|x|) \triangleq x^T x \leq v(t, x), \qquad \psi \in KR,$$

for all $t \geq 0$ and for all $x \in R^2$. Thus, v is positive definite and radially unbounded. It is not decrescent.

(d) For $v: R^+ \times R^2 \to R$ given by $v(t, x) = x_1^2/(1 + t) + x_2^2$, we have

$$v(t, x) \leq x^T x \triangleq \psi(|x|), \qquad \psi \in KR.$$

Hence, v is decrescent and positive semidefinite. It is not positive definite.

(e) The function $v: R^+ \times R^2 \to R$ given by

$$v(t, x) = (x_2 - x_1)^2(1 + t)$$

is positive semidefinite. It is not positive definite nor decrescent.

We close the present section with a discussion of an important class of v functions. Let $x \in R^n$, let $B = [b_{ij}]$ be a real symmetric $n \times n$ matrix, and consider the **quadratic form** $v: R^n \to R$ given by

$$v(x) = x^T B x = \sum_{i, k = 1}^{n} b_{ik} x_i x_k. \tag{7.4}$$

Recall that in this case, B is diagonalizable and all of its eigenvalues are real. We state the following results (which are due to Sylvester) without proof.

Theorem 7.17. Let v be the quadratic form defined in (7.4). Then

(i) v is positive definite (and radially unbounded) if and only if all principal minors of B are positive, i.e., if and only if

$$\det \begin{bmatrix} b_{11} & \cdots & b_{1k} \\ \vdots & & \vdots \\ b_{k1} & \cdots & b_{kk} \end{bmatrix} > 0, \qquad k = 1, \ldots, n.$$

(These inequalities are called the **Sylvester inequalities**.)

(ii) v is negative definite if and only if

$$(-1)^k \det \begin{bmatrix} b_{11} & \cdots & b_{1k} \\ \vdots & & \vdots \\ b_{k1} & \cdots & b_{kk} \end{bmatrix} > 0, \qquad k = 1, \ldots, n.$$

(iii) v is **definite** (i.e., either positive definite or negative definite) if and only if all eigenvalues are nonzero and have the same sign. (Thus, v is positive definite, if and only if all eigenvalues of B are positive.)

(iv) v is **semidefinite** (i.e., either positive semidefinite or negative semidefinite) if and only if the nonzero eigenvalues of B have the same sign.

(v) If $\lambda_1, \ldots, \lambda_n$ denote the eigenvalues of B (not necessarily distinct), if $\lambda_m = \min_i \lambda_i$, if $\lambda_M = \max_i \lambda_i$, and if we use the Euclidean norm ($|x| = (x^T x)^{1/2}$), then

$$\lambda_m |x|^2 \leq v(x) \leq \lambda_M |x|^2 \qquad \text{for all} \quad x \in R^n.$$

(vi) v is indefinite if and only if B possesses both positive and negative eigenvalues.

The reader will find the following example very instructive.

Example 7.18. The purpose of this example is to point out some of the geometric properties of (two-dimensional) quadratic forms. Let B be a real symmetric 2×2 matrix and let

$$v(x) = x^T B x.$$

Assume that both eigenvalues of B are positive so that v is positive definite and radially unbounded. In R^3, let us now consider the surface determined by

$$z = v(x) = x^T B x. \tag{7.5}$$

5.7 Lyapunov Functions

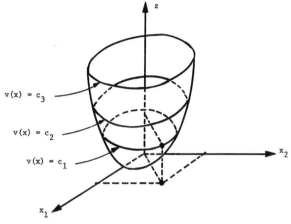

FIGURE 5.12

Equation (7.5) describes a cup-shaped surface as depicted in Fig. 5.12. Note that corresponding to every point on this cup-shaped surface there exists one and only one point in the $x_1 x_2$ plane. Note also that the loci defined by

$$C_i = \{x \in R^2 : v(x) = c_i \geq 0\} \qquad (c_i = \text{const})$$

determine closed curves in the $x_1 x_2$ plane as shown in Fig. 5.13. We call these curves **level curves**. Note that $C_0 = \{0\}$ corresponds to the case when

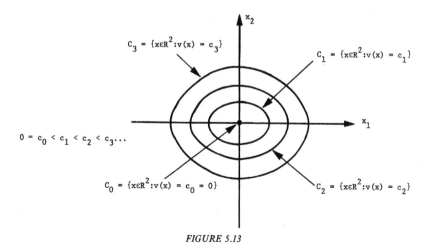

FIGURE 5.13

$z = c_0 = 0$. Note also that this function v can be used to cover the entire R^2 plane with closed curves by selecting for z all values in R^+.

In the case when $v = x^T B x$ is a positive definite quadratic form with $x \in R^n$, the preceding comments are still true; however, in this case, the closed curves C_i must be replaced by closed hypersurfaces in R^n and a simple geometric visualization as in Figs. 5.12 and 5.13 is no longer possible.

5.8 LYAPUNOV STABILITY AND INSTABILITY RESULTS: MOTIVATION

Before we state and prove the principal Lyapunov-type of stability and instability results, we give a geometric interpretation of some of these results in R^2. To this end, we consider the system of equations

$$\begin{aligned} x_1' &= f_1(x_1, x_2), \\ x_2' &= f_2(x_1, x_2), \end{aligned} \quad (8.1)$$

and we assume that f_1 and f_2 are such that for every (t_0, x_0), $t_0 \geq 0$, Eq. (8.1) has a unique solution $\phi(t, t_0, x_0)$ with $\phi(t_0, t_0, x_0) = x_0$. We also assume that $(x_1, x_2)^T = (0, 0)^T$ is the only equilibrium in $B(h)$ for some $h > 0$.

Next, let v be a positive definite, continuously differentiable function with nonvanishing gradient ∇v on $0 < |x| \leq h$. Then

$$v(x) = c \quad (c \geq 0)$$

defines for sufficiently small constants $c > 0$ a family of closed curves C_i which cover the neighborhood $B(h)$ as shown in Fig. 5.14. Note that the origin $x = 0$ is located in the interior of each such curve and in fact $C_0 = \{0\}$.

Now suppose that all trajectories of (8.1) originating from points on the circular disk $|x| \leq r_1 < h$ cross the curves $v(x) = c$ from the exterior toward the interior when we proceed along these trajectories in the direction of increasing values of t. Then we can conclude that these trajectories approach the origin as t increases, i.e., the equilibrium $x = 0$ is in this case asymptotically stable.

In terms of the given v function, we have the following interpretation. For a given solution $\phi(t, t_0, x_0)$ to cross the curve $v(x) = r$, $r = v(x_0)$, the angle between the outward normal vector $\nabla v(x_0)$ and the derivative of $\phi(t, t_0, x_0)$ at $t = t_0$ must be greater than $\pi/2$, i.e.,

$$v'_{(8.1)}(x_0) = \nabla v(x_0) f(x_0) < 0.$$

5.8 Lyapunov Stability and Instability Results: Motivation

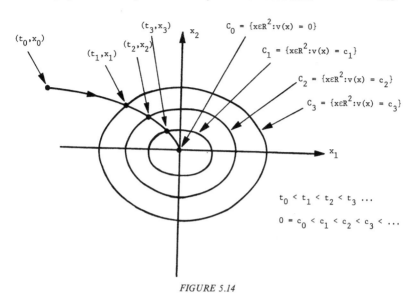

FIGURE 5.14

For this to happen at all points, we must have $v'_{(8.1)}(x) < 0$ for $0 < |x| \le r_1$. The same result can be arrived at from an analytic point of view. The function

$$V(t) = v(\phi(t, t_0, x_0))$$

decreases monotonically as t increases. This implies that the derivative $v'(\phi(t, t_0, x_0))$ along the solution $\phi(t, t_0, x_0)$ must be negative definite in $B(r)$ for $r > 0$ sufficiently small.

Next, let us assume that (8.1) has only one equilibrium (at $x = 0$) and that v is positive definite and radially unbounded. It turns out that in this case, the relation $v(x) = c$, $c \in R^+$, can be used to cover all of R^2 by closed curves of the type shown in Fig. 5.14. If for arbitrary (t_0, x_0), the corresponding solution of (8.1), $\phi(t, t_0, x_0)$, behaves as already discussed, then it follows that the derivative of v along this solution $v'(\phi(t, t_0, x_0))$, will be negative definite in R^2.

Since the foregoing discussion was given in terms of an arbitrary solution of (8.1), we may suspect that the following results are true:

1. If there exists a positive definite function v such that $v'_{(8.1)}$ is negative definite, then the equilibrium $x = 0$ of (8.1) is asymptotically stable.

2. If there exists a positive definite and radially unbounded function v such that $v'_{(8.1)}$ is negative definite for all $x \in R^2$, then the equilibrium $x = 0$ of (8.1) is asymptotically stable in the large.

In the next section we shall state and prove results which include the foregoing conjectures as special cases.

Continuing our discussion by making reference to Fig. 5.15, let us assume that we can find for Eq. (8.1) a continuously differentiable function $v: R^2 \to R$ which is indefinite and which has the properties discussed below. Since v is indefinite, there exist in each neighborhood of the origin points for which $v > 0$, $v < 0$, and $v(0) = 0$. Confining our attention to $B(k)$, where $k > 0$ is sufficiently small, we let $D = \{x \in B(k) : v(x) < 0\}$. The boundary of D, ∂D, which may consist of several subdomains, as shown in Fig. 5.15, consists of points in $\partial B(k)$ and of points determined by $v(x) = 0$. Assume that in the interior of D, v is bounded. Suppose $v'_{(8.1)}(x)$ is negative definite in D and that $x(t)$ is a trajectory of (8.1) which originates somewhere on the boundary of D ($x(t_0) \in \partial D$) with $v(x(t_0)) = 0$. Then this trajectory will penetrate the boundary of D at points where $v = 0$ as t increases and it can never again reach a point where $v = 0$. In fact, as t increases, this trajectory will penetrate the set of points determined by $|x| = k$ (since by assumption, $v'_{(8.1)} < 0$ along this trajectory and $v < 0$ in D). But this indicates that the equilibrium $x = 0$ of (8.1) is unstable. We are once more led to a conjecture (which we shall prove in the next section):

3. Let a function $v: R^2 \to R$ be given which is continuously differentiable and which has the following properties:

(i) There exist points x arbitrarily close to the origin such that $v(x) < 0$; they form the domain D which is bounded by the set of points determined by $v = 0$ and the disk $|x| = k$.

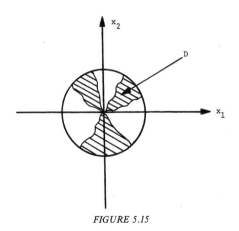

FIGURE 5.15

5.9 Principal Lyapunov Stability and Instability Theorems

(ii) In the interior of D, v is bounded.
(iii) In the interior of D, $v'_{(8.1)}$ is negative.

Then the equilibrium $x = 0$ of (8.1) is unstable.

5.9 PRINCIPAL LYAPUNOV STABILITY AND INSTABILITY THEOREMS

We are now in a position to give precise statements and proofs of some of the more important stability, instability, and boundedness results for the system of equations given by

$$x' = f(t, x). \tag{E}$$

These results comprise the **direct method of Lyapunov**, which is also sometimes called the **second method of Lyapunov**. The reason for this nomenclature is clear: results of the type presented here allow us to make qualitative statements about whole families of solutions of (E), without actually solving this equation.

As already mentioned, in the case of local stability results, we shall require that $x = 0$ is an isolated equilibrium of (E), and in the case of global stability results, we shall require that $x = 0$ is the only equilibrium of (E).

The results given in this section require the existence of functions $v: R^+ \times B(h) \to R$ (resp., $v: R^+ \times R^n \to R$) which are assumed to be *continuously differentiable* with respect to all arguments of v. We emphasize that these results can be generalized to the case where v is only continuous on its domain of definition and where v is required to satisfy locally a Lipschitz condition with respect to x. In this case, $v'_{(E)}$ must be interpreted in the sense of Eq. (7.3).

A. Stability

In our first two results, we concern ourselves with the stability and uniform stability of the equilibrium $x = 0$ of (E).

Theorem 9.1. If there exists a continuously differentiable positive definite function v with a negative semidefinite (or identically zero) derivative $v'_{(E)}$, then the equilibrium $x = 0$ of (E) is **stable**.

Proof. According to Definition 3.1, we fix $\varepsilon > 0$ and $t_0 \geq 0$ and we seek a $\delta > 0$ such that (3.1) and (3.2) are satisfied. Without loss of generality, we can assume that $\varepsilon < h_1$. Since $v(t, x)$ is positive definite, then by Theorem 7.12 there is a function $\psi \in K$ such that $v(t, x) \geq \psi(|x|)$ for $0 \leq |x| \leq h_1$, $t \geq 0$. Pick $\delta > 0$ so small that $v(t_0, x_0) < \psi(\varepsilon)$ if $|x_0| \leq \delta$. Since $v'_{(E)}(t, x) \leq 0$, then $v(t, \phi(t, t_0, x_0))$ is monotone nonincreasing and $v(t, \phi(t, t_0, x_0)) < \psi(\varepsilon)$ for all $t \geq t_0$. Thus, $|\phi(t, t_0, x_0)|$ cannot reach the value ε, since this would imply that $v(t, \phi(t, t_0, x_0)) \geq \psi(|\phi(t, t_0, x_0)|) = \psi(\varepsilon)$. ∎

Theorem 9.2. *If there exists a continuously differentiable, positive definite, decrescent function v with a negative semidefinite derivative $v'_{(E)}$, then the equilibrium $x = 0$ of (E) is* **uniformly stable**.

Proof. By Theorems 7.12 and 7.14, there are two functions ψ_1 and $\psi_2 \in K$ such that $\psi_1(|x|) \leq v(t, x) \leq \psi_2(|x|)$ for all $t \geq 0$ and for all x with $|x| \leq h_1$. Fix ε in the range $0 < \varepsilon < h_1$. Pick $\delta > 0$ so small that $\psi_2(\delta) < \psi_1(\varepsilon)$. If $t_0 \geq 0$ and if $|x_0| \leq \delta$, then $v(t_0, x_0) \leq \psi_2(\delta) < \psi_1(\varepsilon)$. Since $v'_{(E)}$ is nonpositive, then $v(t, \phi(t, t_0, x_0))$ is monotone nonincreasing. Thus $v(t, \phi(t, t_0, x_0)) < \psi_1(\varepsilon)$ for all $t \geq t_0$. Hence, $\psi_1(|\phi(t, t_0, x_0)|) < \psi_1(\varepsilon)$ for all $t \geq t_0$. Since ψ_1 is strictly increasing, then $|\phi(t, t_0, x_0)| < \varepsilon$ for all $t \geq t_0$. ∎

Let us now consider some specific examples.

Example 9.3. Consider the simple pendulum (see Chapter 1 and Example 2.2)

$$\begin{aligned} x'_1 &= x_2, \\ x'_2 &= -k \sin x_1, \end{aligned} \quad (9.1)$$

where $k > 0$ is a constant. As noted before, the system (9.1) has an isolated equilibrium at $x = 0$. The total energy for the pendulum is the sum of the kinetic energy and potential energy, given by

$$v(x) = \tfrac{1}{2} x_2^2 + k \int_0^{x_1} \sin \eta \, d\eta = \tfrac{1}{2} x_2^2 + k(1 - \cos x_1).$$

Note that this function is continuously differentiable, that $v(0) = 0$, and that v is positive definite. Also, note that v is automatically decrescent, since it does not depend on t.

Along the solutions of (9.1) we have

$$v'_{(9.1)}(x) = (k \sin x_1) x'_1 + x_2 x'_2 = (k \sin x_1) x_2 + x_2(-k \sin x_1) = 0.$$

In accordance with Theorem 9.1, the equilibrium $x = 0$ of (9.1) is stable, and in accordance with Theorem 9.2, the equilibrium $x = 0$ of (9.1) is uniformly stable.

Note that since $v'_{(9.1)} = 0$, the total energy in system (9.1) will be constant for a given set of initial conditions for all $t \geq 0$.

5.9 Principal Lyapunov Stability and Instability Theorems

There are in general no specific rules which tell us how to choose a v function in a particular problem. This is perhaps the major shortcoming of the results that comprise the direct method of Lyapunov. The preceding example suggests that a good choice for a v function is the total energy of a system. It turns out that another widely used class of v functions consists of quadratic forms defined by (7.4).

Example 9.4. Consider the second order system

$$x'' + x' + e^{-t}x = 0. \tag{9.2}$$

Letting $x = x_1$, $x' = x_2$, we can express (9.2) equivalently by

$$\begin{aligned} x_1' &= x_2, \\ x_2' &= -x_2 - e^{-t}x_1. \end{aligned} \tag{9.3}$$

This system has an isolated equilibrium at the origin $(x_1, x_2) = (0, 0)$. In studying the stability properties of this equilibrium, let us choose the positive definite function

$$v(x_1, x_2) = x_1^2 + x_2^2.$$

Along the solutions of (9.3), we have

$$v'_{(9.3)}(t, x_1, x_2) = 2x_1 x_2 (1 - e^{-t}) - 2x_2^2.$$

Since for this choice of v function neither of the preceding two theorems is applicable, we can reach no conclusion. So let us choose another v function,

$$v(t, x_1, x_2) = x_1^2 + e^t x_2^2.$$

In this case, we obtain

$$v'_{(9.3)}(t, x_1, x_2) = -e^t x_2^2.$$

This v function is positive definite and $v'_{(9.3)}$ is negative semidefinite. Therefore, Theorem 9.1 is applicable and we conclude that the equilibrium $x = 0$ is stable. However, since v is not decrescent, Theorem 9.2 is not applicable and we cannot conclude that the equilibrium $x = 0$ is uniformly stable.

Example 9.5. Consider a conservative dynamical system with n degrees of freedom, which we discussed in Chapter 1 and which is given by

$$\begin{aligned} q_i' &= \frac{\partial H}{\partial p_i}(p, q), \\ p_i' &= -\frac{\partial H}{\partial q_i}(p, q), \end{aligned} \tag{9.4}$$

where $q^T = (q_1, \ldots, q_n)$ denotes the generalized position vector, $p^T = (p_1, \ldots, p_n)$ the momentum vector, $H(p,q) = T(p) + W(q)$ the Hamiltonian, $T(p)$ the kinetic energy, and $W(q)$ the potential energy. The positions of equilibrium points of (9.4) correspond to the points in R^{2n} where the partial derivatives of H all vanish. In the following, we assume that $(p^T, q^T) = (0^T, 0^T)$ is an isolated equilibrium of (9.4), and without loss of generality we also assume that $H(0,0) = 0$. Furthermore, we assume that H is smooth and that $T(p)$ and $W(q)$ are of the form

$$T(p) = T_2(p) + T_3(p) + \cdots$$

and

$$W(q) = W_k(q) + W_{k+1}(q) + \cdots, \qquad k \geq 2.$$

Here $T_j(p)$ denotes the terms in p of order j and $W_j(q)$ denotes the terms in q of order j. The kinetic energy $T(p)$ is always assumed to be positive definite with respect to p. If the potential energy has an isolated minimum at $q = 0$, then W is positive definite with respect to q. So let us choose as a v function

$$v(p,q) = H(p,q) = T(p) + W(q)$$

which is positive definite. Since

$$v'_{(9.4)}(p,q) = H'(p,q) = 0,$$

Theorem 9.1 is applicable and we conclude that the equilibrium at the origin is stable. Since v is independent of t, it is also decrescent, Theorem 9.2 is also applicable, and we conclude that the equilibrium at the origin is also **uniformly stable**. Note that Example 9.3 (for the simple pendulum) is a special case of the present example.

B. Asymptotic Stability

The next two results address the asymptotic stability of the equilibrium $x = 0$ of (E).

Theorem 9.6. If there exists a continuously differentiable, positive definite, decrescent function v with a negative definite derivative $v'_{(E)}$, then the equilibrium $x = 0$ of (E) is **uniformly asymptotically stable.**

Proof. By Theorem 9.2 the equilibrium $x = 0$ is uniformly stable. It remains to be shown that Definition 3.3(ii) is also satisfied.

The hypotheses of this theorem along with Theorems 7.12–7.14 imply that there are functions ψ_1, ψ_2, and ψ_3 in class K such that

$$\psi_1(|x|) \leq v(t,x) \leq \psi_2(|x|) \qquad \text{and} \qquad v'_{(E)}(t,x) \leq -\psi_3(|x|)$$

5.9 Principal Lyapunov Stability and Instability Theorems

for all $(t, x) \in R^+ \times B(r_1)$ for some $r_1 > 0$. Pick $\delta_1 > 0$ such that $\psi_2(\delta_1) < \psi_1(r_1)$. Choose ε such that $0 < \varepsilon \leq r_1$. Choose δ_2 such that $0 < \delta_2 < \delta_1$ and such that $\psi_2(\delta_2) < \psi_1(\varepsilon)$. Define $T = \psi_1(r_1)/\psi_3(\delta_2)$. Fix $t_0 \geq 0$ and x_0 with $|x_0| < \delta_1$.

We now claim that $|\phi(t^*, t_0, x_0)| < \delta_2$ for some $t^* \in [t_0, t_0 + T]$. For if this were not true, we would have $|\phi(t, t_0, x_0)| \geq \delta_2$ for all $t \in [t_0, t_0 + T]$. Thus

$$0 < \psi_1(\delta_2) \leq v(t, \phi(t, t_0, x_0))$$
$$\leq v(t_0, x_0) + \int_{t_0}^{t} v'_{(E)}(s, \phi(s, t_0, x_0))\, ds$$
$$\leq \psi_2(\delta_1) - \int_{t_0}^{t} \psi_3(\delta_2)\, ds.$$

Now at $t = t_0 + T$ we find that

$$0 < \psi_2(\delta_1) - T\psi_3(\delta_2) = \psi_2(\delta_1) - \psi_1(r_1) < 0.$$

a contradiction. Hence, t^* exists.

Now for $t \geq t^*$ we have

$$\psi_1(|\phi(t, t_0, x_0)|) \leq v(t, \phi(t, t_0, x_0)) \leq v(t^*, \phi(t^*, t_0, x_0))$$
$$\leq \psi_2(|\phi(t^*, t_0, x_0)|) \leq \psi_2(\delta_2) < \psi_1(\varepsilon).$$

Since $\psi_1 \in K$ it follows that $|\phi(t, t_0, x_0)| < \varepsilon$ for all $t \geq t^*$ and hence for all $t \geq t_0 + T$. ∎

Theorem 9.7. If there exists a continuously differentiable, positive definite, decrescent, and radially unbounded function v such that $v'_{(E)}$ is negative definite for all $(t, x) \Pi R^+ \times R^n$, then the equilibrium $x = 0$ of (E) is **uniformly asymptotically stable in the large**.

Proof. The trivial solution of (E) is uniformly asymptotically stable by Theorem 9.6. It remains to be shown that the domain of attraction of $x = 0$ is all of R^n.

Fix $(t_0, x_0) \in R^+ \times R^n$. Then $v(t, \phi(t, t_0, x_0))$ is nonincreasing and so has a limit $\eta \geq 0$. If $|x_0| \leq \alpha$, then

$$\psi_2(\alpha) \geq v(t, \phi(t, t_0, x_0)) \geq \psi_1(|\phi(t, t_0, x_0)|), \qquad \psi_i \in KR,$$

and so $|\phi(t, t_0, x_0)| \leq \alpha_1 = \psi_1^{-1}(\psi_2(\alpha))$.

Suppose that no $T(\alpha, \varepsilon)$ exists. Then for some $x_0, \eta > 0$. By Theorem 7.12, for $|x| \leq \alpha_1$ find $\psi_3 \in K$ such that $v'_{(E)}(t, x) \leq -\psi_3(|x|)$. Thus, for $t \geq t_0$ we have $\psi_2^{-1}(\eta) \leq |\phi(t, t_0, x_0)|$ and

$$\eta \leq v(t, \phi(t, t_0, x_0)) \leq v(t_0, x_0) - \int_{t_0}^{t} \psi_3(|\phi(s, t_0, x_0)|)\, ds$$
$$\leq v(t_0, x_0) - \int_{t_0}^{t} \psi_3(\psi_2^{-1}(\eta))\, ds.$$

Thus, the right-hand side of this inequality becomes negative for t sufficiently large. But this is impossible when $\eta > 0$. Hence, $\eta = 0$. ∎

Example 9.8. Consider the system
$$x_1' = (x_1 - c_2 x_2)(x_1^2 + x_2^2 - 1),$$
$$x_2' = (c_1 x_1 + x_2)(x_1^2 + x_2^2 - 1) \tag{9.5}$$
which has an isolated equilibrium at the origin $x = 0$. Choosing
$$v(x) = c_1 x_1^2 + c_2 x_2^2,$$
we obtain
$$v'_{(9.5)}(x) = 2(c_1 x_1^2 + c_2 x_2^2)(x_1^2 + x_2^2 - 1).$$

If $c_1 > 0$, $c_2 > 0$, then v is positive definite and radially unbounded and $v'_{(9.5)}$ is negative definite in the domain $x_1^2 + x_2^2 < 1$. As such, Theorem 9.6 is applicable and we conclude that the equilibrium $x = 0$ is uniformly asymptotically stable. Theorem 9.7 is not applicable and we cannot conclude that the equilibrium $x = 0$ is uniformly asymptotically stable in the large.

Example 9.9. Consider the system
$$x_1' = x_2 + c x_1 (x_1^2 + x_2^2),$$
$$x_2' = -x_1 + c x_2 (x_1^2 + x_2^2), \tag{9.6}$$
where c is a real constant. Note that $x = 0$ is the only equilibrium. Choosing
$$v(x) = x_1^2 + x_2^2,$$
we obtain
$$v'_{(9.6)}(x) = 2c(x_1^2 + x_2^2)^2.$$

If $c = 0$, then Theorems 9.1 and 9.2 are applicable and the equilibrium $x = 0$ of (9.6) is uniformly stable. If $c < 0$, then Theorem 9.7 is applicable and the equilibrium $x = 0$ of (9.6) is uniformly asymptotically stable in the large.

C. Exponential Stability

The next two results deal with the exponential stability of the equilibrium $x = 0$ of (E).

Theorem 9.10. If there exists a continuously differentiable function v and three positive constants c_1, c_2, and c_3 such that
$$c_1 |x|^2 \leq v(t, x) \leq c_2 |x|^2, \qquad v'_{(E)}(t, x) \leq -c_3 |x|^2$$

5.9 Principal Lyapunov Stability and Instability Theorems

for all $t \in R^+$ and for all $x \in B(r)$ for some $r > 0$, then the equilibrium $x = 0$ of (E) is **exponentially stable**.

Proof. Given any $(t_0, x_0) \in R^+ \times B(r)$, let $\phi_0(t) = \phi(t, t_0, x_0)$ and let $V(t) = v(t, \phi_0(t))$. Then $V(t)$ satisfies the differential inequality

$$V'(t) \leq -c_3|\phi(t)|^2 \leq -(c_3/c_2)V(t).$$

By Lemma 2.8.2 it follows that $V(t) \leq V(t_0)\exp[(-c_3/c_2)(t - t_0)]$. Thus

$$c_1|\phi_0(t)|^2 \leq V(t_0)\exp[(-c_3/c_2)(t - t_0)]$$
$$\leq c_2|x_0|^2 \exp[(-c_3/c_2)(t - t_0)].$$

Hence, the conditions of Definition 3.4 are fulfilled with $\alpha = c_3/(2c_2)$ and $\delta(\varepsilon) = (c_2/c_1)^{1/2}\varepsilon$. ∎

Theorem 9.11. If there exists a continuously differentiable function v and three positive constants c_1, c_2, and c_3 such that

$$c_1|x|^2 \leq v(t, x) \leq c_2|x|^2, \qquad v'_{(E)}(t, x) \leq -c_3|x|^2$$

for all $t \in R^+$ and for all $x \in R^n$, then the equilibrium $x = 0$ of (E) is **exponentially stable in the large**.

Proof. Similarly as in the proof of Theorem 9.10, we have for any $(t_0, x_0) \in R^+ \times R^n$ the estimate

$$|\phi(t, t_0, x_0)| \leq (c_2/c_1)^{1/2}|x_0|\exp[(-c_3/2c_2)(t - t_0)]. \qquad ∎$$

Let us consider a specific example.

Example 9.12. Consider the system described by the equations

$$\begin{aligned} x'_1 &= -a(t)x_1 - bx_2, \\ x'_2 &= bx_1 - c(t)x_2, \end{aligned} \qquad (9.7)$$

where b is a real constant and where a and c are real and continuous functions defined for $t \geq 0$ satisfying $a(t) \geq \delta > 0$ and $c(t) \geq \delta > 0$ for all $t \geq 0$. We assume that $x = 0$ is the only equilibrium for (9.7).

If we choose

$$v(x) = \tfrac{1}{2}(x_1^2 + x_2^2),$$

then

$$v'_{(9.7)}(t, x) = -a(t)x_1^2 - c(t)x_2^2 \leq -\delta(x_1^2 + x_2^2).$$

Since all hypotheses of Theorem 9.11 are satisfied, we conclude that the equilibrium $x = 0$ of (9.7) is exponentially stable in the large.

D. Boundedness of Solutions

The next two results are concerned with the boundedness of the solutions of (E). The assumption that $x = 0$ is an isolated equilibrium of (E) is not needed in these results.

Theorem 9.13. If there exists a continuously differentiable function v defined on $|x| \geq R$ (where R may be large) and $0 \leq t < \infty$, and if there exist $\psi_1, \psi_2 \in KR$ such that

$$\psi_1(|x|) \leq v(t, x) \leq \psi_2(|x|), \qquad v'_{(E)}(t, x) \leq 0$$

for all $|x| \geq R$ and for all $0 \leq t < \infty$, then the solutions of (E) are **uniformly bounded**.

Proof. Fix $k > R$, let $(t_0, x_0) \in R^+ \times B(k)$ with $|x_0| > R$, let $\phi_0(t) = \phi(t, t_0, x_0)$, and let $V(t) = v(t, \phi_0(t))$ for as long as $|\phi_0(t)| > R$. Since $v'_{(E)}(t, x) \leq 0$, it follows that

$$\psi_1(|\phi_0(t)|) \leq V(t) \leq V(t_0) \leq \psi_2(k).$$

Since $\psi_1 \in KR$, its inverse exists and $|\phi_0(t)| \leq \beta \triangleq \psi_1^{-1}(\psi_2(k))$ for as long as $|\phi_0(t)| > R$.

If $|\phi_0(t)|$ starts at a value less than R or if it reaches a value less than R for some $t > t_0$, then $\phi_0(t)$ can remain in $B(k)$ for all subsequent t or else it may leave $B(k)$ over some interval $t_1 < t < t_2 \leq +\infty$. On the interval $I = (t_1, t_2)$, the foregoing argument yields $|\phi_0(t)| \leq \beta$ over I. Thus $|\phi(t)| \leq \max\{R, \beta\}$ for all $t \geq t_0$. ∎

Theorem 9.14. If there exists a continuously differentiable function v defined on $|x| \geq R$ (where R may be large) and $0 \leq t < \infty$, and if there exist $\psi_1, \psi_2 \in KR$ and $\psi_3 \in K$ such that

$$\psi_1(|x|) \leq v(t, x) \leq \psi_2(|x|), \qquad v'_{(E)}(t, x) \leq -\psi_3(|x|)$$

for all $|x| \geq R$ and $0 \leq t < \infty$, then the solutions of (E) are **uniformly ultimately bounded**.

Proof. Fix $k_1 > R$ and choose $B > k_1$ such that $\psi_2(k_1) < \psi_1(B)$. This is possible since $\psi_1 \in KR$. Choose $k > B$ and let $T = [\psi_2(k)/\psi_3(k_1)] + 1$. With $B < |x_0| \leq k$ and $t_0 \geq 0$, let $\phi_0(t) = \phi(t, t_0, x_0)$ and $V(t) = v(t, \phi_0(t))$. Then $|\phi_0(t)|$ must satisfy $|\phi_0(t^*)| \leq k_1$ for some $t^* \in (t_0, t_0 + T)$, for otherwise

$$V(t) \leq V(t_0) - \int_{t_0}^{t} \psi_3(k_1) \, ds \leq \psi_2(k) - \psi_3(k_1)(t - t_0).$$

5.9 Principal Luapunov Stability and Instability Theorems

The right-hand side of the preceding expression is negative when $t = T$. Hence, t^* must exist.

Suppose now that $|\phi_0(t^*)| = k_1$ and $|\phi_0(t)| > k_1$ for $t \in (t^*, t_1)$, where $t_1 \leq +\infty$. Since $V(t)$ is nondecreasing in t, we have

$$\psi_1(|\phi_0(t)|) \leq V(t) \leq V(t^*) \leq \psi_2(|\phi_0(t^*)|) = \psi_2(k_1) < \psi_1(B)$$

for all $t \geq t^*$. Hence, $|\phi_0(t)| < B$ for all $t \geq t^*$. ∎

Let us now consider a specific case.

Example 9.15. Consider the system

$$\begin{align} x' &= -x - \sigma, \\ \sigma' &= -\sigma - f(\sigma) + x, \end{align} \quad (9.8)$$

where $f(\sigma) = \sigma(\sigma^2 - 6)$. Note that there are isolated equilibrium points at $x = \sigma = 0$, $x = -\sigma = 2$, and $x = -\sigma = -2$.

Choosing

$$v(x, \sigma) = \tfrac{1}{2}(x^2 + \sigma^2),$$

we obtain

$$v'_{(9.8)}(x, \sigma) = -x^2 - \sigma^2(\sigma^2 - 5) \leq -x^2 - (\sigma^2 - \tfrac{5}{2})^2 + \tfrac{25}{4}.$$

Note that v is positive definite and radially unbounded and that $v'_{(9.8)}$ is negative for all (x, σ) such that $x^2 + \sigma^2 > R^2$, where, e.g., $R = 10$ will do. It follows from Theorem 9.13 that all solutions of (9.8) are uniformly bounded, and in fact, it follows from Theorem 9.14 that the solutions of (9.8) are uniformly ultimately bounded.

E. Instability

In the final three results of this section, we present conditions for the instability of the equilibrium $x = 0$ of (E).

Theorem 9.16. The equilibrium $x = 0$ of (E) is **unstable** (at $t = t_0 \geq 0$) if there exists a continuously differentiable, decrescent function v such that $v'_{(E)}$ is positive definite (negative definite) and if in every neighborhood of the origin there are points x such that $v(t_0, x) > 0$ ($v(t_0, x) < 0$).

Proof. Pick a function $\psi_2 \in K$ and ε such that $0 < \varepsilon \leq h$ and such that $|v(t, x)| \leq \psi_2(|x|)$ for all $(t, x) \in R^+ \times B(\varepsilon)$. Pick $\psi_3 \in K$ such that $v'_{(E)}(t, x) \geq \psi_3(|x|)$ on $R^+ \times B(\varepsilon)$. (If $v'_{(E)}$ is negative definite, we replace v by $-v$.) Let $\{x_m\}$ be a sequence of points satisfying $0 < |x_m| < \varepsilon$, $x_m \to 0$ as $m \to \infty$, and $v(t_0, x_m) > 0$. Let $\phi_m(t) = \phi(t, t_0, x_m)$ and $V_m(t) = v(t, \phi_m(t))$. Then

$\phi_m(t)$ must reach the sphere $|x| = \varepsilon$ in finite time. For otherwise we have for all $t \geq t_0$,

$$\psi_2(|\phi_m(t)|) \geq V_m(t) \geq V_m(t_0)$$

or

$$|\phi_m(t)| \geq \psi_2^{-1}(V_m(t_0)) = \alpha_m > 0.$$

Thus,

$$\psi_2(\varepsilon) > \psi_2(|\phi_m(t)|) \geq V_m(t) \geq V_m(t_0) + \int_{t_0}^t \psi_3(\alpha_m)\, ds$$
$$\geq V_m(t_0) + \psi_3(\alpha_m)(t - t_0)$$

for all $t \geq t_0$. But this is impossible. Hence $x = 0$ is unstable. ∎

Theorem 9.16 is called **Lyapunov's first instability theorem**.

In the special case when in Theorem 9.16 v and $v'_{(E)}$ are both positive definite (or negative definite), the equilibrium $x = 0$ of (E) is said to be **completely unstable** (since in this case all trajectories tend away from the origin).

Example 9.17. If in Example 9.9 $c > 0$, then we have $v(x) = x_1^2 + x_2^2$ and $v'_{(9.6)}(x) = 2c(x_1^2 + x_2^2)^2$ and we can conclude from Theorem 9.16 that the equilibrium $x = 0$ of (E) is unstable, in fact, it is completely unstable.

Example 9.18. Consider the system

$$\begin{aligned} x_1' &= c_1 x_1 + x_1 x_2, \\ x_2' &= -c_2 x_2 + x_1^2, \end{aligned} \qquad (9.9)$$

where $c_1 > 0$, $c_2 > 0$ are constants. Choosing

$$v(x) = x_1^2 - x_2^2,$$

we obtain

$$v'_{(9.9)}(x) = 2(c_1 x_1^2 + c_2 x_2^2).$$

Since v is indefinite and $v'_{(9.9)}$ is positive definite, Theorem 9.16 is applicable and the equilibrium $x = 0$ of (9.9) is unstable.

Example 9.19. Let us now return to the conservative system considered in Example 9.5. This time we assume that $W(0) = 0$ is an isolated maximum. This is ensured by assuming that W_k is a negative definite homogeneous polynomial of degree k. (Clearly k must be an even integer.) Recall that we also assumed that T_2 is positive definite. So let us now choose as a

5.9 Principal Lyapunov Stability and Instability Theorems

v function

$$v(p, q) = p^T q = \sum_{i=1}^{n} p_i q_i.$$

Then

$$v'_{(9.4)}(p,q) = \sum_{i=1}^{n} p_i \frac{\partial T_2}{\partial p_i} + \sum_{i=1}^{n} p_i \frac{\partial T_3}{\partial p_i} + \cdots$$
$$- \sum_{i=1}^{n} q_i \frac{\partial W_k}{\partial q_i} - \sum_{i=1}^{n} q_i \frac{\partial W_{k+1}}{\partial q_i} - \cdots$$
$$= 2T_2(p) + 3T_3(p) + \cdots - kW_k(q) - (k+1)W_{k+1}(q) - \cdots.$$

In a sufficiently small neighborhood of the origin, the sign of $v'_{(9.4)}$ is determined by the sign of the term $2T_2(p) - kW_k(q)$, and thus, $v'_{(9.4)}$ is positive definite. Since v is indefinite, Theorem 9.16 is applicable and we conclude that the equilibrium $(p^T, q^T) = (0^T, 0^T)$ is unstable.

The next result is known as **Lyapunov's second instability theorem**.

Theorem 9.20. Let there exist a bounded and continuously differentiable function $v: D \to R$, $D = \{(t, x): t \geq t_0, x \in B(h)\}$, with the following properties:

(i) $v'_{(E)}(t, x) = \lambda v(t, x) + w(t, x)$, where $\lambda > 0$ is a constant and $w(t, x)$ is either identically zero or positive semidefinite;

(ii) in the set $D_1 = \{(t, x): t = t_1, x \in B(h_1)\}$ for fixed $t_1 \geq 0$ and with arbitrarily small h_1, there exist values x such that $v(t_1, x) > 0$.

Then the equilibrium $x = 0$ of (E) is **unstable**.

Proof. Fix $h_1 > 0$ and then pick $x_1 \in B(h_1)$ with $v(t_1, x_1) > 0$. Let $\phi_1(t) = \phi(t, t_1, x_1)$ and $V(t) = v(t, \phi_1(t))$ so that

$$V'(t) = \lambda V(t) + w(t, \phi_1(t)).$$

Hence

$$V(t) = V(t_1)e^{\lambda(t-t_1)} + \int_{t_1}^{t} e^{\lambda(t-s)} w(s, \phi_1(s))\, ds \geq V(t_1)e^{\lambda(t-t_1)}$$

for all t such that $t \geq t_1$ and $\phi_1(t)$ exists and satisfies $|\phi_1(t)| \leq h$. If $|\phi_1(t)| \leq h$ for all $t \geq t_1$, then we see that $V(t) \to \infty$ as $t \to \infty$. But $V(t) = v(t, \phi(t))$ is bounded since v is bounded on $R^+ \times B(h)$, a contradiction. Hence $\phi_1(t)$ must reach $|x| = h$ in finite time. ∎

Let us now consider a specific example.

Example 9.21. Consider the system

$$x_1' = x_1 + x_2 + x_1 x_2^4,$$
$$x_2' = x_1 + x_2 - x_1^2 x_2, \tag{9.10}$$

which has an isolated equilibrium at $x = 0$. Choosing

$$v(x) = (x_1^2 - x_2^2)/2,$$

we obtain

$$v'_{(9.10)}(x) = \lambda v(x) + w(x),$$

where $w(x) = x_1^2 x_2^4 + x_1^2 x_2^2$ and $\lambda = 2$. It follows from Theorem 9.20 that the equilibrium $x = 0$ of (9.10) is unstable.

Our last result of the present section is called **Chetaev's instability theorem**.

Theorem 9.22. Let there exist a continuously differentiable function v having the following properties:

(i) For every $\varepsilon > 0$ and for every $t \geq t_0$, there exist points $\bar{x} \in B(\varepsilon)$ such that $v(t, \bar{x}) < 0$. We call the set of all points (t, x) such that $x \in B(h)$ and such that $v(t, x) < 0$ the "domain $v < 0$." It is bounded by the hypersurfaces which are determined by $|x| = h$ and by $v(t, x) = 0$ and it may consist of several component domains.

(ii) In at least one of the component domains D of the domain $v < 0$, v is bounded from below and $0 \in \partial D$ for all $t \geq 0$.

(iii) In the domain D, $v'_{(E)} \leq -\psi(|v|)$, where $\psi \in K$.

Then the equilibrium $x = 0$ of (E) is **unstable**.

Proof. Let $M > 0$ be a number such that $-M \leq v(t, x)$ on D. Given any $h_1 > 0$ choose $(0, x_0) \in R^+ \times B(h_1) \cap D$. Then the solution $\phi_0(t) = \phi(t, 0, x_0)$ must leave $B(h)$ in finite time. Indeed, $|\phi_0(t)|$ must become equal to h in finite time. To see this, assume the contrary. Let $V(t) = v(t, \phi(t))$. Since $V(0) < 0$ and $v'_{(E)}(t, x) \leq -\psi(|v(t, x)|)$, we have $V(t) \leq V(0) < 0$ for all $t \geq 0$. Thus

$$V(t) \leq V(0) - \int_0^t \psi(|V(0)|)\, ds \to -\infty$$

as $t \to \infty$. This contradicts the bound $V(t) \geq -M$. Hence there is a $t^* > 0$ such that $(t^*, \phi_0(t^*)) \in \partial D$. But $V(t^*) < 0$, so the only part of ∂D which $(t, \phi_0(t))$ can penetrate is that part where $|\phi_0(t)| = h$. Since this can happen for arbitrarily small $|x_0|$, the instability of $x = 0$ is proved. ∎

5.9 Principal Lyapunov Stability and Instability Theorems

A simpler version of this instability theorem can be proved for the autonomous system (A).

Theorem 9.23. Let there exist a continuously differentiable function $v: \overline{B(h)} \to R$ having the following properties:

(i) The set $\{x \in B(h): v(x) < 0\}$ is called the "domain $v < 0$." We assume that this domain contains a component D for which $0 \in \partial D$.

(ii) $v'_{(A)}(x) < 0$ for all $x \in D$, $x \neq 0$.

Then the equilibrium $x = 0$ of (A) is **unstable**.

The proof of Theorem 9.23 is similar to the proof of Theorem 9.22 and will be left as an exercise for the reader. We now consider two specific examples.

Example 9.24. Consider the system

$$x'_1 = x_1 + x_2,$$
$$x'_2 = x_1 - x_2 + x_1 x_2, \qquad (9.11)$$

which has an isolated equilibrium at the origin $x = 0$. Choosing

$$v(x) = -x_1 x_2,$$

we obtain

$$v'_{(9.11)}(x) = -x_1^2 - x_2^2 - x_1^2 x_2.$$

Let $D = \{x \in R^2 : x_1 > 0, x_2 > 0 \text{ and } x_1^2 + x_2^2 < 1\}$. Then for all $x \in D$, $v < 0$ and $v'_{(9.11)} < 2v$. We see that Theorem 9.22 is applicable and conclude that the equilibrium $x = 0$ of (9.11) is unstable.

Example 9.25. Returning to the conservative system considered in Example 9.5, let us assume that $W(0) = 0$ is not a local minimum of the potential energy. Then there are points q arbitrarily near the origin such that $W(q) < 0$. Since $H(0, q) = W(q)$, there are points (p, q) arbitrarily near the origin where $H(p, q) < 0$ for all p sufficiently near the origin. Therefore, there are points (p, q) arbitrarily close to the origin where $p^T q > 0$ and $-H(p, q) > 0$. Now let U be some neighborhood of the origin, and let U_1 be the region of points in U where both of the inequalities

$$p^T q > 0 \quad \text{and} \quad -H(p, q) > 0$$

are satisfied. The origin is then a boundary point of U_1. So let us now choose the v function

$$v(p, q) = p^T q H(p, q).$$

Since $H'(p, q) = 0$, we have, as before,

$$v'_{(9.4)}(p, q) = -H(p, q)[-2T_2(p) - 3T_3(p) - \cdots + kW_k(q) + \cdots]. \qquad (9.12)$$

If we select U sufficiently small, then $T(p) > 0$ within U and therefore $W_k(q) < 0$ within U_1. Hence, for U sufficiently small, the term in brackets in (9.12) is negative within U_1 and $v'_{(9.4)}$ is negative within U_1. On the boundary points of U_1 that are in $B(h)$ it must be that either $p^T q = 0$ or $H(p, q) = 0$ and at these points $v = 0$. Thus, all conditions of Theorem 9.23 are satisfied and we conclude that the equilibrium $(p^T, q^T) = (0^T, 0^T)$ of (9.4) is unstable.

Henceforth, we shall call any function v which satisfies any one of the results of the present section (as well as Section 11) a **Lyapunov function**.

We conclude this section by observing that frequently the theorems of the present section yield more than just stability (resp., instability and boundedness) information. For example, suppose that for the system

$$x' = f(x), \qquad (A)$$

there exists a continuously differentiable function v and three positive constants c_1, c_2, c_3 such that

$$c_1 |x|^2 \le v(x) \le c_2 |x|^2, \qquad v'_{(A)}(x) \le -c_3 |x|^2, \qquad (9.13)$$

for all $x \in R^n$. Then, clearly, the equilibrium $x = 0$ of (A) is exponentially stable in the large. However, as noted in the proof of Theorem 9.11, the condition (9.13) yields also the estimate

$$|\phi(t, t_0, \xi)| \le \sqrt{c_2/c_1}\, |\xi| e^{-[c_3/(2c_2)](t-t_0)}.$$

5.10 LINEAR SYSTEMS REVISITED

Most of the important questions concerning the stability of linear systems

$$x' = Ax \qquad (L)$$

were answered in Section 5. However, further investigation is necessary to construct a suitable Lyapunov function for (L), since by modifying such a function, we can find appropriate Lyapunov functions for a large class of nonlinear equations, consisting of a "linear and nonlinear part." Such systems, which are sometimes called "nearly linear systems," are treated in the next chapter.

We begin by considering as a Lyapunov function the quadratic form

$$v(x) = x^T B x, \qquad B = B^T, \qquad (10.1)$$

5.10 Linear Systems Revisited

where $x \in R^n$ and B is a real $n \times n$ matrix. If we evaluate the derivative of v with respect to t along the solutions of (L), we obtain

$$v'_{(L)}(x) = -x^T C x, \tag{10.2}$$

where

$$-C = A^T B + BA, \qquad C = C^T. \tag{10.3}$$

Our objective will now be to determine the as yet unknown matrix B in such a way that $v'_{(L)}$ becomes a preassigned negative definite quadratic form (i.e., in such a way that C is a pressigned positive definite matrix).

Equation (10.3) constitutes a system of $n(n+1)/2$ linear equations. We need to determine under what conditions we can solve for the $n(n+1)/2$ elements, b_{ik}, given C and A. To this end, we choose a similarity transformation P such that

$$PAP^{-1} = \bar{A}, \tag{10.4}$$

or equivalently,

$$A = P^{-1}\bar{A}P, \tag{10.5}$$

where $\bar{A}$ is similar to A and P is a real $n \times n$ nonsingular matrix. From (10.5) and (10.3), we obtain

$$(\bar{A})^T (P^{-1})^T B P^{-1} + (P^{-1})^T B P^{-1} \bar{A} = -(P^{-1})^T C P^{-1} \tag{10.6}$$

or

$$(\bar{A})^T \bar{B} + \bar{B}\bar{A} = -\bar{C}, \qquad \bar{B} = (P^{-1})^T B P^{-1}, \qquad \bar{C} = (P^{-1})^T C P^{-1}. \tag{10.7}$$

In (10.6), B and C are subjected to a congruence transformation and $\bar{B}$ and $\bar{C}$ have the same definiteness properties as B and C, respectively. Since every real $n \times n$ matrix can be triangularized, we can choose P in such a fashion that $\bar{A} = [\bar{a}_{ij}]$ is triangular, i.e., $\bar{a}_{ij} = 0$ for $i < j$. Note that in this case the eigenvalues of A, $\lambda_1, \ldots, \lambda_n$, appear in the main diagonal of $\bar{A}$. To simplify our notation, we rewrite (10.7) [in the form of (10.3)] by dropping the bars, i.e.,

$$A^T B + BA = -C, \qquad C = C^T, \tag{10.8}$$

and we assume that $A = [a_{ij}]$ has been triangularized, i.e., $a_{ij} = 0$ for $i < j$.

Since the eigenvalues $\lambda_1, \ldots, \lambda_n$ appear in the diagonal of A, we can rewrite (10.8) as

$$\begin{aligned} 2\lambda_1 b_{11} &= -c_{11}, \\ a_{21}b_{11} + (\lambda_1 + \lambda_2)b_{12} &= -c_{12}, \\ &\vdots \end{aligned} \tag{10.9}$$

Since this system is triangular and since its determinant is equal to

$$2^n \lambda_1 \lambda_2 \cdots \lambda_n \prod_{i<j}(\lambda_i + \lambda_j), \tag{10.10}$$

the matrix B can be determined (uniquely) if and only if this determinant is not zero. This is true when all eigenvalues of A are nonzero and no two of them are such that $\lambda_i + \lambda_j = 0$. This condition is not affected by a similarity transformation and is therefore also valid for the original system of equations (10.3).

The foregoing construction assures that $v'_{(L)}$ is negative definite. We must still check for the definiteness of the matrix B. This can be accomplished in a purely algebraic way. However, it is much easier to apply the results of Section 9 to make the following observations:

(a) If all the eigenvalues λ_i have negative real parts, then the equilibrium $x = 0$ of (L) is asymptotically stable and B must be positive definite. Indeed, if B were not positive definite, then for δ positive and sufficiently small, $(B - \delta E)$ will have at least one negative eigenvalue while the function $V(x) = x^T(B - \delta E)x$ has negative definite derivative, i.e.,

$$V'_{(L)}(x) = x^T[(B - \delta E)A + A^T(B - \delta E)]x = x^T[-C - \delta(A + A^T)]x^T < 0$$

for all $x \neq 0$. By Theorem 9.16, $x = 0$ would be unstable, a contradiction. Hence B must be positive definite.

(b) If at least one of the eigenvalues λ_k has positive real part and none of the eigenvalues has zero real part, then B cannot be positive definite. Otherwise, we could use an argument similar to (a), apply Theorem 9.7, and arrive at a contradiction. Furthermore, if the real parts of all eigenvalues are positive, then B must be negative definite.

If A has eigenvalues with positive real parts and if we have the case in which one of the sums $(\lambda_k + \lambda_l)$ vanishes, then B cannot be constructed in the foregoing manner. However, in this case we can use a transformation $x = Py$ so that $P^{-1}AP$ is a block diagonal matrix of the form $\mathrm{diag}(A_1, A_2)$, all eigenvalues of A_1 have positive real parts, and all eigenvalues of A_2 have nonpositive real parts. By the results already proved, given any positive definite matrices C_1 and C_2 there exist two symmetric matrices B_1 and B_2 with B_1 positive definite such that

$$A_1^T B_1 + B_1 A_1 = C_1, \qquad A_2^T B_2 + B_2 A_2 = -C_2.$$

Thus $w(y) = y^T By$, with $B = \mathrm{diag}(-B_1, B_2)$, is a Lyapunov function for $y' = P^{-1}APy$ which satisfies the hypotheses of Theorem 9.16.

We can now summarize the foregoing discussion in the following theorem.

5.11 Invariance Theory

Theorem 10.1. Assume $\det A \neq 0$. If all the eigenvalues of the matrix A have negative real parts or if at least one of the eigenvalues have a positive real part, then there exists a Lyapunov function v of the form (10.1) whose derivative $v'_{(L)}$ is definite (i.e., negative definite or positive definite).

The preceding result shows that if A is a stable matrix, then for (L), the conditions of Theorem 9.7 are also necessary conditions for asymptotic stability. Also, if A is an unstable matrix, then for (L), the conditions of Theorem 9.16 are also necessary conditions for instability.

When all the eigenvalues of A have negative real parts, we can solve Eq. (10.3) in closed form. We have:

Theorem 10.2. If all the eigenvalues of A have negative real parts, we can solve (10.3) in closed form; in fact, we have

$$B = \int_0^\infty e^{A^T s} C e^{As} \, ds.$$

Proof. Let $\phi(t)$ be the solution of (L) satisfying $\phi(0) = x_0$, i.e., $\phi(t) = [\exp(At)]x_0$. Define $v(x) = x^T B x$ and compute

$$v(\phi(t)) = x_0^T e^{A^T t} \left(\int_0^\infty e^{A^T s} C e^{As} \, ds \right) e^{At} x_0$$

$$= x_0^T \left(\int_0^\infty e^{A^T (t+s)} C e^{A(t+s)} \, ds \right) x_0$$

$$= x_0^T \left(\int_t^\infty e^{A^T s} C e^{As} \, ds \right) x_0.$$

Thus

$$v'_{(L)}(\phi(t)) = x_0^T (-e^{A^T t} C e^{At}) x_0 = \phi(t)^T (-C) \phi(t)$$

and at $t = 0$ we see that $v'_{(L)}(x_0) = -x_0^T C x_0$ for all $x_0 \in R^n$. ∎

5.11 INVARIANCE THEORY

In the present section, we extend some of the results of Section 5.9 for autonomous systems

$$x' = f(x). \tag{A}$$

Here we assume that f and $\partial f/\partial x_i$, $i = 1, \ldots, n$, are continuous in a region $D \subset R^n$ (where D may be all of R^n) and we assume that $x = 0$ is in the interior of D. As usual, we assume that $x = 0$ is an isolated equilibrium. Note that our assumptions are sufficient for the local existence, uniqueness, continuability, and continuity with respect to parameters of solutions of (A). Note also that the solution $\phi(t, t_0, x_0)$ of (A) satisfying $x(t_0) = x_0$ must satisfy $\phi(t, t_0, x_0) = \phi(t - t_0, 0, x_0)$. Thus, the initial time is not important and it will be suppressed, i.e. we shall write $\phi(t, x_0)$ for the solution of (A) satisfying $x(0) = x_0$, i.e., $\phi(t, x_0) \triangleq \phi(t, 0, x_0)$.

Recall that the solutions of (A) describe in $R \times R^n$ the **motions** for (A) and that the projections of the motions into R^n determine the **trajectories** for (A). The trajectory $C(x_0)$ is the set of points $\phi(t, x_0)$ over $-\infty < t < \infty$ when $\phi(t, x_0)$ exists on all of R. Similarly, the **positive semitrajectory** $C^+(x_0)$ is the set of points $\phi(t, x_0)$ for $t \geq 0$ and the negative semitrajectory $C^-(x_0)$ is the set of all points $\phi(t, x_0)$ for $t \leq 0$. If the point x_0 is understood or unimportant, we shall write C^+ in place of $C^+(x_0)$ and C^- in place of $C^-(x_0)$. Note that if $C(x_0)$ exists, then $C(x_0) = C^+(x_0) \cup C^-(x_0)$.

Definition 11.1. A set Γ of points in R^n is **invariant** (respectively, **positively invariant**) **with respect to** (A) if every solution of (A) starting in Γ remains in Γ for all time (respectively, for all $t \geq 0$).

Note that if $x_0 = \phi(t_0)$ is a point of the set Γ and if Γ is positively invariant with respect to (A), then the semitrajectory $C^+(x_0)$ lies in Γ, for all $t_0 \in R^+$. Similarly, if Γ is invariant with respect to (A), then the trajectory $C(x_0)$ lies in Γ for all $t_0 \in R$.

Example 11.2. A set consisting of any equilibrium point of (A) is an invariant set with respect to (A).

Example 11.3. Consider the nonlinear spring problem discussed in Chapter 1.

$$x_1' = x_2,$$
$$x_2' = -g(x_1), \tag{11.1}$$

where g is continuously differentiable and where $x_1 g(x_1) > 0$ for all $x_1 \neq 0$. This has only one equilibrium which is located at the origin. The total energy for this system is given by

$$v(x) = \tfrac{1}{2} x_2^2 + \int_0^{x_1} g(\eta)\, d\eta = \tfrac{1}{2} x_2^2 + G(x_1),$$

where $G(x_1) = \int_0^{x_1} g(\eta)\, d\eta$. Note that v is a positive definite function and that

$$v'_{(11.1)}(x) = 0.$$

5.11 Invariance Theory

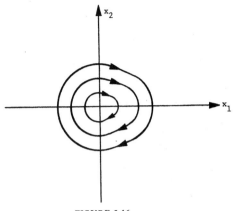

FIGURE 5.16

Therefore, (11.1) is a conservative dynamical system and $x = 0$ is a stable equilibrium. Since $v'_{(11.1)} = 0$, it follows that

$$\tfrac{1}{2}x_2^2 + G(x_1) = c, \qquad (11.2)$$

where c is determined by the initial conditions (x_{10}, x_{20}). For different values of c, we obtain different trajectories, as shown, e.g., in Fig. 5.16. The exact shapes of these trajectories depend, of course, on the nature of the function G. Note, however, that the curves determined by (11.2) will always be symmetric about the x_1 axis. If $G(x) \to \infty$ as $|x| \to \infty$, then each one of these closed trajectories is an invariant set with respect to (11.1). ∎

Example 11.4. Let us now consider a somewhat more complicated and general situation. Assume that all solutions $\phi(t, x_0)$ of (A) exist for all $t \geq 0$ and for all $x_0 \in R^n$. Let $v: R^n \to R$ be continuously differentiable and not necessarily positive definite. Assume that v is such that along the solutions of (A) we have $v'_{(A)}(x) \leq 0$ for all $x \in R^n$. Now let

$$S_k = \{x \in R^n : v(x) \leq k\}.$$

As depicted in Fig. 5.17 for the case $n = 2$, such a set may consist of several components, say, $S_{k1}, S_{k2}, \ldots$. We now show that for every k, the set S_k, and in fact, each of the components of S_k, is a positively invariant set with respect to (A).

To show this, assume that $\phi(t, x_0)$ is the solution of (A) such that $\phi(0, x_0) = x_0$. Then $v'(\phi(t, x_0)) \leq 0$ and thus $v(\phi(t, x_0)) \leq v(x_0)$. Since by assumption $x_0 \in S_k$, it follows that $v(\phi(t, x_0)) \leq k$ for all $t \geq 0$ and thus $\phi(t, x_0) \in S_k$ for all $t \geq 0$, which shows that S_k is positively invariant with

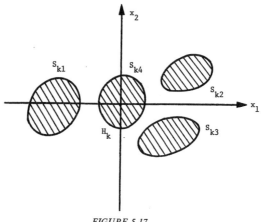

FIGURE 5.17

respect to (A). Furthermore, if x_0 belongs to a particular component of S_k, then since $C^+(x_0)$ is a connected set, $\phi(t, x_0)$ will remain in the same component for all $t \geq 0$. This shows that each component of S_k is a positively invariant set.

We note that if solutions of (A) are not assumed to exist for all time $t \geq 0$, then the foregoing argument still works for any *bounded* component of S_k. However, the preceding conclusions are not necessarily true for unbounded components of S_k as the following example shows. For $x' = x^2$, let $v(x) = -x$ so that $v'_{(E)}(x) = -x^2 \leq 0$. The set S_k is not invariant for any value of $k \in R$, since there exist x_0 for which $\phi(t, x_0)$ has **finite escape time** (i.e., there exist x_0 for which $\phi(t, x_0)$ does not exist for all t in the future).

Since in the preceding discussion, no conditions were imposed on S_k, the solution $\phi(t, x_0)$ that lies in S_k can become "infinite" (i.e., $|\phi(t, x_0)|$ can become arbitrarily large) whenever any one of the components of S_k is unbounded. Now if in particular, v is positive definite, then the set S_k must have at least one component, say H_k, which for sufficiently small $k > 0$ contains the origin. Note that H_k will become an arbitrarily small neighborhood containing the origin as $k \to 0$. Since H_k is a positively invariant set, every solution starting in H_k will remain in H_k for sufficiently small k. This, of course, means that the equilibrium $x = 0$ of (A) is stable.

To establish the main results of this section, we also require the following concept.

Definition 11.5. A point $a \in R^n$ is said to lie in the **positive limit set** $\Omega(C^+)$ (or to be an Ω-**limit point** of the semitrajectory C^+) of the

5.11 Invariance Theory

solution $\phi(t)$ of (A) if there exists a sequence $\{t_m\} \to \infty$ as $m \to \infty$ such that

$$\lim_{m \to \infty} \phi(t_m) = a.$$

Thus, $\Omega(C^+)$ is the set of all accumulation points of the semi-trajectory C^+.

Example 11.6. In the case of a stable focus (see Fig. 5.10), $\Omega(C^+) = \{0\}$.

Example 11.7. Referring to Fig. 5.16, we see that (for $|\xi|$ small) every trajectory C^+ of system (11.1) has the property that $\Omega(C^+) = C^+ = C$.

Example 11.8. Every equilibrium point x_e of (A) is the limit set $\Omega(C^+)$ of the semitrajectory $C^+(x_e)$.

In the proofs of the main results of this section, we require some preliminary results which we now state and prove.

Lemma 11.9. *If the solution $\phi(t, x_0)$ of (A) remains in a compact set K for $0 \le t < \infty$, then its positive limit set $\Omega(C^+)$ is a nonempty, compact, invariant set with respect to (A). Moreover, $\phi(t, x_0)$ approaches the set $\Omega(C^+)$ as $t \to \infty$ (i.e., for every $\varepsilon > 0$ there exists a $T > 0$ such that for every $t > T$ there exists a point $a \in \Omega(C^+)$ (possibly depending on t) such that $|\phi(t, x_0) - a| < \varepsilon$).*

Proof. We claim that

$$\Omega(C^+) = \bigcap \{\overline{[C^+(\phi(t, x_0))]} : t \ge 0\}, \qquad (11.3)$$

where $\overline{[B]}$ denotes the closure in R^n of the set B. Clearly if $y \in \Omega(C^+)$, then there is a sequence $\{t_m\}$, with $t_m \to \infty$ as $m \to \infty$, such that $\phi(t_m, x_0) \to y$ as $m \to \infty$. For any $t \ge 0$ we can delete the $t_m < t$ and see that $y \in \overline{[C^+(\phi(t, x_0))]}$. Conversely, if y is a member of the set on the right-hand side of (11.3), then for any integer m, there is a point $y_m \in C^+(\phi(m, x_0))$ such that $|y - y_m| < 1/m$. But y_m has the form $y_m = \phi(t_m, x_0)$, where $t_m > m$. Thus $t_m \to \infty$ and $\phi(t_m, x_0) \to y$, i.e., $y \in \Omega(C^+)$. The right-hand side of (11.3) is the intersection of a decreasing family of compact sets. Hence $\Omega(C^+)$ is compact (i.e., closed and bounded). By the Bolzano–Weierstrass theorem, $\Omega(C^+)$ is also nonempty. The invariance of $\Omega(C^+)$ is a direct consequence of Corollary 2.5.3.

Suppose now that $\phi(t, x_0)$ does not approach $\Omega(C^+)$ as $t \to \infty$. Then there is an $\varepsilon > 0$ and a sequence $\{t_m\}$, such that $t_m \to \infty$ as $m \to \infty$, and such that the distance from $\phi(t_m, x_0)$ to $\Omega(C^+)$ is at least $\varepsilon > 0$. Since the sequence $\{\phi(t_m, x_0)\}$ is bounded, then by the Bolzano–Weierstrass theorem a subsequence will tend to a limit y. Clearly $y \in \Omega(C^+)$ and at the same time the distance from y to $\Omega(C^+)$ must be at least ε, a contradiction. ∎

Lemma 11.10. Let v be a continuously differentiable function defined on a domain D containing the origin and let $v'_{(A)}(x) \leq 0$ for all $x \in D$. Let $x_0 \in D$ and let $\phi(t, x_0)$ be a bounded solution of (A) whose positive semitrajectory C^+ lies in D for all $t \geq 0$ and let the positive limit set $\Omega(C^+)$ of $\phi(t, x_0)$ lie in D. Then

$$v'_{(A)} = 0$$

at all points of $\Omega(C^+)$.

Proof. Let $y_0 \in \Omega(C^+)$. By Corollary 2.5.3, there is a sequence $\{t_m\}$, such that $t_m \to \infty$ as $m \to \infty$, and such that $\phi(t + t_m, x_0) \to \phi(t, y_0)$ as $m \to \infty$ uniformly for t on compact subsets of R. Since $v(\phi(t, x_0))$ is nonincreasing and bounded, it tends to a limit v_0. Thus for any $t \in R$ we have

$$\lim_{m \to \infty} v(\phi(t + t_m, x_0)) = v(\phi(t, y_0)) \equiv v_0.$$

Since $v(\phi(t, y_0))$ is constant, its derivative is zero. ∎

We are now in a position to present the main results of this section.

Theorem 11.11. Let v be a continuously differentiable, real valued function defined on some domain $D \subset R^n$ containing the origin and assume that $v'_{(A)} \leq 0$ on D. Assume that $v(0) = 0$. For some real constant $k \geq 0$, let H_k be the component of the set $S_k = \{x : v(x) \leq k\}$ which contains the origin. Suppose that H_k is a closed and bounded subset of D. Let $E = \{x \in D : v'_{(A)}(x) = 0\}$. Let M be the largest invariant subset of E with respect to (A). Then every solution of (A) starting in H_k at $t = 0$ approaches the set M as $t \to \infty$.

Proof. Let $x_0 \in H_k$. Since H_k is compact and invariant, then by the remarks in Example 11.4, it is positively invariant. By Lemma 11.9, $\phi(t, x_0) \to \Omega(C^+)$ as $t \to \infty$, where $C^+ = C^+(x_0)$. By Lemmas 11.9 and 11.10, the set $\Omega(C^+)$ is an invariant set and $v'_{(A)}(x) = 0$ on $\Omega(C^+)$. Hence $\Omega(C^+) \subset M$. ∎

Using Theorem 11.11, we can now establish the following stability results.

Corollary 11.12. Assume that for system (A) there exists a continuously differentiable, real valued, positive definite function v defined on some set $D \subset R^n$ containing the origin. Assume that $v'_{(A)} \leq 0$ on D. Suppose that the origin is the only invariant subset with respect to (A) of the set $E = \{x \in D : v'_{(A)} = 0\}$. Then the equilibrium $x = 0$ of (A) is **asymptotically stable**.

Proof. In Example 11.4 we have already remarked that $x = 0$ is stable. By Theorem 11.11 any solution starting in H_k will tend to the origin as $t \to \infty$. ∎

5.11 Invariance Theory

If by some method we can show that all solutions of (A) remain bounded as $t \to \infty$, then the following result concerning bounded solutions of (A) will be useful.

Corollary 11.13. Let $v: R^n \to R^+$ be a continuously differentiable function and let $v(0) = 0$. Suppose that $v'_{(A)} \leq 0$ for all $x \in R^n$. Let $E = \{x \in R^n : v'_{(A)}(x) = 0\}$. Let M be the largest invariant subset of E. Then all bounded solutions of (A) approach M as $t \to \infty$.

Proof. The proof of this result is essentially the same as the proof of Theorem 11.11. By Lemma 11.9, a bounded solution $\phi(t, x_0)$ will tend to $\Omega(C^+)$ as $t \to \infty$. By Lemmas 11.9 and 11.10, $\Omega(C^+) \subset M$. ∎

From Corollaries 11.12 and 11.13, we know that if v is positive definite for all $x \in R^n$, if $v'_{(E)} \leq 0$ for all $x \in R^n$ and if in the set $E = \{x \in R^n : v'_{(E)}(x) = 0\}$ the origin is the only invariant subset, then the origin of (A) is asymptotically stable and all bounded solutions of (A) approach 0 as $t \to \infty$. Therefore, if we can provide additional conditions which ensure that all solutions of (A) are bounded, then we have shown that the equilibrium $x = 0$ of (A) is asymptotically stable in the large. However, this follows immediately from our boundedness result given in Theorem 9.13. We therefore have the following result.

Theorem 11.14. Assume that there exists a continuously differentiable, positive definite, and radially unbounded function $v: R^n \to R$ such that

  (i) $v'_{(A)}(x) \leq 0$ for all $x \in R^n$, and
  (ii) the origin is the only invariant subset of the set
  $$E = \{x \in R^n : v'_{(A)}(x) = 0\}.$$

Then the equilibrium $x = 0$ of (A) is **asymptotically stable in the large**.

Sometimes it may be difficult to find a v function which satisfies all of the conditions of Theorem 11.14. In such cases, it is often useful to prove the boundedness of solutions first, and separately show that all bounded solutions approach zero. We shall demonstrate this by means of an example (see Example 11.16).

Example 11.15. Let us consider the Lienard equation discussed in Chapter 1, given by

$$x'' + f(x)x' + g(x) = 0, \tag{11.4}$$

where f and g are continuously differentiable for all $x \in R$, where $g(x) = 0$ if and only if $x = 0$, $xg(x) > 0$ for all $x \neq 0$ and $x \in R$, $\lim_{|x| \to \infty} \int_0^x g(\eta)\, d\eta = \infty$, and $f(x) > 0$ for all $x \in R$. Letting $x_1 = x$, $x_2 = x'$, (11.4) is equivalent to the system of equations

$$x_1' = x_2,$$
$$x_2' = -f(x_1)x_2 - g(x_1). \quad (11.5)$$

Note that the only equilibrium of (11.5) is the origin $(x_1, x_2) = (0, 0)$.

If for the moment, we were to assume that the damping term $f \equiv 0$, then (11.5) would reduce to the conservative system of Example 11.3. Recall that the total energy for this conservative system is given by

$$v(x_1, x_2) = \tfrac{1}{2}x_2^2 + \int_0^{x_1} g(\eta)\,d\eta \quad (11.6)$$

which is positive definite and radially unbounded.

Returning to our problem at hand, let us choose the v function (11.6) for the system (11.5). Along the solutions of this system, we have

$$v'_{(11.5)}(x_1, x_2) = -x_2^2 f(x_1) \le 0 \quad \text{for all} \quad (x_1, x_2) \in R^2.$$

The set E in Theorem 11.14 is the x_1 axis. Let M be the largest invariant subset of E. If $x = (x_1, 0) \in M$, then at the point x the differential equation is $x_1' = 0$ and $x_2' = -g(x_1) \ne 0$ if $x_1 \ne 0$. Hence the solution emanating from x must cross the x_1 axis. This means that $(x_1, 0) \notin M$ if $x_1 \ne 0$. If $x_1 = 0$, then x is the trivial solution and does remain on the x_1 axis. Thus $M = \{(0, 0)^T\}$. By Theorem 11.14 the origin $x = 0$ is asymptotically stable in the large.

Example 11.16. Let us reconsider the Lienard equation of Example 11.15,

$$x_1' = x_2,$$
$$x_2' = -f(x_1)x_2 - g(x_1). \quad (11.7)$$

This time we assume that $x_1 g(x_1) > 0$ for all $x_1 \ne 0$, $f(x_1) > 0$ for all $x_1 \in R$, and $\lim_{|x_1| \to \infty} \left| \int_0^{x_1} f(\sigma)\,d\sigma \right| = \infty$. This is the case if, e.g., $f(\sigma) = k > 0$. We choose again as a v function

$$v(x) = \tfrac{1}{2}x_2^2 + \int_0^{x_1} g(\eta)\,d\eta$$

so that

$$v'_{(11.6)}(x) = -f(x_1)x_2^2.$$

Since we no longer assume that $\lim_{|x_1| \to \infty} \int_0^{x_1} g(\eta)\,d\eta = \infty$, we cannot apply Theorem 11.14, for in this case, v is not necessarily radially unbounded. However, since the hypotheses of Corollary 11.12 are satisfied, we can still conclude that the equilibrium $(x_1, x_2) = (0, 0)$ of (11.7) is asymptotically stable. Furthermore, by showing that all solutions of (11.7) are bounded, we

5.11 Invariance Theory

can conclude from Corollary 11.13 that the equilibrium $x = 0$ of the Lienard equation is still asymptotically stable in the large.

To this end, let l and a be arbitrary given positive numbers and consider the region U defined by the inequalities

$$v(x) < l \quad \text{and} \quad \left(x_2 + \int_0^{x_1} f(\eta)\, d\eta\right)^2 < a^2.$$

For each pair of numbers (l, a), U is a bounded region as shown, e.g., in Fig. 5.18.

Now let $x_0^T = (x_{10}, x_{20}) = (x_1(0), x_2(0))$ be *any* point in R^2. If we choose (l, a) properly, x_0 will be in the interior of U. Now let $\phi(t, x_0)$ be a solution of (11.7) such that $\phi(0, x_0) = x_0$. We shall show that $\phi(t, x_0)$ cannot leave the bounded region U. This in turn will show that all solutions of (11.7) are bounded, since $\phi(t, x_0)$ is arbitrary.

In order to leave U, the solution $\phi(t, x_0)$ must either cross the locus of points determined by $v(x) = l$ or one of the loci determined by $x_2 + \int_0^{x_1} f(\eta)\, d\eta = \pm a$. Here we choose, without loss of generality, $a > 0$ so large that the part of the curve determined by $x_2 + \int_0^{x_1} f(\eta)\, d\eta = a$ that is also the boundary of U corresponds to $x_1 > 0$ and the part of the curve determined by $x_2 + \int_0^{x_1} f(\eta)\, d\eta = -a$ corresponds to $x_1 < 0$. Now since $v'(\phi(t, x_0)) \leq 0$, the solution $\phi(t, x_0)$ cannot cross the curve determined by $v(x) = l$. To show that it does not cross either of the curves determined by

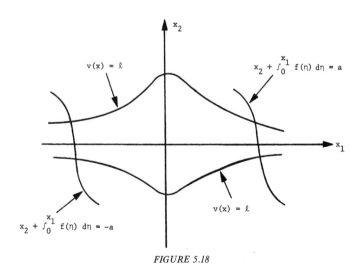

FIGURE 5.18

$x_2 + \int_0^{x_1} f(\eta)\,d\eta = \pm a$, we consider the function

$$w(t) = \left[\phi_2(t, x_0) + \int_0^{\phi_1(t,x_0)} f(\eta)\,d\eta\right]^2,$$

where $\phi(t, x_0)^T = [\phi_1(t, x_0), \phi_2(t, x_0)]$. Then

$$w'(t) = -2\left[\phi_2(t, x_0) + \int_0^{\phi_1(t,x_0)} f(\eta)\,d\eta\right]g(\phi_1(t, x_0)).$$

Now suppose that $\phi(t, x_0)$ reaches the boundary determined by the equation $x_2 + \int_0^{x_1} f(\eta)\,d\eta = a$, $x_1 > 0$. Then along this part of the boundary $w'(t) = -2ag(\phi(t, x_0)) < 0$ because $x_1 > 0$ and $a > 0$. Therefore, the solution $\phi(t, x_0)$ cannot cross outside of U through that part of the boundary determined by $x_2 + \int_0^{x_1} f(\eta)\,d\eta = a$. We apply the same argument to the part of the boundary determined by $x_2 + \int_0^{x_1} f(\eta)\,d\eta = -a$.

Therefore, every solution of (11.7) is bounded and the equilibrium $x = 0$ of (11.7) is asymptotically stable in the large. ∎

5.12 DOMAIN OF ATTRACTION

Many practical systems possess more than one equilibrium point. In such cases, the concept of asymptotic stability in the large is no longer applicable and one is usually very interested in knowing the extent of the domain of attraction of an asymptotically stable equilibrium. In this section, we briefly address the problem of obtaining estimates of the domain of attraction of the equilibrium $x = 0$ of the autonomous system

$$x' = f(x). \tag{A}$$

As in Section 11, we assume that f and $\partial f/\partial x_i$, $i = 1, \ldots, n$, are continuous in a region $D \subset R^n$ and we assume that $x = 0$ is in the interior of D. As usual, we assume that $x = 0$ is an isolated equilibrium point. Again we let $\phi(t, x_0)$ be the solution of (A) satisfying $x(0) = x_0$.

Let us assume that there exists a continuously differentiable and positive definite function v such that

$$v'_{(A)}(x) \leq 0 \quad \text{for all} \quad x \in D.$$

Let $E = \{x \in D : v'_{(A)}(x) = 0\}$ and suppose that $\{0\}$ is the only invariant subset of E with respect to (A). In view of Corollary 11.12 we might conclude that the set D is contained in the domain of attraction of $x = 0$. However, this conjecture is false, as can be seen from the following. Let $n = 2$ and suppose that $S_l = \{x : v(x) \leq l\}$ is a closed and bounded subset of D. Let H_l be the component of S_l which contains the origin for $l \geq 0$. (Note that when $l = 0$,

5.12 Domain of Attraction

$H_l = \{0\}$.) Referring to Fig. 5.19, we note that for small $l > 0$, the level curves $v = l$ determine closed bounded regions which are contained in D and which contain the origin. However, for l sufficiently large, this may no longer be true, for in this case the sets H_l may extend outside of D and they may even be unbounded.

Note however, that from Theorem 11.11 we can say that every closed and bounded region H_l which is contained in D, will also be contained in the domain of attraction of the origin $x = 0$. Thus, we can compute $l = l_m$, so that H_{l_m} has this property, as the largest value of l for which the component of $S_{l_m} = \{x : v(x) = l_m\}$ containing the origin, actually meets the boundary of D. Note that even when D is unbounded, all sets H_l which are completely contained in D are positively invariant sets with respect to (A). Thus, every bounded solution of (A) which starts in H_l tends to the origin by Corollary 11.13.

Example 12.1. Consider the system

$$\begin{aligned} x_1' &= x_2 - \varepsilon(x_1 - \tfrac{1}{3}x_1^3), \\ x_2' &= -x_1, \end{aligned} \quad (12.1)$$

where $\varepsilon > 0$. This system has an isolated equilibrium at the origin.

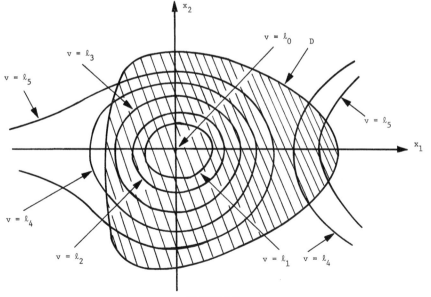

FIGURE 5.19

Choose $v(x) = \frac{1}{2}(x_1^2 + x_2^2)$. Then

$$v'_{(12.1)}(x) = -\varepsilon x_1^2(1 - \tfrac{1}{3}x_1^2),$$

and $v'_{(12.1)} \leq 0$ when $|x_1| \leq \sqrt{3}$. By Corollary 11.12, the equilibrium $x = 0$ is asymptotically stable. Furthermore, the region $\{x \in R^2 : x_1^2 + x_2^2 < 3\}$ is contained in the domain of attraction of the equilibrium $x = 0$.

There are also results which determine the domain of attraction of the origin $x = 0$ precisely. In the following, we let $G \subset D$ and we assume that G is a simply connected domain containing a neighborhood of the origin. The following result is called **Zubov's theorem**.

Theorem 12.2. Suppose there exist two functions $v: G \to R$ and $h: R^n \to R$ with the following properties:

(i) v is continuously differentiable and positive definite in G and satisfies in G the inequality $0 < v(x) < 1$ when $x \neq 0$. For any $b \in (0, 1)$ the set $\{x \in G : v(x) \leq b\}$ is bounded.

(ii) h is continuous on R^n, $h(0) = 0$, and $h(x) > 0$ for $x \neq 0$.

(iii) For $x \in G$, we have

$$v'_{(A)}(x) = -h(x)[1 - v(x)][1 + |f(x)|^2]^{1/2}. \tag{12.2}$$

(iv) As $x \in G$ approaches a point on the boundary of G, or in case of an unbounded region G, as $|x| \to \infty$, $\lim v(x) = 1$.

Then G is exactly the domain of attraction of the equilibrium $x = 0$.

Proof. Under the given hypotheses, it follows from Theorem 9.6 that $x = 0$ is uniformly asymptotically stable. Note also that if we introduce the change of variables

$$ds = (1 + |f(\phi(t))|^2)^{1/2} dt,$$

then (12.2) reduces to

$$dv/ds = -h(x)(1 - v(x)),$$

but the stability properties of (A) remain unchanged. Let $V(s) = v(\phi(s))$ for a given function $\phi(s)$ such that $\phi(0) = x_0$. Then

$$\frac{d}{ds}\log[1 - V(s)] = h(\phi(s)),$$

or

$$1 - V(s) = (1 - V(0))\exp\left[\int_0^s h(\phi(u))\,du\right]. \tag{12.3}$$

5.12 Domain of Attraction

Let $x_0 \in G$ and assume that x_0 is not in the domain of attraction of the trivial solution. Then $h(\phi(s)) \geq \delta > 0$ for some fixed δ and for all $s \geq 0$. Hence, in (12.3) as $s \to \infty$ the term on the left is at most one, while the term on the right tends to infinity. This is impossible. Thus x_0 is in the domain of attraction of $x = 0$.

Suppose x_1 is in the domain of attraction but $x_1 \notin G$. Then $\phi(s, x_1) \to 0$ as $s \to \infty$, so there must exist s_1 and s_2 such that $\phi(s_1, x_1) \in \partial G$ and $\phi(s_2, x_1) \in G$. Let $x_0 = \phi(s_2, x_1)$ in (12.3). Take the limit in (12.3) as $s \to s_1^+$. We see that

$$\lim_{s \to s_1^+} [1 - V(s)] = 1 - 1 = 0,$$

while the limit on the right-hand side is

$$[1 - v(x_0)] \exp\left[\int_{s_2}^{s_1} h(\phi(s, x_1))\, ds\right] > 0.$$

This is impossible. Hence x_1 must be in G. ∎

An immediate consequence of Theorem 12.2 is the following result.

Corollary 12.3. Suppose there is a function h which satisfies the hypotheses of Theorem 12.2 and suppose there is a continuously differentiable, positive definite function $v: G \to R$ which satisfies the inequality $0 \leq v(x) \leq 1$ for all $x \in G$ as well as the differential equation

$$\nabla v^T(x) f(x) = -h(x)[1 - v(x)][1 + |f(x)|^2]^{1/2}. \tag{12.4}$$

Then the boundary of the domain of attraction is defined by the equation

$$v(x) = 1. \tag{12.5}$$

If the domain of attraction G is all of R^n, then we have asymptotic stability in the large. The condition on v in this case is

$$v(x) \to 1 \quad \text{as} \quad |x| \to \infty.$$

In the foregoing results, we can also work with a different v function. For example, if we let

$$w(x) = -\log[1 - v(x)], \tag{12.6}$$

then (12.2) assumes the form

$$w'_{(A)}(x) = -h(x)[1 + |f(x)|^2]^{1/2}$$

and the condition (12.5) defining the boundary becomes $w(x) \to \infty$.

Note that the function $h(x)$ in the preceding results is arbitrary. In applications, it is chosen in a fashion which makes the solution of the partial differential equations easy.

From the proofs of the preceding results, we can also conclude that the relation

$$v(x) = l, \quad 0 < l < 1,$$

defines a family of closed hypersurfaces which covers the domain G. The origin corresponds to $l = 0$ and the boundary of G corresponds to $l = 1$.

Example 12.4. Consider the system

$$x'_1 = 2x_1 \frac{1 - x_1^2 + x_2^2}{(x_1 + 1)^2 + x_2^2} + x_1 x_2 = f_1(x_1, x_2),$$

$$x'_2 = \frac{1 - x_1^2 + x_2^2}{2} - \frac{4x_1^2 x_2}{(x_1 + 1)^2 + x_2^2} = f_2(x_1, x_2). \quad (12.7)$$

The equilibrium is at $x_1 = 1$, $x_2 = 0$. The partial differential equation is

$$v_x f_1(x_1, x_2) + v_y f_2(x_1, x_2) = -2 \frac{(x_1 - 1)^2 + x_2^2}{(x_1 + 1)^2 + x_2^2}(1 - v),$$

where

$$h(x_1, x_2) = 2 \frac{(x_1 - 1)^2 + x_2^2}{(x_1 + 1)^2 + x_2^2}(1 + f_1^2 + f_2^2)^{-1/2}.$$

It is easily verified that a solution is

$$v(x_1, x_2) = \frac{(x_1 - 1)^2 + x_2^2}{(x_1 + 1)^2 + x_2^2}.$$

Since $v(x_1, x_2) = 1$ if and only if $x_1 = 0$, the domain of attraction is the set $\{(x_1, x_2): 0 < x_1 < \infty, -\infty < x_2 < \infty\}$.

5.13 CONVERSE THEOREMS

It turns out that for virtually every result of Section 9 there is a converse theorem. That is, in virtually every case, the hypotheses of the results of Section 9 constitute necessary and sufficient conditions for some appropriate stability, instability, or boundedness statement. (See the books

5.13 Converse Theorems

by Hahn [17, Chapter 6] and Yoshizawa [46, Chapter 5].) To establish these necessary and sufficient conditions, one needs to prove the so-called converse Lyapunov theorems. Results of this type are important, since they frequently allow us to establish additional qualitative results; however, they are not useful in constructing Lyapunov functions in a given situation. For this reason, we shall confine ourselves to presenting only one sample result. We first prove two preliminary results.

Lemma 13.1. Let $f, f_x \in C(R^+ \times \overline{B(h)})$. Then there is a function $\psi \in C^1(R^+)$ such that $\psi(0) = 0$, $\psi'(t) > 0$ and such that $s = \psi(t)$ transforms (E) into

$$dx/ds = f^*(s, x), \qquad (E^*)$$

where $|\partial f^*(s, x)/\partial x| \leq 1$ on $R^+ \times B(h)$. Moreover, if $v(s, x)$ is a C^1-smooth function such that $v'_{(E^*)}(s, x)$ is negative definite, then $v(\psi(t), x)$ has a derivative with respect to (E) which is negative definite.

Proof. Pick a positive and continuous function F such that $|\partial f(t, x)/\partial x| \leq F(t)$ for all $(t, x) \in R^+ \times B(h)$. We can assume that $F(t) \geq 1$ for all $t \geq 0$. Define

$$\psi(t) = \int_0^t F(v)\,dv$$

and define Ψ as the inverse function $\Psi = \psi^{-1}$. Define $s = \psi(t)$ so that (E) becomes (E*) with

$$f^*(s, x) = f(\Psi(s), x)/F(\Psi(s)).$$

Clearly, for all $(t, x) \in R^+ \times B(h)$ we have

$$\left|\frac{\partial f^*}{\partial x}(s, x)\right| = \left|\frac{\partial f}{\partial x}(\Psi(s), x)\right|/F(\Psi(s)) \leq \frac{F(\Psi(s))}{F(\Psi(s))} = 1.$$

If $v(s, x)$ has negative definite derivative with respect to system (E*), then define $V(t, x) = v(\psi(t), x)$. There is a function $\psi_1 \in K$ such that $v'_{(E^*)}(s, x) \leq -\psi_1(|x|)$. Thus

$$V'_{(E)}(t, x) = v_s(\psi(t), x)\psi'(t) + \nabla v(\psi(t), x) f(t, x)$$

$$= v_s(\psi(t), x) F(t) + \nabla v(\psi(t), x) \frac{f(t, x)}{F(t)} F(t)$$

$$= F(t) v'_{(E^*)}(\psi(t), x)$$

$$\leq v'_{(E^*)}(\psi(t), x) \leq -\psi_1(|x|).$$

Thus $V'_{(E)}(t, x)$ is also negative definite. ∎

Lemma 13.2. Let $g(t)$ be a positive, continuous function defined for all $t \geq 0$ and satisfying $g(t) \to 0$ as $t \to \infty$. Let $h(t)$ be a positive, continuous, monotone nondecreasing function defined for all $t \geq 0$. Then there exists a function $G(u)$ defined for $u \geq 0$, positive for $u > 0$, continuous, increasing, having an increasing, continuous derivative G', and such that $G(0) = G'(0) = 0$, and such that for any $a > 0$ and any continuous function $g^*(t)$ which satisfies $0 < g^*(t) \leq ag(t)$ the integrals

$$\int_0^\infty G(g^*(t))\,dt \quad \text{and} \quad \int_0^\infty G'(g^*(t))h(t)\,dt \tag{13.1}$$

converge uniformly in g^*.

Proof. We first construct a function $u(t)$ defined for $t > 0$, continuous, decreasing, $u(t) \to 0$ as $t \to \infty$, and $u(t) \to \infty$ as $t \to 0^+$ such that for any $a > 0$ there exists a $T(a)$ with the property that if $t \geq T(a)$ then $ag(t) \leq u(t)$.

Pick a sequence $\{t_m\}$ such that $t_1 \geq 1$, $t_{m+1} \geq t_m + 1$ and such that if $t \geq t_m$ then $g(t) \leq (m+1)^{-2}$. Define $u(t_m) = m^{-1}$, $u(t)$ linear between the t_m's and such that $u(t) = (t_1/t)^p$ on $0 < t < t_1$, where p is chosen so large that $u'(t_1^-) < u'(t_1^+)$. For $t_m \leq t \leq t_{m+1}$ we have

$$ag(t) \leq a(m+1)^{-2} \quad \text{and} \quad u(t) \geq (m+1)^{-1}$$

so that

$$ag(t) \leq u(t)a(m+1)^{-1} \leq u(t)$$

as soon as m is larger than $[a]$, the integer part of a. Thus we can take $T(a) = [a]$.

Define $F(u)$ to be the inverse function of $u(t)$ and define

$$G(u) = \int_0^u \{e^{-F(s)}/[h(F(s))]\}\,ds. \tag{13.2}$$

Since F is continuous and h is positive, the integrand in (13.2) is continuous on $0 < u < \infty$ while $F(u) \to \infty$ as $u \to 0^+$. Hence the integral exists and defines a function $G \in C^1(R^+)$.

Fix $a > 0$ and choose a continuous function g^* such that $0 < g^*(t) < ag(t)$. For $t \geq T(a)$ we have $0 < g^*(t) \leq u(t)$ or $F(g^*(t)) \geq t$. Thus

$$G'(g^*(t)) = \frac{e^{-F(g^*(t))}}{h(F(g^*(t)))} \leq \frac{e^{-t}}{h(t)}, \qquad t \geq T(a).$$

Hence the uniform convergence of the second integral in (13.1) is clear.

5.13 Converse Theorems

The tail of the first integral in (13.1) can be estimated by

$$\int_{T(a)}^{\infty} \left(\int_0^{u(t)} \frac{e^{-F(s)}}{h(0)} \, ds \right) dt.$$

Since $u(t)$ is piecewise C^1 on $0 < t < \infty$, we can change variables from u to s in the inner integral to compute

$$\int_{T(a)}^{\infty} \left(\int_{+\infty}^{t} [u'(s)e^{-s}/h(s)] \, ds \right) dt \leq \int_{T(a)}^{\infty} \left(\int_{+\infty}^{t} [u'(s)e^{-s}/h(0)] \, ds \right) dt$$

$$\leq h(0)^{-1} \int_{T(a)}^{\infty} \left(\int_t^{\infty} e^{-s} \, ds \right) dt < \infty$$

since $0 > u'(t) > -1$. Hence the uniform convergence of the first integral in (13.1) is also clear. ∎

We now state and prove the main result of this section.

Theorem 13.3. *If f and f_x are in $C(R^+ \times B(r))$ and if the equilibrium $x = 0$ of (E) is uniformly asymptotically stable, then there exists a Lyapunov function $v \in C^1(R^+ \times B(r_1))$ for some $r_1 > 0$ such that v is positive definite and decrescent and such that $v'_{(E)}$ is negative definite.*

Proof. By Lemma 13.1 we can assume without loss of generality that $|\partial f/\partial x| \leq 1$ on $R^+ \times B(r)$. Thus by Theorem 2.4.4 we have

$$|\phi(t, \tau, x) - \phi(t, \tau, y)| \leq |x - y|e^{(t-\tau)}$$

for all $x, y \in B(r)$, $\tau \geq 0$, and all $t \geq \tau$ for which the solutions exist. Define $h(t) = e^t$.

Pick r_1 such that $0 < r_1 \leq r$ and such that if $(\tau, x) \in R^+ \times B(r_1)$, then $\phi(t, \tau, x) \in B(r)$ for all $t \geq \tau$ and such that

$$\lim_{t \to \infty} \phi(t + \tau, \tau, x) = 0$$

uniformly for $(\tau, x) \in R^+ \times B(r_1)$. This is possible since $x = 0$ is uniformly asymptotically stable. Let $g(s)$ be a positive, continuous function such that $g(s) \to 0$ as $s \to \infty$, and such that $|\phi(s + t, t, x)|^2 \leq g(s)$ on $s \geq 0$, $t \geq 0$, $x \in B(r_1)$.

Let G be the function given by Lemma 13.2 and define

$$v(t, x) = \int_0^{\infty} G(|\phi(s + t, t, x)|^2) \, ds,$$

where $|\phi|$ denotes the Euclidean norm of ϕ. Clearly v is defined on $R^+ \times B(r_1)$. Since the integral converges uniformly in $(t, x) \in R^+ \times B(r_1)$, then v is also continuous. If $D = \partial/\partial x_1$, then $D\phi(s + t, t, x)$ must satisfy the linear equation

$$dy/ds = f_x(s, \phi(s + t, t, x))y, \qquad y(t) = (1, 0, \ldots, 0)^T$$

(by Theorem 2.7.1). Thus $|D\phi(s + t, t, x)| \leq ke^s$ for some constant $k \geq 1$. Thus

$$\frac{\partial v}{\partial x_1}(t, x) = \int_0^\infty G'(|\phi(s + t, t, x)|^2)\left[2\phi(s + t, t, x)\frac{\partial \phi}{\partial x_1}(s + t, t, x)\right]ds$$

exists and is continuous while

$$\left|\frac{\partial v}{\partial x_1}(t, x)\right| \leq \int_0^\infty G'(g(s))k_1 e^s\, ds < \infty$$

for some constant $k_1 > 0$. A similar argument can be used on the other partial derivatives. Hence $v \in C^1(R^+ \times B(r_1))$.

Since v_x exists and is bounded by some number B while $v(t, 0)$ is zero, then clearly

$$0 \leq v(t, x) = v(t, x) - v(t, 0) \leq B|x|.$$

Thus v is decrescent. To see that v is positive definite, first find $M_1 > 0$ such that $|f(t, x)| \leq M_1 |x|$ for all $(t, x) \in R^+ \times B(r_1)$. Thus, for $M = M_1 r_1$ we have

$$|\phi(t + s, t, x) - x| \leq \int_t^{t+s} |f(u, \phi(u, t, x))|\, du \leq Ms.$$

Thus for $0 \leq s \leq |x|/(2M)$ we have $|\phi(t + s, t, x)| \geq |x|/2$ and

$$v(t, x) \geq \int_0^{|x|/(2M)} G(|\phi(t + s, t, x)|^2)\, ds \geq (|x|/(2M))G(|x|^2/4).$$

This proves that v is positive definite.

To compute $v'_{(E)}$ we replace x by a solution $\phi(t, t_0, x_0)$. Since by uniqueness $\phi(t + s, t, \phi(t, t_0, x_0)) = \phi(t + s, t_0, x_0)$, then

$$v(t, \phi(t, t_0, x_0)) = \int_0^\infty G(|\phi(t + s, t_0, x_0)|^2)\, ds = \int_t^\infty G(|\phi(s, t_0, x_0)|^2)\, ds,$$

and

$$\frac{d}{dt} v(t, \phi(t, t_0, x_0)) = -G(|\phi(t, t_0, x_0)|^2).$$

Thus $v'_{(E)}(t_0, x_0) = -G(|x_0|^2)$. ■

5.14 COMPARISON THEOREMS

In the present section, we state and prove several comparison theorems for the system

$$x' = f(t, x) \tag{E}$$

which are the basis of the **comparison principle** in the stability analysis of the isolated equilibrium $x = 0$ of (E). In this section, we shall assume that $f: R^+ \times B(r) \to R^n$ for some $r > 0$, and that f is continuous there.

We begin by considering a scalar ordinary differential equation of the form

$$y' = G(t, y), \tag{$\tilde{C}$}$$

where $y \in R$, $t \in R^+$, and $G: R^+ \times [0, r) \to R$ for some $r > 0$. Assume that G is continuous on $R^+ \times [0, r)$ and that $G(t, 0) = 0$ for all $t \geq 0$. Recall that under these assumptions Eq. ($\tilde{C}$) possesses solutions $\phi(t, t_0, y_0)$ for every $\phi(t_0, t_0, y_0) = y_0 \in [0, r)$, $t_0 \in R^+$, which are not necessarily unique. These solutions either exist for all $t \in [t_0, \infty)$ or else must leave the domain of definition of G at some finite time $t_1 > t_0$. Also, under the foregoing assumptions, Eq. ($\tilde{C}$) admits the trivial solution $y = 0$ for all $t \geq t_0$. We assume that $y = 0$ is an isolated equilibrium. For the sake of brevity, we shall frequently write $\phi(t)$ in place of $\phi(t, t_0, y_0)$ to denote solutions, with $\phi(t_0) = y_0$.

We also recall that under the foregoing assumptions, Eq. ($\tilde{C}$) has both a maximal solution $p(t)$ and a minimal solution $q(t)$ for any $p(t_0) = q(t_0) = y_0$. Furthermore, each of these solutions either exists for all $t \in [t_0, \infty)$ or else must leave the domain of definition of G at some finite time $t_1 > t_0$.

Theorem 14.1. Let f and G be continuous on their respective domains of definition. Let $v: R^+ \times B(r) \to R$ be a continuously differentiable, positive definite function such that

$$v'_{(E)}(t, x) \leq G(t, v(t, x)). \tag{14.1}$$

Then the following statements are true.

(i) If the trivial solution of Eq. ($\tilde{C}$) is stable, then the trivial solution of system (E) is **stable**.

(ii) If v is decrescent and if the trivial solution of Eq. ($\tilde{C}$) is uniformly stable, then the trivial solution of system (E) is **uniformly stable**.

(iii) If v is decrescent and if the trivial solution of Eq. ($\tilde{C}$) is uniformly asymptotically stable, then the trivial solution of system (E) is **uniformly asymptotically stable**.

(iv) If there are constants $a > 0$ and $b > 0$ such that $a|x|^b \le v(t,x)$, if v is decrescent, and if the trivial solution of Eq. $(\tilde{C})$ is exponentially stable, then the trivial solution of system (E) is **exponentially stable**.

(v) If $f: R^+ \times R^n \to R^n$, $G: R^+ \times R \to R$, $v: R^+ \times R^n \to R$ is decrescent and radially unbounded, if (14.1) holds for all $t \in R^+$, $x \in R^n$, and if the solutions of Eq. $(\tilde{C})$ are uniformly bounded (uniformly ultimately bounded), then the solutions of system (E) are also **uniformly bounded (uniformly ultimately bounded)**.

Proof. We make use of the *comparison theorem*, which was proved in Chapter 2 (Theorem 2.8.4), in the following fashion. Given a solution $\phi(t, t_0, x_0)$ of (E) define $v_0 = v(t_0, x_0)$ and let $y(t, t_0, v_0)$ be the maximal solution of $(\tilde{C})$ which satisfies $y(t_0) = v_0$. By (14.1) and Theorem 2.8.4 it follows that

$$v(t, \phi(t, t_0, x_0)) \le y(t, t_0, v_0) \tag{14.2}$$

for as long as both solutions exist and $t \ge t_0$.

(i) Assume that the trivial solution of $(\tilde{C})$ is stable. Fix $\varepsilon > 0$. Since $v(t,x)$ is positive definite, there is a function $\psi_1 \in K$ such that $\psi_1(|x|) \le v(t,x)$. Let $\eta = \psi_1(\varepsilon)$ so that $v(t,x) < \eta$ implies $|x| < \varepsilon$. Since $y = 0$ is stable, there is a $v > 0$ such that if $|v_0| < v$ then $y(t, t_0, v_0) < \eta$ for all $t \ge t_0$. Since $v(t_0, 0) = 0$, there is a $\delta = \delta(t_0, \varepsilon) > 0$ such that $v(t_0, x_0) < v$ if $|x_0| < \delta$. Take $|x_0| < \delta$ so that by the foregoing chain of reasoning we know that (14.2) implies $v(t, \phi(t, t_0, x_0)) < \eta$ and thus $|\phi(t, t_0, x_0)| < \varepsilon$ for all $t \ge t_0$. This proves that $x = 0$ is stable.

(ii) Let $\psi_1, \psi_2 \in K$ be such that $\psi_1(|x|) \le v(t,x) \le \psi_2(|x|)$. Let $\eta = \psi_1(\varepsilon)$ and choose $v = v(\eta) > 0$ such that $|v_0| < v$ implies $y(t, t_0, v_0) < \eta$ for all $t \ge t_0$. Choose $\delta > 0$ such that $\psi_2(\delta) < v$. Take $|x_0| < \delta$ so that by the foregoing chain of reasoning, we again have $|\phi(t, t_0, x_0)| < \varepsilon$ for all $t \ge t_0$.

(iii) We note that $x = 0$ is uniformly stable by part (ii). Let $\psi_1(|x|) \le v(t,x) \le \psi_2(|x|)$, as before. Fix $\varepsilon > 0$ and let $\eta = \psi_1(\varepsilon)$. Since $y = 0$ is asymptotically stable, there is a $v > 0$ and a $T(\eta) > 0$ such that

$$|y(t + t_0, t_0, v_0)| \le \eta \quad \text{for} \quad t \ge T(\eta), \quad |v_0| < v.$$

Choose $\delta > 0$ so that $\psi_2(\delta) \le v$. For $|x_0| \le \delta$ we have $v(t_0, x_0) \le v$, so by (14.2) $v(t + t_0, \phi(t + t_0, t_0, x_0)) \le \eta$ for $t \ge T(\eta)$ or $|\phi(t + t_0, t_0, x_0)| \le \varepsilon$ when $t \ge T(\eta)$.

(iv) There is an $\alpha > 0$ such that for any $\eta > 0$ there is a $v(\eta) > 0$ such that when $|v_0| < v$, then $|y(t, t_0, v_0)| \le \eta e^{-\alpha(t - t_0)}$ for all $t \ge t_0$.

5.14 Comparison Theorems

Let $a|x|^b \leq v(t,x) \leq \psi_2(|x|)$ as before. Fix $\varepsilon > 0$ and choose $\eta = a\varepsilon^b$. Choose δ such that $\psi_2(\delta) < v(\eta)$. If $|x_0| < \delta$, then $v(t_0, x_0) \leq \psi_2(\delta) < v$ so $y(t, t_0, v_0) \leq \eta e^{-\alpha(t-t_0)}$. So for $t \geq t_0$ we have

$$a|\phi(t, t_0, x_0)|^b \leq \eta e^{-\alpha(t-t_0)},$$

or

$$|\phi(t, t_0, x_0)| \leq (\eta/a)^{1/b} e^{-(\alpha/b)(t-t_0)}.$$

But $(\eta/a)^{1/b} = \varepsilon$ which completes the proof of this part.

(v) Assume that the solutions of $(\tilde{C})$ are uniformly bounded. (Uniform ultimate boundedness is proved in a similar way.) Let $\psi_1 \in KR$, $\psi_2 \in KR$ be such that $\psi_1(|x|) \leq v(t, x) \leq \psi_2(|x|)$. If $|x_0| \leq \alpha$, then $v_0 = v(t_0, x_0) \leq \psi_2(\alpha) \triangleq \alpha_1$. Since the solutions of $(\tilde{C})$ are uniformly bounded and since (14.2) is true, it follows that $v(t, \phi(t, t_0, x_0)) \leq \beta_1(\alpha_1)$ for $t \geq t_0$. So $|\phi(t, t_0, x_0)| \leq \psi_1^{-1}(\beta_1(\alpha_1)) = \beta(\alpha)$. ∎

In practice, the special case $G(t, y) \equiv 0$ is most commonly used in parts (i) and (ii), and the special case $G(t, y) = -\alpha y$ for some constant $\alpha > 0$ is most commonly used in parts (iii) and (iv) of the preceding theorem. An instability theorem can also be proved using this method. For further details, refer to the problems to the end of this chapter.

When applicable, the foregoing results are very useful in applications because they enable us to deduce the qualitative properties of a high-dimensional system [system (E)] from those of a simpler one-dimensional comparison system [system $(\tilde{C})$]. The generality and effectiveness of the preceding comparison technique can be improved and extended by considering *vector valued comparison equations and vector Lyapunov functions*. This will be accomplished in some of the problems given at the end of this chapter.

Example 14.2. A large class of time-varying capacitor–linear resistor networks can be described by equations of the form

$$x'_i = -\sum_{j=1}^{n} \{a_{ij}d_{1j}(t) + b_{ij}d_{2j}(t)\}x_j, \qquad i = 1, \ldots, n, \qquad (14.3)$$

where a_{ij} and b_{ij} are real constants and where $d_{1j}: R^+ \to R$ and $d_{2j}: R^+ \to R$ are continuous functions. It is assumed that $a_{ii} > 0$ and $b_{ii} > 0$ for all i, that $d_{1j}(t) \geq 0$ and $d_{2j}(t) \geq 0$ for all $t \geq 0$ and for all j, and that $d_{1j}(t) + d_{2j}(t) \geq \delta > 0$ for all $t \geq 0$ and for all j.

Now choose as a v function

$$v(x) = \sum_{i=1}^{n} \lambda_i |x_i|,$$

where it is assumed that $\lambda_i > 0$ for all i. Assume that there exists an $\varepsilon > 0$ such that

$$a_{jj} - \sum_{i=1, i \neq j}^{n} \frac{\lambda_i}{\lambda_j} |a_{ij}| \geq \varepsilon > 0, \quad j = 1, \ldots, n,$$

$$b_{jj} - \sum_{i=1, i \neq j}^{n} \frac{\lambda_i}{\lambda_j} |b_{ij}| \geq \varepsilon > 0, \quad j = 1, \ldots, n.$$

(14.4)

For this Lyapunov function, we shall need the more general definition of $v'_{(E)}$ given in (7.3). Note that if D denotes the right-hand Dini derivative, then for any $y \in C^1(R)$ we have $D|y(t)| = y'(t)$ when $y(t) > 0$, $D|y(t)| = -y'(t)$ when $y(t) < 0$, and $D|y(t)| = |y'(t)|$ when $y(t) = 0$. Thus $D|y(t)| = [\text{sgn } y(t)]y'(t)$, except possibly at isolated points. Hence,

$$v'_{(14.3)}(x) \leq \sum_{i=1}^{n} \lambda_i \{ -(a_{ii}d_{1i} + b_{ii}d_{2i})|x_i|$$

$$+ \sum_{j=1, j \neq i}^{n} (|a_{ij}|d_{1j} + |b_{ij}|d_{2j})|x_j| \}$$

$$\leq -\sum_{j=1}^{n} \lambda_j (a_{jj}d_{1j} + b_{jj}d_{2j})|x_j|$$

$$+ \sum_{j=1}^{n} \sum_{i=1, j \neq i}^{n} \lambda_i (|a_{ij}|d_{1j} + |b_{ij}|d_{2j})|x_j|.$$

We want

$$(a_{jj}d_{1j} + b_{jj}d_{2j}) - \sum_{i=1, i \neq j}^{n} \frac{\lambda_i}{\lambda_j} (|a_{ij}|d_{1j} + |b_{ij}|d_{2j}) \geq c > 0$$

for some $c > 0$. But conditions (14.4) and the condition $d_{1i} + d_{2i} \geq \delta > 0$ are sufficient to ensure this for $c = \varepsilon\delta$. Hence we find that

$$v'_{(14.3)}(x) \leq -cv(x),$$

from which we obtain the comparison equation

$$y' = -cy, \quad c > 0. \tag{14.5}$$

Since the equilibrium $y = 0$ of (14.5) is exponentially stable in the large, it follows from Theorem 14.1 (iv) (and from Theorem 5.3) that if there exist constants $\lambda_1, \ldots, \lambda_n$ such that the inequalities (14.4) are true, then the equilibrium $x = 0$ of system (14.3) is exponentially stable in the large.

5.15 APPLICATIONS: ABSOLUTE STABILITY OF REGULATOR SYSTEMS

An important class of problems in applications are regulator systems which can be described by equations of the form

$$x' = Ax + b\eta, \quad \sigma = c^T x + d\eta, \quad \eta = -\phi(\sigma), \tag{15.1}$$

where A is a real $n \times n$ matrix, b, c, and x are real n vectors, and d, σ, and η are real scalars. Also, $\phi(0) = 0$ and $\phi: R \to R$ is continuous. We shall assume that ϕ is such that the system (15.1) has unique solutions for all $t \geq 0$ and for every $x(0) \in R^n$, which depend continuously on $x(0)$.

We can represent system (15.1) symbolically by means of the block diagram of Fig. 5.20. An inspection of this figure indicates that we may view (15.1) as an interconnection of a linear system component (with "input" η and "output" σ) and a nonlinear component. In Fig. 5.20, r denotes a "reference input." Since we are interested in studying the stability properties of the equilibrium $x = 0$ of (15.1), we take $r \equiv 0$.

If we assume for the time being that $x(0) = 0$ and if we take the Laplace transform of both sides of the first two equations in (15.1), we obtain

$$(sE - A)\hat{x}(s) = b\hat{\eta}(s) \quad \text{and} \quad \hat{\sigma}(s) = c^T \hat{x}(s) + d\hat{\eta}(s).$$

Solving for $\hat{\sigma}(s)/\hat{\eta}(s) \triangleq \hat{g}(s)$, we obtain the **transfer function** of the linear component as

$$\hat{g}(s) = c^T(sE - A)^{-1}b + d. \tag{15.2}$$

This enables us to represent system (15.1) symbolically as shown in Fig. 5.21.

Systems of this type have been studied extensively and several monographs have appeared on this subject. See, e.g., the books by LaSalle and Lefschetz [27], Lefschetz [29], Hahn [17], Narendra and Taylor [34], and Vidyasagar [42].

We now list several assumptions that we shall have occasion to use in the subsequent results.

(A1) A is a Hurwitz matrix.

(A2) A has a simple eigenvalue equal to zero and the remaining eigenvalues of A have negative real parts.

(A3) rank $[b | Ab | \cdots | A^{n-1}b] = n$.

(A4) $\sigma\phi(\sigma) \geq 0$ for all $\sigma \in R$.

(A5) there exist constants $k_2 \geq k_1 \geq 0$ such that $k_1\sigma^2 \leq \sigma\phi(\sigma) \leq k_2\sigma^2$ for all $\sigma \in R$.

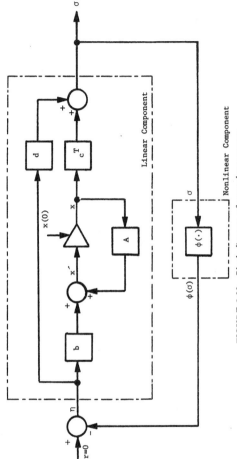

FIGURE 5.20 *Block diagram of a regulator system.*

5.15 Applications: Absolute Stability of Regulator Systems

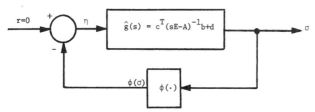

FIGURE 5.21 Block diagram of the regulator system.

When (A3) holds, we say the pair (A, b) is **controllable** and when (A5) holds, we say that ϕ **belongs to the sector** $[k_1, k_2]$. Similarly, if we require that $k_1 \sigma^2 < \sigma \phi(\sigma) < k_2 \sigma^2$, we say that ϕ belongs to the sector (k_1, k_2). Other sectors, such as $(k_1, k_2]$ and $[k_1, k_2)$, are defined in the obvious way.

If we let $d = 0$ and if we replace $\phi(\sigma)$ by $k\sigma$, $k_1 \leq k \leq k_2$, then we can associate with (15.1) the *linear system*

$$x' = (A - kbc^T)x. \tag{15.3}$$

One might conjecture (as was done by M. A. Aizerman in 1949) that if $d = 0$, if ϕ belongs to the sector $[k_1, k_2]$, and if for each $k \in [k_1, k_2]$ the matrix $(A - kbc^T)$ is a Hurwitz matrix [so that system (15.3) is exponentially stable in the large], then the equilibrium $x = 0$ of the nonlinear system (15.1) is asymptotically stable in the large. This conjecture, called **Aizerman's conjecture**, turns out to be false. However, this conjecture is still useful, for it enables us to determine how conservative some of the subsequent results are in a particular application.

In the sequel, we shall address the following problem, which is called the **absolute stability problem** for (15.1): Find conditions on A, b, c, d (involving assumptions of the type given in (A1)–(A3)) which ensure that the equilibrium $x = 0$ of system (15.1) is asymptotically stable in the large for *any* nonlinearity ϕ satisfying either (A4) or (A5). A system (15.1) satisfying this property is said to be **absolutely stable**.

We shall address the absolute stability problem by different methods which will result in (a) *Lure's criterion* and (b) *Popov's criterion*. There are several ways of establishing (b), some of which depend heavily on results from functional analysis. In the present approach, we shall make use of the Yacubovich–Kalman lemma, given in Theorem 15.1. The reader should consult the book by Lefschetz [29, pp. 114–118] for a proof of this result.

Theorem 15.1. Given is a Hurwitz matrix A, a vector b such that the pair (A, b) is controllable, a real vector w, real scalars $\gamma \geq 0$ and $\varepsilon > 0$, and a positive definite matrix Q. Then there exist a positive definite matrix P and a vector q satisfying the equations

$$A^T P + PA = -qq^T - \varepsilon Q \tag{15.4}$$

and

$$Pb - w = \sqrt{\gamma} q \tag{15.5}$$

if and only if ε is small enough and

$$\gamma + 2 \operatorname{Re} w^T (i\omega E - A)^{-1} b > 0 \tag{15.6}$$

for all $\omega \in R$.

A. Lure's Result

In our first result, we let $d = 0$, we assume that A is Hurwitz and that ϕ belongs to the sector $[0, \infty)$ [i.e., ϕ satisfies (A4)], and we use a Lyapunov function of the form

$$v(x) = x^T P x + \beta \int_0^\sigma \phi(\xi) \, d\xi, \tag{15.7}$$

where P is a positive definite matrix and $\beta \geq 0$. This result will require that P be a solution of the Lyapunov matrix equation

$$A^T P + PA = -Q, \tag{15.8}$$

where, as usual, Q is a positive definite matrix of our choice. We have:

Theorem 15.2. Suppose that A is Hurwitz, that ϕ belongs to the sector $[0, \infty)$, and that $d = 0$. Let Q be a positive definite matrix, and let P be the corresponding solution of (15.8). Let

$$w = Pb - (\beta/2) A^T c, \tag{15.9}$$

where $\beta \geq 0$ is some constant [see (15.7)]. Then the system (15.1) is absolutely stable if

$$\beta c^T b - w^T Q^{-1} w > 0. \tag{15.10}$$

Proof. Let $\phi: R \to R$ be a continuous function which satisfies assumption (A4). We must show that the trivial solution of (15.1) is asymptotically stable in the large. To this end, define v by (15.7). Computing the

5.15 Applications: Absolute Stability of Regulator Systems

derivative of v with respect to t along the solutions of (15.1), we obtain

$$\begin{aligned} v'_{(15.1)}(x) &= x^T P(Ax - b\phi(\sigma)) + (x^T A^T - b^T \phi(\sigma))Px + \beta\phi(\sigma)\sigma' \\ &= x^T(A^T P + PA)x - 2x^T Pb\phi(\sigma) + \beta\phi(\sigma)c^T(Ax - b\phi(\sigma)) \\ &= -x^T Qx - 2x^T Pb\phi(\sigma) + \beta x^T A^T c\phi(\sigma) - \beta(c^T b)\phi(\sigma)^2 \\ &= -x^T Qx - 2\phi(\sigma)x^T w - \beta(c^T b)\phi(\sigma)^2 \\ &= -(x + Q^{-1}w\phi(\sigma))^T Q(x + Q^{-1}w\phi(\sigma)) - (\beta c^T b - w^T Q^{-1}w)\phi(\sigma)^2. \end{aligned}$$

In the foregoing calculation, we have used (15.8) and (15.9). By (15.10) and the choice of Q, we see that the derivative of v with respect to (15.1) is negative definite. Indeed if $v'_{(15.1)}(x) = 0$, then $\phi(\sigma) = 0$ and

$$x + Q^{-1}w\phi(\sigma) = x + Q^{-1}w \cdot 0 = x = 0.$$

Clearly v is positive definite and $v(0) = 0$. Hence $x = 0$ is uniformly asymptotically stable. ∎

B. Popov Criterion

In this case, we consider systems described by equations of the form

$$x' = Ax - b\phi(\sigma), \qquad \sigma = c^T x + d\xi, \qquad \xi' = -\phi(\sigma), \qquad (15.11)$$

where A is assumed to be a Hurwitz matrix. We assume that $d \neq 0$ [for otherwise (15.11) would be essentially the same as (15.1) with $d = 0$]. System (15.11) can be rewritten as

$$\begin{bmatrix} x' \\ \xi' \end{bmatrix} = \begin{bmatrix} A & 0 \\ 0 & 0 \end{bmatrix} \begin{bmatrix} x \\ \xi \end{bmatrix} + \begin{bmatrix} b \\ 1 \end{bmatrix} \eta, \qquad \sigma = \begin{bmatrix} c^T & d \end{bmatrix} \begin{bmatrix} x \\ \xi \end{bmatrix}, \qquad \eta = -\phi(\sigma). \quad (15.12)$$

Equation (15.12) is clearly of the same form as Eq. (15.1). However, note that in the present case, the matrix of the linear system component is given by

$$\tilde{A} = \begin{bmatrix} A & 0 \\ 0 & 0 \end{bmatrix}$$

and satisfies assumption (A2), i.e., it has an eigenvalue equal to zero since matrix A satisfies assumption (A1).

Theorem 15.3. System (15.11) with (A1) true and $d > 0$ is absolutely stable for all nonlinearities ϕ belonging to the sector $(0, k)$ if (A3) holds and if there exists a nonnegative constant δ such that

$$\text{Re}[(1 + i\omega\delta)\hat{g}(i\omega)] + k^{-1} > 0 \quad \text{for all} \quad \omega \in R, \quad \omega \neq 0, \quad (15.13)$$

where

$$\hat{g}(s) = (d/s) + c^T(sE - A)^{-1}b. \quad (15.14)$$

Proof. In proving this result, we make use of Theorem 15.1. Choose $\alpha > 0$ and $\beta \geq 0$ such that $\delta = \beta(2\alpha d)^{-1}$. Also, choose $\gamma = \beta(c^T b + d) + (2\alpha d)/k$ and $w = \alpha dc + \frac{1}{2}\beta A^T c$. We must show that $\gamma \geq 0$ and that (15.6) is true.

Note that by (15.13) we have

$$0 < \text{Re}(1 + i\omega\delta)\hat{g}(i\omega) + k^{-1}$$
$$= k^{-1} + \delta d + \text{Re}\, c^T[i\omega(i\omega E - A)^{-1}\delta + (i\omega E - A)^{-1}]b$$
$$= k^{-1} + d\delta + \text{Re}\, c^T[\delta E + \delta A(i\omega E - A)^{-1} + (i\omega E - A)^{-1}]b$$

for all $\omega > 0$. In the limit as $\omega \to \infty$ we have

$$0 \leq k^{-1} + \delta(d + c^T b) = k^{-1} + \beta/(2\alpha d)(d + c^T b) = \gamma/(2\alpha d).$$

Thus $\gamma \geq 0$.

To verify (15.6), we note that since $\delta = \beta(2\alpha d)^{-1}$, then (15.13) is equivalent to

$$\beta/(2\alpha d)d + \text{Re}\{c^T[1 + i\omega(\beta/(2\alpha d))](i\omega E - A)^{-1}b\} + k^{-1} > 0$$

or

$$\beta d + \text{Re}\{c^T[2\alpha d + i\omega\beta](i\omega E - A)^{-1}b\} + (2\alpha d/k) > 0.$$

Since $s(sE - A)^{-1} = E + A(sE - A)^{-1}$ for any complex number s, then (15.13) is equivalent to

$$\beta d + \text{Re}\{c^T[i\omega\beta(i\omega E - A)^{-1} + 2\alpha d(i\omega E - A)^{-1}]b\} + (2\alpha d/k) > 0$$

or

$$\beta d + \text{Re}\{c^T[\beta E + \beta A(i\omega E - A)^{-1} + 2\alpha d(i\omega E - A)^{-1}]b + (2\alpha d/k) > 0.$$

This can be further rearranged to

$$\beta(c^T b + d) + \text{Re}\{c^T(2\alpha d + \beta A)(i\omega E - A)^{-1}b\} + (2\alpha d/k) > 0$$

or

$$[\beta(c^T b + d) + (2\alpha d/k)] + 2\,\text{Re}\{[\alpha dc + \tfrac{1}{2}\beta A^T c]^T(i\omega E - A)^{-1}b\} > 0$$

for all $\omega \neq 0$. For choice of γ and w, this is (15.6).

5.15 Applications: Absolute Stability of Regulator Systems

Use Theorem 15.1 to pick P, q, and $\varepsilon > 0$. Define

$$v(x, \xi) = x^T P x + \alpha d^2 \xi^2 + \beta \int_0^\sigma \phi(s)\, ds$$

for the given values of P, α, and β. The derivative of v with respect to t along the solutions of (15.11) is computed as

$$\begin{aligned}
v'_{(15.11)}(x, \xi) &= x^T P(Ax - b\phi(\sigma)) + (x^T A^T - b^T \phi(\sigma))Px - 2d^2 \alpha \xi \phi(\sigma) + \beta \phi(\sigma) \sigma' \\
&= x^T(PA + A^T P)x - 2x^T P b \phi(\sigma) - 2\alpha d^2 \xi \phi(\sigma) \\
&\quad + \beta \phi(\sigma)[c^T(Ax - b\phi(\sigma)) - d\phi(\sigma)] \\
&= x^T(-qq^T - \varepsilon Q)x - 2x^T P b \phi(\sigma) - 2\alpha d \phi(\sigma)(\sigma - c^T x) \\
&\quad + \beta x^T A^T c \phi(\sigma) - \beta(c^T b + d)\phi(\sigma)^2 \\
&= x^T(-qq^T - \varepsilon Q)x - 2x^T(Pb - w)\phi(\sigma) \\
&\quad - [\beta(c^T b + d) + (2\alpha d/k)]\phi(\sigma)^2 - 2\alpha d[\sigma - (\phi(\sigma)/k)]\phi(\sigma).
\end{aligned}$$

We have used (15.4) and the definition of w. Define

$$R(\sigma) = 2\alpha d(\sigma - \phi(\sigma)/k)\phi(\sigma).$$

Since ϕ is in the sector $(0, k)$, it follows that $R(\sigma) \geq 0$ for all $\sigma \in R$. The definition of $R(\sigma)$, Eq. (15.5), and the choice of γ can now be used to see that

$$\begin{aligned}
v'_{(15.11)}(x, \xi) &= -\varepsilon x^T Q x - R(\sigma) - [x^T qq^T x + 2x^T \sqrt{\gamma} q \phi(\sigma) + \gamma \phi(\sigma)^2] \\
&\leq -\varepsilon x^T Q x - [x^T q + \sqrt{\gamma} \phi(\sigma)]^2.
\end{aligned}$$

If $x \neq 0$, then since Q is positive definite, it follows that $v'_{(15.11)}(x, \xi) < 0$. If $x = 0$, but $\xi \neq 0$, then $\sigma = d\xi \neq 0$ and so $\phi(\sigma) \neq 0$. Hence $v'_{(15.11)}(x, \xi) < 0$ in this case. This show that $v'_{(15.11)}$ is negative definite along solutions of (15.11) for any continuous function ϕ in the sector $(0, k)$. ∎

Theorem 15.3 has a very useful geometric interpretation. If we plot in the complex plane, $\operatorname{Re} \hat{g}(i\omega)$ versus $\omega \operatorname{Im} \hat{g}(i\omega)$, with ω as a parameter (such a plot is called a *Popov plot* or a *modified Nyquist plot*), then condition (15.13) states that there exists a number $\delta \geq 0$ such that the Popov plot of $\hat{g}$ lies to the right of a straight line with slope $1/\delta$ and passing through the point $-1/k + i0$. A typical situation for which Theorem 15.3 is satisfied, using this interpretation, is given in Fig. 5.22. Note that it suffices to consider only $\omega \geq 0$ in generating a Popov plot, since both $\operatorname{Re} \hat{g}(i\omega)$ and $\omega \operatorname{Im} \hat{g}(i\omega)$ are even functions of ω. In Fig. 5.22, the arrow indicates the direction of increasing ω.

We conclude by noting that results of the form given in Theorem 15.3 can also be established for other system configurations [e.g., when in (15.11) $d = 0$].

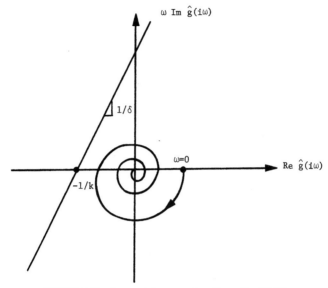

FIGURE 5.22 Geometric interpretation of inequality (15.13).

PROBLEMS

1. Show that the trivial solution of (E) is stable if and only if for some fixed $t_0 \in R^+$ it is true that for any $\varepsilon > 0$, there is a $\delta(\varepsilon) > 0$ such that when $|\xi| < \delta(\varepsilon)$ then $|\phi(t, t_0, \xi)| < \varepsilon$ for all $t \geq t_0$.

2. Show that the trivial solution of (E) is unstable if and only if for any $t_0 \in R^+$ there are sequences $\{\xi_m\}$ and $\{t_m\}$ such that $\xi_m \to 0$ and $t_m \to \infty$ as $m \to \infty$ while $|\phi(t_m + t_0, t_0, \xi_m)| \geq \varepsilon$.

3. Show that if the trivial solution of (E) is uniformly asymptotically stable in the large, then solutions of (E) are uniformly bounded, provided either that f is periodic in t or that $|f(t, x)| \leq K_1|x| + K_2$ for some constants K_1 and K_2.

4. Show that if solutions of (E) are uniformly ultimately bounded, then they are uniformly bounded, provided that $|f(t, x)| \leq K_1|x| + K_2$ for some constants K_1 and K_2.

5. Show that if solutions of (P) are uniformly ultimately bounded, then the solutions of (P) are uniformly bounded.

Problems

6. Prove Theorem 5.2.

7. Prove that if the trivial solution of (LH) is uniformly stable, $B(t) \in C[0, \infty)$ and $\int_0^\infty |B(t)|\, dt < \infty$, then the trivial solution of

$$x' = [A(t) + B(t)]x \qquad (16.1)$$

is uniformly stable. (You may assume that $x = 0$ is an isolated equilibrium point for both systems.)

8. Prove that if the trivial solution of (LH) is uniformly asymptotically stable and if $B(t) \in C[0, \infty)$ with $\sup\{|B(t)| : t \geq 0\} \leq M$, then the trivial solution of (16.1) is uniformly stable when M is sufficiently small.

9. Let B be a real $2n$-dimensional, symmetric matrix, and let

$$J = \begin{bmatrix} 0 & E_n \\ -E_n & 0 \end{bmatrix}$$

and define $A = JB$. Show that whenever λ is an eigenvalue of A, then so is $-\lambda$.

10. Show that the trivial solution of a linear, autonomous Hamiltonian system with Hamiltonian $H(q, p) = q^T S_1 q + p^T S_2 p$ can never be asymptotically stable.

11. Prove that if the equilibrium $x = 0$ of (L) is stable, then all eigenvalues of A have nonpositive real parts and in the Jordon canonical form $J = \text{diag}(J_0, J_1, \ldots, J_s)$, all eigenvalues with zero real part occur only in J_0 and not in a block $J_k = \lambda_{k+i} E_i + N_i$, where $1 \leq k \leq s$.

12. Show that the trivial solution of an nth order, linear autonomous equation

$$a_n y^{(n)} + a_{n-1} y^{(n-1)} + \cdots + a_1 y' + a_0 y = 0, \qquad a_n \neq 0,$$

is stable if and only if all roots of

$$p(\lambda) = a_n \lambda^n + \cdots + a_1 \lambda + a_0 \qquad (16.2)$$

have nonpositive real parts and all roots with zero real parts are simple roots.

13. Prove Theorem 5.9.

14. Use Theorem 5.10 to prove Corollary 5.11. *Hint*: Let M be the matrix whose determinant is D_n. Use row operations to show that M is similar to a triangular matrix whose diagonal elements are the c_{1k}'s.

15. Assume that $a_j > 0$ for $j = 0, 1, \ldots, n$. Find necessary and sufficient conditions that all roots of (16.2) have negative real part in case $n = 2, 3, 4$.

16. Let $a(t) \not\equiv 0$ be a continuous, T-periodic function and let ϕ_1 and ϕ_2 be solutions of

$$y'' + a(t)y = 0 \qquad (16.3)$$

such that $\phi_1(0) = \phi'_2(0) = 1$, $\phi'_1(0) = \phi_2(0) = 0$. Define $\alpha = -(\phi_1(T) + \phi'_2(T))$. For what values of α can you be sure that the trivial solution of (16.3) is stable?

17. In Problem 16, let $a(t) = a_0 + \varepsilon \sin t$ and $T = 2\pi$. Find values $a_0 > 0$ for which the trivial solution of (16.3) is stable for $|\varepsilon|$ sufficiently small.

18. Repeat Problem 17 for $a_0 < 0$.

19. Verify (7.3), i.e., show that if $v(t, x)$ is continuous in (t, x) and is locally Lipschitz continuous in x, then

$$\limsup_{\theta \to 0^+} \left[\frac{v(t+\theta, x+\theta f(t,x)) - v(t,x)}{\theta} \right] = \limsup_{\theta \to 0^+} \frac{v(t+\theta, \phi(t+\theta, t, x)) - v(t,x)}{\theta}.$$

20. Prove Theorem 7.13.

21. Prove Theorem 7.14.

22. In Theorem 7.17, (a) show that (i) implies (ii), and (b) prove parts (iii)–(vi).

23. Let $v: R^+ \times B(h) \to R^+$, let $v \in C^1(R^+ \times R^n)$, let v be positive definite, and let $v'_{(E)}$ be negative definite. Prove the following statements.

    (a) If $f(t, x)$ is bounded on $R^+ \times B(h)$, then the trivial solution of (E) is asymptotically stable.

    (b) If $f(t, x)$ is T-periodic in t, then the trivial solution of (E) is uniformly asymptotically stable.

    (c) If $x = 0$ is uniformly stable and if $v(t, x)$ is bounded on $R^+ \times B(h)$, then the equilibrium $x = 0$ of (E) is uniformly asymptotically stable.

24. Suppose there is a C^1 function $v: R^+ \times B(h) \to R^+$ which is positive definite and which satisfies $v(t, x) \geq k|x|^a$ for some $k > 0$ and $a > 0$ such that $v'_{(E)}(t, x) \leq -bv(t, x)$ for some $b > 0$. Show that the trivial solution of (E) is exponentially stable.

25. Let $v \in C^1(R^+ \times R^n)$, $v(t, x) \geq 0$, and $v'_{(E)}(t, x) \leq 0$. Let $v(t, x)$ be ultimately radially unbounded, i.e., there is an $R_0 > 0$ and a $\psi \in KR$ such that $v(t, x) \geq \psi(|x|)$ for all $t \geq 0$ and for all $x \in R^n$ with $|x| \geq R_0$. Prove the following statements:

    (a) System (E) possesses Lagrange stability.

    (b) If for any $h > 0$, $v(t, x)$ is bounded on $R^+ \times B(h)$, then solutions of (E) are uniformly bounded.

    (c) If for any $h > 0$, $v(t, x)$ is bounded on $R^+ \times B(h)$ and if $-v'_{(E)}(t, x)$ is ultimately radially unbounded, then solutions of (E) are uniformly ultimately bounded.

26. Suppose $v \in C^1(R^+ \times R^n)$ is positive definite, decrescent, and radially unbounded and $v'_{(E)}(t, x)$ is negative definite. Show that for any $r > 0$ and $\delta \in (0, r)$ there is a $T > 0$ such that if $(t_0, x_0) \in R^+ \times B(r)$, then $|\phi(t, t_0, x_0)|$ must be less than δ before $t - t_0 = T$.

27. Let $G \in C^1(R^+ \times R)$ with $G(t, y) = G(t, -y)$ and $G(t, 0) = 0$ and let $v \in C^1(R^+ \times B(h))$ be a positive definite and decrescent function such that
$$v'_{(E)}(t, x) \geq G(t, v(t, x))$$
on $R^+ \times B(h)$. If the trivial solution of $y' = G(t, y)$ is unstable, show that the trivial solution of (E) is also unstable.

28. Let $v \in C(R^+ \times B(h))$, let $v(t, x)$ satisfy a Lipschitz condition in x with Lipschitz constant k, and let $v'_{(E)}(t, x) \leq -w(t, x) \leq 0$. Show that for the system
$$x' = f(t, x) + h(t, x) \qquad (16.4)$$
we have $v'_{(16.4)}(t, x) \leq -w(t, x) + k|h(t, x)|$.

29. In Theorem 13.3 show that:

  (a) If f is periodic in t with period T, then v will be periodic in t with period T.

  (b) If f is independent of t, then so is v.

30. Let $f \in C^1(R^n)$ with $f(0) = 0$, $h \in C^1(R^+ \times R^n)$ with $|h(t, x)|$ bounded on sets of the form $R^+ \times B(r)$ for every $r > 0$, and let the trivial solution of (A) be asymptotically stable. Show that for any $\varepsilon > 0$ there is a $\delta > 0$ such that if $|\xi| < \delta$ and if $|\alpha| < \delta$, then the solution $\psi(t, \xi)$ of
$$x' = f(x) + \alpha h(t, x), \qquad x(0) = \xi$$
will satisfy $|\psi(t, \xi)| < \varepsilon$ for all $t \geq 0$. *Hint*: Use the converse theorem and Problem 29.

31. If in addition, in Problem 30, we have $\lim_{t \to \infty} |h(t, x)| = 0$ uniformly for x on compact subsets of R^n, show that there exists a $\delta > 0$ such that if $|\xi| < \delta$ and $|\alpha| < \delta$ then $\psi(t, \xi) \to 0$ as $t \to \infty$. *Hint*: Use Corollary 2.5.3.

32. Show that if a positive semiorbit C^+ of (A) is bounded, then its positive limit set $\Omega(C^+)$ is connected.

33. Let $f \in C^1(R^n)$ with $f(0) = 0$ and let the equilibrium $x = 0$ of (A) be asymptotically stable with a bounded domain of attraction G. Show that ∂G is an invariant set with respect to (A).

34. Find all equilibrium points for the following equations (or systems of equations). Determine the stability of the trivial solution by finding an appropriate Lyapunov function.

  (a) $y' = \sin y$,
  (b) $y' = y^2(y^2 - 3y + 2)$,
  (c) $x'' + (x^2 - 1)x' + x = 0$,
  (d) system (1.2.44) as shown in Fig. 1.23 with $f_1(t) \equiv f_2(t) \equiv 0$,
  (e) $x'_1 = x_2 + x_1 x_2$, $x'_2 = -x_1 + 2x_2$,
  (f) $x'' + x' + \sin x = 0$,
  (g) $x'' + x' + x(x^2 - 4) = 0$,
  (h) $x' = a(1 + t^2)^{-1}x$, where $a > 0$ or $a < 0$.

35. Analyze the stability properties of the trivial solution of the following systems.

(a) $x' = -a_0 f(x) - \sum_{i=1}^{n} a_i z_i,$

$z_i' = -\lambda_i z_i + b_i f(x) \quad (1 \leq i \leq n)$

where a_i, λ_i, and b_i are all positive and $xf(x) > 0$ if $x \neq 0$. *Hint*: Choose $v(x, z) = \int_0^x f(s)\, ds + \frac{1}{2}\sum_{i=1}^{n}(a_i/b_i) z_i^2.$

(b) $x' = -a_0 y - \sum_{i=1}^{n} a_i z_i,$

$y' = f(x),$

$z_i' = -\lambda_i z_i + b_i f(x) \quad (1 \leq i \leq n),$

where $xf(x) > 0$ if $x \neq 0$ and $a_i/b_i > 0$ for all i.

36. Check for boundedness, uniform boundedness, or uniform ultimate boundedness in each of the following:

(a) $x'' + x' + x(x^2 - 4) = 0,$
(b) $x'' + x' + x^3 = \sin t,$
(c) $x_1' = x_2 + \dfrac{x_1 x_2}{1 + x_1^2 + x_2^2},\quad x_2' = -2x_1 + 2x_2 +$ arctan $x_1,$
(d) $x_1' = x_2^3 + x_1(x_3^2 + 1),\quad x_2' = -x_1^3 + x_2(x_3^2 + 2),$
$x_3' = -(x_3)^3.$

Hint: Choose $v = x_1^4 + x_2^4.$

37. Analyze the stability properties of the trivial solution of

$$x^{(n)} + g(x) = 0,$$

when $n > 2$, n is odd and $xg(x) > 0$ if $x \neq 0$. For $n = 2m + 1$, use

$$v = \sum_{k=1}^{m}(-1)^k x_k x_{2m+2-k} + (-1)^{m+1} x_{m+1}^2/2.$$

38. Check the stability of the trivial solution of $x' = -Ax$ for the following cases.

(a) $A = \begin{bmatrix} 1 & 1 & 1 \\ 1 & 1 & 1 \\ 1 & 1 & 0 \end{bmatrix},$ (b) $A = \begin{bmatrix} 2 & 1 & -1 \\ 1 & 2 & 0 \\ -1 & 0 & 2 \end{bmatrix}.$

Check by applying Sylvester's theorem and also by direct computation of the eigenvalues.

39. For each of the following polynomials, determine whether or not all roots have negative real parts.
  (a) $3s^3 - 7s^2 + 4s + 1$,
  (b) $s^4 + s^3 + 2s^2 + 2s + 5$,
  (c) $s^5 + 2s^4 + 3s^3 + 4s^2 + 7s + 5$,
  (d) $s^3 + 2s^2 + s + k$, k any real number.

40. Let $f \in C^1(R \times R^n)$ with $f(t,0) = 0$ and suppose that the eigenvalues of the symmetric matrix
$$J(t,x) = \tfrac{1}{2}[f_x(t,x) + f_x(t,x)^T]$$
satisfy $\lambda_i(t,x) \leq -\mu$ for $i = 1, 2, \ldots, n$ and for all (t,x) in $R \times R^n$.
  (a) If $\mu = 0$, show that the trivial solution of (E) is stable and that solutions of (E) are uniformly bounded.
  (b) If $\mu > 0$, show that the trivial solution of (E) is exponentially stable in the large.
  (c) Find h_0 such that if $h > h_0$, then the trivial solution of
$$x' = y - (x^5 + 3x^3 + x), \qquad y' = -hy + (x + x^3/3)$$
is uniformly asymptotically stable.

41. Let $y \in R^n$ and let $B(y) = [b_{ij}(y)]$ be an $n \times n$ matrix valued function in $C(R^n)$. Consider the system
$$y' = B(y)y. \qquad (16.5)$$
Show that if for all $y \in R^n - \{0\}$ we have
  (a) $\max_i(b_{ii}(y) - \sum_{j \neq i}|b_{ij}(y)|) \triangleq -c(y) < 0$, or
  (b) $\max_j(b_{jj}(y) - \sum_{i \neq j}|b_{ij}(y)|) \triangleq -d(y) < 0$, or
  (c) $\max_i(b_{ii}(y) - \tfrac{1}{2}\sum_{j \neq i}|b_{ij}(y) + b_{ji}(y)|) \triangleq -e(y) < 0$,
then the trivial solution of (16.5) is globally uniformly asymptotically stable. *Hint*: Let $v_1(y) = \max_i|y_i|$, $v_2(y) = \sum_{i=1}^n |y_i|$, and $v_3(y) = \sum_{i=1}^n y_i^2$. Compute $v_1'(y) \leq -c(y)v_1(y)$.

42. Let $B(y)$ be as in Problem 41, let $p: R \to R^n$ be a continuous, 2π-periodic function and let
$$\limsup_{|y| \to \infty} \max\left\{ b_{ii}(y) + \sum_{j \neq i} |b_{ij}(y)| \right\} < 0.$$
Show that solutions of $y' = B(y)y + p(t)$ are uniformly ultimately bounded.

43. (*Comparison principle*) Consider the vector comparison system
$$y' = G(t,y), \qquad (C_v)$$
where $G: R^+ \times R^l \to R^l$, G is continuous, $G(t,0) \equiv 0$, and $G(t,y)$ is quasimonotone in y (see Chapter 2, Problem 17 for the definition of quasimonotone).

Let $w: R^+ \times R^n \to R^l$, $l \le n$, be a C^1 function such that $|w(t, x)|$ is positive definite, $w(t, x) \ge 0$ and such that

$$w'_{(E)}(t, x) \le G(t, w(t, x)),$$

where $w'_{(E)} = (w'_{1(E)}, \ldots, w'_{l(E)})^T$ is defined componentwise. Prove the following.

(i) If the trivial solution of (C_v) is stable, then so is the trivial solution of (E).

(ii) If $|w(t, x)|$ is decrescent and if the trivial solution of (C_v) is uniformly stable, then so is the trivial solution of (E).

(iii) If $|w(t, x)|$ is decrescent and if the trivial solution of (C_v) is uniformly asymptotically stable, then so is the trivial solution of (E).

(iv) If there are constants $a > 0$ and $b > 0$ such that $a|x|^b \le |w(t, x)|$, $|w(t, x)|$ is decrescent, and if the trivial solution of (C_v) is exponentially stable, then so is the trivial solution of (E).

Hint: Use problem 2.18.

44. Let $A = [a_{ij}]$ be an $l \times l$ matrix such that $a_{ij} \ge 0$ for $i, j = 1, 2, \ldots, n$ and $i \ne j$. Suppose for $j = 1, 2, \ldots, l$

$$a_{jj} - \sum_{i=1, i \ne j}^{l} |a_{ij}| < 0.$$

Show that the trivial solution of $x' = Ax$ is exponentially stable.

45. Show that the trivial solution of the system

$$\begin{aligned} x'_1 &= -x_1 - 2x_2^2 + 2kx_4, \\ x'_2 &= -x_2 + 2x_1 x_2, \\ x'_3 &= -3x_3 + x_4 + kx_1, \\ x'_4 &= -2x_4 - x_3 - kx_2 \end{aligned}$$

is uniformly asymptotically stable when $|k|$ is small. *Hint*: Choose $v_1 = x_1^2 + x_2^2$ and $v_2 = x_3^2 + x_4^2$.

46. For the predator–prey model (cf. Example 1.2.12)

$$\begin{aligned} x'_1 &= a_1 x_1 - b_1 x_1 x_2, \\ x'_2 &= -a_2 x_2 + b_2 x_1 x_2 \end{aligned}$$

with a_1, a_2, b_1, and b_2 all positive, find all equilibrium points and determine their stability properties. *Hint*: The function

$$w \triangleq ((x_1 b_2/a_2)e^{-x_1 b_2/a_2})^{1/a_1}((x_2 b_1/a_1)e^{-x_2 b_1/a_1})^{1/a_2}$$

may be of use.

47. For (E) and (C̃) let F and G be C^1 functions and let $v: R^+ \times B(r) \to R$ be a positive definite and decrescent C^1 function such that $G(t, 0) \equiv 0$, $G(t, -y) = G(t, y)$ and

$$v'_{(E)}(t, x) \geq G(t, v(t, x)).$$

If the trivial solution of (C̃) is unstable, then the trivial solution of (E) is also unstable.

6 | PERTURBATIONS OF LINEAR SYSTEMS

In this chapter, we study the effects of perturbations on the properties of trajectories in a neighborhood of a fixed critical point or in a neighborhood of a periodic solution. Throughout, the analysis is accomplished by arranging matters so that the system of interest can be considered as a perturbation of a linear equation with constant coefficients.

In Section 1 we provide some preliminaries. In Section 2 we analyze the case in which the linear part of the equation has a noncritical coefficient matrix. In this section we also show how, in certain situations, the problem of stability of the trivial solution of periodic nonlinear systems can be reduced to this noncritical case. In Section 3 we study conditional stability of the trivial solution of nonlinear autonomous systems, and in Section 4 we study stability and instability of perturbed linear periodic systems. Finally, in Section 5 we define and study the notion of asymptotic equivalence of systems.

6.1 PRELIMINARIES

We recall that for a function $g: R^l \to R^k$, the notation $g(x) = O(|x|^\beta)$ as $|x| \to \alpha$ means that

$$\limsup_{|x| \to \alpha} \frac{|g(x)|}{|x|^\beta} < \infty,$$

6.1 Preliminaries

and that the interesting cases include $\alpha = 0$ and $\alpha = \infty$. (Here $\beta \geq 0$ and $|\cdot|$ denotes any of the equivalent norms on R^l.) Furthermore, when $g: R \times R^l \to R^k$, then $g(t, x) = O(|x|^\beta)$ as $|x| \to \alpha$ uniformly for t in an interval I means that

$$\limsup_{|x| \to \alpha} \left(\sup_{t \in I} \frac{|g(t, x)|}{|x|^\beta} \right) < \infty.$$

Also, we recall that $g(x) = o(|x|^\beta)$ as $|x| \to \alpha$ means that

$$\lim_{|x| \to \alpha} \frac{|g(x)|}{|x|^\beta} = 0.$$

Further variations of the foregoing [such as, e.g., $g(t, x) = o(|x|^\beta)$ as $|x| \to \alpha$ uniformly for $t \in I$, or $g(x) = o(x^\beta)$ as $x \to 0^+$] are defined in the obvious way.

In this chapter as well as in a subsequent chapter, we shall also require the **implicit function theorem** which we present next. To this end we consider a system of functions

$$g_i(x, y) = g_i(x_1, \ldots, x_n, y_1, \ldots, y_r), \quad 1 \leq i \leq r,$$

and we assume that these functions have continuous first derivatives in an open set containing a point (x_0, y_0). We recall that the matrix

$$g_y \triangleq \frac{\partial g}{\partial y} \triangleq \begin{bmatrix} \partial g_1/\partial y_1 & \partial g_1/\partial y_2 & \cdots & \partial g_1/\partial y_r \\ \partial g_2/\partial y_1 & \partial g_2/\partial y_2 & \cdots & \partial g_2/\partial y_r \\ \vdots & \vdots & & \vdots \\ \partial g_r/\partial y_1 & \partial g_r/\partial y_2 & \cdots & \partial g_r/\partial y_r \end{bmatrix}$$

is called the **Jacobian matrix** of $(g_1, \ldots, g_r)$ with respect to $(y_1, \ldots, y_r)$. Also, the determinant of this matrix is called the **Jacobian** of $(g_1, \ldots, g_r)$ with respect to $(y_1, \ldots, y_r)$ and is denoted by

$$J = \det(\partial g/\partial y).$$

The implicit function theorem is as follows:

Theorem 1.1. Let $g_1, \ldots, g_r$ have continuous first derivatives in a neighborhood of a point (x_0, y_0). Assume that $g_i(x_0, y_0) = 0$, $1 \leq i \leq r$, and that $J \neq 0$ at (x_0, y_0). Then there is a δ neighborhood U of x_0 and a γ neighborhood S of y_0 such that for any x in U there is a unique solution y of $g_i(x, y) = 0, 1 \leq i \leq r$, in S. The vector valued function $y(x) = (y_1(x), \ldots, y_r(x))$ defined in this way has continuous first derivatives in R. If $g \in C^k$ or if g is analytic, then so is $y(x)$.

6.2 STABILITY OF AN EQUILIBRIUM POINT

In this section, we consider systems of n real nonlinear first order ordinary differential equations of the form

$$x' = Ax + F(t, x), \tag{PE}$$

where $F: R^+ \times B(h) \to R^n$ for some $h > 0$ and A is a real $n \times n$ matrix. Here we assume that Ax constitutes the **linear part** of the right-hand side of (PE) and $F(t, x)$ represents the remaining terms which are of order higher than one in the various components of x. Such systems may arise in the process of linearizing nonlinear equations of the form

$$x' = g(t, x), \tag{G}$$

or they may arise in some other fashion during the modeling process of a physical system.

To be more specific, let $g: R \times D \to R^n$ where D is some domain in R^n. If $g \in C^1(R \times D)$ and if ϕ is a given solution of (E) defined for all $t \geq t_0 \geq 0$, then we can **linearize** (G) **about** ϕ in the following manner. Define $y = x - \phi(t)$ so that

$$y' = g(t, x) - g(t, \phi(t)) = g(t, y + \phi(t)) - g(t, \phi(t))$$
$$= \frac{\partial g}{\partial x}(t, \phi(t))y + G(t, y).$$

Here

$$G(t, y) \triangleq [g(t, y + \phi(t)) - g(t, \phi(t))] - \frac{\partial g}{\partial x}(t, \phi(t))y$$

is $o(|y|)$ as $|y| \to 0$ uniformly in t on compact subsets of $[t_0, \infty)$.

Of special interest is the case when g is independent of t [i.e., when $g(t, x) \equiv g(x)$] and $\phi(t) = \xi_0$ is a constant (equilibrium point). Under these conditions we have

$$y' = Ay + G(y),$$

where $A = \partial g(\xi_0)/\partial x$.

Also of special interest is the case in which $g(t, x)$ is T periodic in t (or is independent of t) and $\phi(t)$ is T periodic. We shall consider this case in some detail in Section 4.

6.2 Stability of an Equilibrium Point

Theorem 2.1. Let A be a real, constant, and stable $n \times n$ matrix and let $F: R^+ \times B(h) \to R^n$ be continuous in (t, x) and satisfy

$$F(t, x) = o(|x|) \quad \text{as} \quad |x| \to 0, \tag{2.1}$$

uniformly in $t \in R^+$. Then the trivial solution of (PE) is uniformly asymptotically stable.

Since this type of result is very important in applications, we shall give two different proofs of this theorem. Each proof provides insight into the qualitative behavior of perturbations of the associated linear system given by

$$y' = Ay. \tag{L}$$

Proof 1. Since (L) is an autonomous linear system, Theorem 5.10.1 applies. In view of that theorem, there exists a symmetric, real, positive definite $n \times n$ matrix B such that $BA + A^T B = -C$, where C is positive definite. Consider the Lyapunov function $v(x) = x^T B x$. The derivative of v with respect to t along the solutions of (PE) is given by

$$v'_{(PE)}(t, x) = -x^T C x + 2 x^T B F(t, x). \tag{2.2}$$

Now pick $\gamma > 0$ such that $x^T C x \geq 3\gamma |x|^2$ for all $x \in R^n$. By (2.1) there is a δ with $0 < \delta < h$ such that if $|x| \leq \delta$, then $|BF(t, x)| \leq \gamma |x|$ for all $(t, x) \in R^+ \times \overline{B(\delta)}$. For all (t, x) in $R^+ \times \overline{B(\delta)}$ we obtain, in view of (2.2), the estimate

$$v'_{(PE)}(t, x) \leq -3\gamma |x|^2 + 2\gamma |x|^2 = -\gamma |x|^2.$$

It follows that $v'_{(PE)}(t, x)$ is negative definite in a neighborhood of the origin. By Theorem 5.9.6 it follows that the trivial solution of (PE) is uniformly asymptotically stable. ∎

Proof 2. A fundamental matrix for (L) is e^{At}. Moreover, since A is stable, there are positive constants M and σ such that

$$|e^{At}| \leq M e^{-\sigma t} \quad \text{for all} \quad t \geq 0.$$

Given a solution ϕ of (PE), for as long as ϕ exists we can use the variation of constants formula (3.3.3) to write ϕ in the form

$$\phi(t) = e^{A(t-t_0)} \phi(t_0) + \int_{t_0}^{t} e^{A(t-s)} F(s, \phi(s)) \, ds.$$

Hence, for all $t \geq t_0$ we have

$$|\phi(t)| \leq M |\phi(t_0)| e^{-\sigma(t-t_0)} + M \int_{t_0}^{t} e^{-\sigma(t-s)} |F(s, \phi(s))| \, ds.$$

Given ε with $0 < \varepsilon < \sigma$, by (2.1) there is a δ with $0 < \delta < h$ such that $|F(t,x)| \le \varepsilon|x|/M$ for all pairs (t,x) in $R^+ \times B(\delta)$. Thus, if $|\phi(t_0)| < \delta$, then for as long as $|\phi(t)|$ remains less than δ, we have

$$|\phi(t)| \le M|\phi(t_0)|e^{-\sigma(t-t_0)} + \varepsilon \int_{t_0}^t e^{-\sigma(t-s)}|\phi(s)|\,ds$$

and

$$|\phi(t+t_0)| \le M|\phi(t_0)|e^{-\sigma t} + \varepsilon \int_0^t e^{-\sigma(t-s)}|\phi(s+t_0)|\,ds$$

or

$$e^{\sigma t}|\phi(t+t_0)| \le M|\phi(t_0)| + \varepsilon \int_0^t e^{\sigma s}|\phi(s+t_0)|\,ds. \tag{2.3}$$

Applying the Gronwall inequality (Theorem 2.1.6) to the function $e^{\sigma t}|\phi(t+t_0)|$ in (2.3), we obtain

$$e^{\sigma t}|\phi(t+t_0)| \le M|\phi(t_0)|e^{\varepsilon t},$$

or

$$|\phi(t)| \le M|\phi(t_0)|e^{-(\sigma-\varepsilon)(t-t_0)} \tag{2.4}$$

for as long as $|\phi(t)| < \delta$. Choose $\gamma < \delta/M$ and pick ϕ so that $|\phi(t_0)| \le \gamma$. Since $\sigma - \varepsilon > 0$, inequality (2.4) implies that $|\phi(t)| \le M\gamma < \delta$. Hence, ϕ exists for all $t \ge t_0$ and satisfies (2.4). It follows that the trivial solution of (PE) is exponentially stable, and hence, also asymptotically stable. ∎

We now consider a specific case.

Example 2.2. Recall that the Lienard equation is given by

$$x'' + f(x)x' + x = 0, \tag{2.5}$$

where $f: R \to R$ is a continuous function with $f(0) > 0$. We can rewrite (2.5) as

$$x' = y, \qquad y' = -x - f(0)y + [f(0) - f(x)]y,$$

and we can apply Theorem 2.1 with $x = (x_1, x_2)^T$,

$$A = \begin{bmatrix} 0 & 1 \\ -1 & -f(0) \end{bmatrix}, \qquad F(t,x) = \begin{bmatrix} 0 \\ [f(0) - f(x_1)]x_2 \end{bmatrix}.$$

Noting that A is a stable matrix and that F satisfies (2.1), we conclude that the trivial solution $(x, x') = (0, 0)$ of (2.5) is uniformly asymptotically stable. We emphasize that this is a local property, i.e., it is true even if $f(x)$ becomes negative for some or all x with $|x|$ large.

In the next result, we consider the case in which A has an eigenvalue with positive real part.

6.2 Stability of an Equilibrium Point

Theorem 2.3. Assume that A is a real nonsingular $n \times n$ matrix which has at least one eigenvalue with positive real part. If $F: R^+ \times B(h) \to R^n$ is continuous and satisfies (2.1), then the trivial solution of (PE) is unstable.

Proof. We use Theorem 5.10.1 to choose a real, symmetric $n \times n$ matrix B such that $BA + A^TB = -C$ is negative definite. The matrix B is not positive definite or even positive semidefinite. Hence, the function $v(x) \triangleq x^TBx$ is negative at points arbitrarily close to the origin. Evaluating the derivative of v with respect to t along the solutions of (PE), we obtain

$$v'_{(PE)}(t,x) = -x^TCx + 2x^TBF(t,x).$$

Pick $\gamma > 0$ such that $x^TCx \geq 3\gamma|x|^2$ for all $x \in R^n$. In view of (2.1) we can pick δ such that $0 < \delta < h$ and $|BF(t,x)| \leq \gamma|x|$ for all $(t,x) \in R^+ \times B(\delta)$. Thus, for all (t,x) in $R^+ \times B(\delta)$, we obtain

$$v'_{(PE)}(t,x) \leq -3\gamma|x|^2 + 2|x|\gamma|x| = -\gamma|x|^2,$$

so that $v'_{(PE)}$ is negative definite. By Theorem 5.9.16 the trivial solution of (PE) is unstable. ∎

Let us consider another specific case.

Example 2.4. Consider the simple pendulum (see Example 1.2.9) described by

$$x'' + a\sin x = 0,$$

where a is a positive constant. Note that $x_e = \pi$, $x'_e = 0$ is an equilibrium of this equation. Let $y = x - x_e$ so that

$$y'' + a\sin(y + \pi) = y'' - ay + a(\sin(y + \pi) + y) = 0.$$

This equation can be put into the form (PE) with

$$A = \begin{bmatrix} 0 & 1 \\ a & 0 \end{bmatrix}, \quad F(t,x) = \begin{bmatrix} 0 \\ a(\sin(y + \pi) + y) \end{bmatrix}.$$

Applying Theorem 2.3, we conclude that the equilibrium point $(\pi, 0)$ is unstable.

Next, we consider periodic systems described by equations of the form

$$x' = P(t)x + F(t,x), \tag{2.6}$$

where P is a real $n \times n$ matrix which is continuous on R and which is periodic with period $T > 0$, and where F has the properties enumerated above.

Systems of this type may arise in the process of linearizing equations of the form (E) or they may arise in the process of modeling a physical system. For such systems, we establish the following result.

Corollary 2.5. Let P be defined as above and let F satisfy the hypotheses of Theorem 2.1.

(i) If all characteristic exponents of the linear system

$$z' = P(t)z \tag{2.7}$$

have negative real parts, then the trivial solution of (2.6) is uniformly asymptotically stable.

(ii) If at least one characteristic exponent of (2.7) has positive real part, then the trivial solution of (2.6) is unstable.

Proof. By Theorem 3.4.2 the fundamental matrix Φ for (2.7) satisfying $\Phi(0) = E$ has the form

$$\Phi(t) = U(t)e^{tA},$$

where $U(t)$ is a continuous, periodic, and nonsingular matrix. (From the results in Section 3.4, we see that A is uniquely defined up to $2n\pi i E$. Hence we can assume that A is nonsingular.) Now define $x = U(t)y$, where x solves (2.6), so that

$$U'(t)y + U(t)y' = P(t)U(t)y + F(t, U(t)y),$$

while

$$U' = PU - UA.$$

Thus y solves the equation

$$y' = Ay + U^{-1}(t)F(t, U(t)y),$$

and $U^{-1}(t)F(t, U(t)y)$ satisfies (2.1). Now apply Theorem 2.1 or 2.3 to determine the stability of the equilibrium $y = 0$. Since $U(t)$ and $U^{-1}(t)$ are both bounded on R, the trivial solutions $y = 0$ and $x = 0$ have the same stability properties. ∎

We see from Theorems 2.1 and 2.3 that the stability properties of the trivial solution of many nonlinear systems can be determined by checking the stability of a linear approximation, called a "first approximation." This technique is called *determining stability by "linearization"* or *determining stability from the first approximation*. Also, Theorem 2.1 together with Theorem 2.3 are sometimes called **Lyapunov's first method** or **Lyapunov's indirect method** of stability analysis of an equilibrium point.

6.3 THE STABLE MANIFOLD

We reconsider the system of equations

$$x' = Ax + F(t, x) \qquad \text{(PE)}$$

under the assumption that the matrix A is noncritical. We wish to study in detail the properties of the solutions in a neighborhood of the origin $x = 0$. In doing so, we shall need to strengthen hypothesis (2.1) and we shall be able to prove the existence of stable and unstable manifolds for (PE). The precise definition of these manifolds is given later.

We begin by making the following assumption:

$F: R \times B(h) \to R^n$, F is continuous on $R \times B(h)$, $F(t, 0) = 0$ for all $t \in R$ and for any $\varepsilon > 0$ there is a δ with $0 < \delta < h$ such that if (t, x) and $(t, y) \in R \times B(\delta)$, then $|F(t, x) - F(t, y)| \leq \varepsilon |x - y|$. (3.1)

This hypothesis is satisfied for example if $F(t, x)$ is periodic in t (or independent of t), if $F \in C^1(R \times B(h))$ and both $F(t, 0) = 0$ and $F_x(t, 0) = 0$ for all $t \in R$.

In order to provide motivation and insight for the main results of the present section, we recall the phase portraits of the two-dimensional systems considered in Section 5.6. We are especially interested in the noncritical cases. Specifically, let us consider Fig. 5.7b which depicts the qualitative behavior of the trajectories in the neighborhood of a saddle. There is a one-dimensional linear subspace S^* such that the solutions starting in S^* tend to the origin as $t \to \infty$ (see Fig. 6.1). This set S^* is called the stable manifold. There is also an unstable manifold U^* consisting of those trajectories which tend to the origin as $t \to -\infty$. If time is reversed, S^* and U^* change roles. What we shall prove in the following is that if the linear system is perturbed by terms which satisfy hypothesis (3.1), then the resulting phase portrait (see, e.g., Fig. 6.2) remains essentially unchanged. The stable manifold S and the unstable manifold U may become slightly distorted but they persist (see Fig. 6.2).

Our analysis is local, i.e., it is valid in a small neighborhood of the origin. For n-dimensional systems, we shall allow k eigenvalues with negative real parts and $n - k$ eigenvalues with positive real parts. We allow $k = 0$ or $k = n$ as special cases and, of course, we shall allow F to depend on time t. In (t, x) space, we show that there is a $(k + 1)$-dimensional stable manifold and an $(n - k + 1)$-dimensional unstable manifold in a sufficiently small neighborhood of the line determined by $(t, 0)$, $t \in R$.

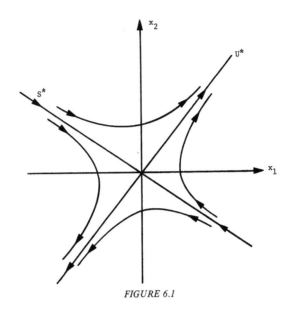

FIGURE 6.1

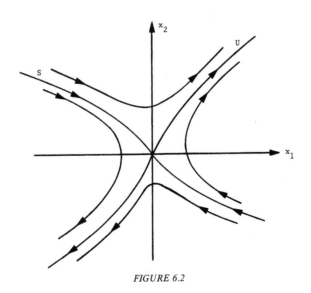

FIGURE 6.2

6.3 The Stable Manifold

Definition 3.1. A **local hypersurface** S **of dimension** $k + 1$ **located along a curve** $v(t)$ is determined as follows. There is a neighborhood V of the origin in R^n and there are $(n - k)$ functions $H_i \in C^1(R \times V)$ such that

$$S = \{(t, x) : t \in R, x - v(t) \in V \text{ and } H_i(t, x + v(t)) = 0 \text{ for } i = k + 1, \ldots, n\}.$$

Here $H_i(t, v(t)) = 0$ for $i = k + 1, \ldots, n$ and for all $t \in R$. Moreover, if ∇ denotes the gradient with respect to x, then for each $t \in R$, $\{\nabla H_i(t, v(t)) : k + 1 \leq i \leq n\}$ is a set of $n - k$ linearly independent vectors. A **tangent hypersurface** to S at a point (t, x) is determined by $\{y \in R^n : \langle y, \nabla H_i(t, v(t)) \rangle = 0, i = k + 1, \ldots, n\}$. We say that S is C^m **smooth** if the functions v and H_i are in C^m and we say that S is **analytic** if v and the H_i are holomorphic in t and in (t, x).

In the typical situation in the present chapter, $v(t)$ will be a constant [usually $v(t) \equiv 0$] or it will be a periodic function. Moreover, typically there will be a constant $n \times n$ matrix Q, a neighborhood U of the origin in $\hat{y} = (y_1, \ldots, y_k)^T$ space, and a C^1 function $G : R \times U \to R^{n-k}$ such that $G(t, 0) \equiv 0$ and such that

$$S = \{(t, x) : y = Q(x - v) \in U \text{ and } (y_{k+1}, \ldots, y_n)^T = G(t, y_1, \ldots, y_k)\}.$$

The functions $H_i(t, x)$ can be determined immediately from $G(t, y)$ and Q.

We are now in a position to prove a qualitative result for a noncritical linear system with k-dimensional stable manifold.

Theorem 3.2. Let the function F satisfy hypothesis (3.1) and let A be a real, constant $n \times n$ matrix which has k eigenvalues with negative real parts and $(n - k)$ eigenvalues with positive real parts. Then there exists a $(k + 1)$-dimensional local hypersurface S, located at the origin, called the **stable manifold** of (PE), such that S is positively invariant with respect to (PE), and for any solution ϕ of (PE) and any τ such that $(\tau, \phi(\tau)) \in S$, we have $\phi(t) \to 0$ as $t \to \infty$. Moreover, there is a $\delta > 0$ such that if $(\tau, \phi(\tau)) \in R \times B(\delta)$ for some solution ϕ of (PE) but $(\tau, \phi(\tau)) \notin S$, then $\phi(t)$ must leave the ball $B(\delta)$ at some finite time $t_1 > \tau$.

If $F \in C^l(R \times B(h))$ for $l = 1, 2, 3, \ldots$ or $l = \infty$ or if F is holomorphic in (t, x), then S has the same degree of smoothness as F. Moreover, S is tangent at the origin to the stable manifold S^* for the linear system (L).

Proof. Pick a linear transformation $x = Qy$ such that (PE) becomes

$$y' = By + g(t, y), \qquad \text{(PE')}$$

where $B = Q^{-1}AQ = \text{diag}(B_1, B_2)$ and $g(t, y) = Q^{-1}F(t, Qy)$. The matrix Q can be chosen so that B_1 is a $k \times k$ stable matrix and $-B_2$ is an $(n - k) \times (n - k)$ stable matrix. Clearly g will satisfy (3.1). Moreover, if we define

$$U_1(t) = \begin{bmatrix} e^{B_1 t} & 0 \\ 0 & 0 \end{bmatrix}, \qquad U_2(t) = \begin{bmatrix} 0 & 0 \\ 0 & e^{B_2 t} \end{bmatrix},$$

then $e^{Bt} = U_1(t) + U_2(t)$ and for some positive constants K and σ we have

$$|U_1(t)| \leq K e^{-2\sigma t}, \quad t \geq 0, \quad \text{and} \quad |U_2(t)| \leq K e^{\sigma t}, \quad t \leq 0.$$

Let ϕ be a bounded solution of (PE') with $\phi(\tau) = \xi$. Then by the variation of constants formula we have

$$\phi(t) = e^{B(t-\tau)}\xi + \int_\tau^t e^{B(t-s)} g(s, \phi(s)) \, ds$$

$$= U_1(t - \tau)\xi + \int_\tau^t U_1(t - s) g(s, \phi(s)) \, ds + U_2(t - \tau)\xi$$

$$+ \int_\tau^\infty U_2(t - s) g(s, \phi(s)) \, ds - \int_t^\infty U_2(t - s) g(s, \phi(s)) \, ds.$$

Since $U_2(t - s) = U_2(t) U_2(-s)$, the bounded solution ϕ of (PE') must satisfy

$$\phi(t) = U_1(t - \tau)\xi + \int_\tau^t U_1(t - s) g(s, \phi(s)) \, ds - \int_t^\infty U_2(t - s) g(s, \phi(s)) \, ds$$

$$+ U_2(t)\left[U_2(-\tau)\xi + \int_\tau^\infty U_2(-s) g(s, \phi(s)) \, ds\right]. \tag{3.2}$$

Conversely, any solution ϕ of (3.2) which is bounded and continuous on $[\tau, \infty)$ must solve (PE').

In order to satisfy (3.2) it is sufficient to find bounded and continuous solutions of the integral equation

$$\psi(t, \tau, \xi) = U_1(t - \tau)\xi + \int_\tau^t U_1(t - s) g(s, \psi(s, \tau, \xi)) \, ds$$

$$- \int_t^\infty U_2(t - s) g(s, \psi(s, \tau, \xi)) \, ds \tag{3.3}$$

which also satisfy the side condition

$$U_2(-\tau)\xi + \int_\tau^\infty U_2(-s) g(s, \psi(s, \tau, \xi)) \, ds = 0. \tag{S}$$

Successive approximation will be used to solve (3.3) starting with $\psi_0(t, \tau, \xi) \equiv 0$. Pick $\varepsilon > 0$ such that $4\varepsilon K < \sigma$, pick $\delta = \delta(\varepsilon)$ using (3.1), and pick ξ with $|\xi| < \delta/(2K)$. Define

$$\|\psi\| \triangleq \sup\{|\psi(t)| : t \geq \tau\}.$$

6.3 The Stable Manifold

If $\|\psi_j\| \leq \delta$, then ψ_{j+1} must satisfy

$$|\psi_{j+1}(t,\tau,\xi)| \leq K|\xi| + \int_\tau^t Ke^{-\sigma(t-s)}\varepsilon\|\psi_j\|\,ds + \int_t^\infty Ke^{\sigma(t-s)}\varepsilon\|\psi_j\|\,ds$$
$$\leq \tfrac{1}{2}\delta + (2\varepsilon K/\sigma)\|\psi_j\| \leq \delta.$$

Since $\psi_0 \equiv 0$, then the ψ_j are well defined and satisfy $\|\psi_j\| \leq \delta$ for all j. Thus

$$|\psi_{j+1}(t,\tau,\xi) - \psi_j(t,\tau,\xi)|$$
$$\leq \int_\tau^t Ke^{-\sigma(t-s)}\varepsilon\|\psi_j - \psi_{j-1}\|\,ds + \int_t^\infty Ke^{\sigma(t-s)}\varepsilon\|\psi_j - \psi_{j-1}\|\,ds$$
$$\leq (2\varepsilon K/\sigma)\|\psi_j - \psi_{j-1}\| \leq \tfrac{1}{2}\|\psi_j - \psi_{j-1}\|.$$

By induction, we have $\|\psi_{k+l+1} - \psi_{k+l}\| \leq 2^{-l}\|\psi_{k+1} - \psi_k\|$ and

$$\|\psi_{k+j} - \psi_k\| \leq \|\psi_{k+j} - \psi_{k+j-1}\| + \cdots + \|\psi_{k+1} - \psi_k\|$$
$$\leq (2^{-j+1} + \cdots + 2^{-1} + 1)\|\psi_{k+1} - \psi_k\|$$
$$\leq 2\|\psi_{k+1} - \psi_k\| \leq 2^{-k+1}\|\psi_1\|.$$

From this estimate, it follows that $\{\psi_j\}$ is a Cauchy sequence uniformly in (t,τ,ξ) over $\tau \in R$, $t \in [\tau,\infty)$, and $\xi \in B(\delta/(2K))$. Thus $\psi_j(t,\tau,\xi)$ tends to a limit $\psi(t,\tau,\xi)$ uniformly for (t,τ,ξ) on compact subsets of $(\tau,\xi) \in R \times B(\delta/(2K))$, $t \in [\tau,\infty)$. The limit function ψ must be continuous in (t,τ,ξ) and it must satisfy $\|\psi\| \leq \delta$.

The limit function ψ must satisfy (3.3). This is argued as follows. Note first that

$$\left|\int_t^\infty U_2(t-s)g(s,\psi(s,\tau,\xi))\,ds - \int_t^\infty U_2(t-s)g(s,\psi_j(s,\tau,\xi))\,ds\right|$$
$$\leq \int_t^\infty Ke^{\sigma(t-s)}\varepsilon|\psi(s,\tau,\xi) - \psi_j(s,\tau,\xi)|\,ds$$
$$\leq (K\varepsilon/\sigma)\|\psi - \psi_j\| \to 0, \qquad j \to \infty.$$

A similar procedure applies to the other integral term in (3.3). Thus we can take the limit as $j \to \infty$ in the equation

$$\psi_{j+1}(t,\tau,\xi) = U_1(t-\tau)\xi + \int_\tau^t U_1(t-s)g(s,\psi_j(s,\tau,\xi))\,ds$$
$$- \int_t^\infty U_2(t-s)g(s,\psi_j(s,\tau,\xi))\,ds$$

to obtain (3.3). Note that the solution of (3.3) is unique for given τ and ξ since a second solution $\bar{\psi}$ would have to satisfy $\|\psi - \bar{\psi}\| \leq \tfrac{1}{2}\|\psi - \bar{\psi}\|$.

The stable manifold S is the set of all points (τ,ξ) such that Eq. (S) is true. It will be clear that S is a local hypersurface of dimension $(k+1)$. If $\xi = 0$, then by uniqueness $\psi(t,\tau,0) \equiv 0$ for $t \geq \tau$ and so

$g(t, \psi(t, \tau, 0)) \equiv 0$. Hence $(\tau, 0) \in S$ for all $\tau \in R$. To see that S is positively invariant, let $(\tau, \xi) \in S$. Then $\psi(t, \tau, \xi)$ will solve (3.2), and hence it will solve (PE′). For any $\tau_1 > \tau$ let $\xi_1 = \psi(\tau_1, \tau, \xi)$ and define $\phi(t, \tau_1, \xi_1) \triangleq \psi(t, \tau, \xi)$. Then $\phi(t, \tau_1, \xi_1)$ solves (PE′) and hence it also solves (3.2) with (τ, ε) replaced by (τ_1, ξ_1). Hence

$$\left| U_2(t)\left[U_2(-\tau_1)\xi_1 + \int_{\tau_1}^{\infty} U_2(-s)g(s, \phi(s, \tau_1, \xi_1))\,ds \right] \right|$$

$$= \left| \phi(t, \tau_1, \xi_1) - U_1(t - \tau_1)\xi_1 - \int_{\tau_1}^{t} U_1(t - s)g(s, \phi(s, \tau_1, \xi_1))\,ds \right.$$

$$\left. + \int_t^{\infty} U_2(t - s)g(s, \phi(s, \tau_1, \xi_1))\,ds \right|$$

$$\leq \delta + Ke^{-\sigma(t-\tau_1)}|\xi_1| + \int_{\tau_1}^{t} Ke^{-\sigma(t-s)}\varepsilon\delta\,ds + \int_t^{\infty} Ke^{\sigma(t-s)}\varepsilon\delta\,ds$$

$$\leq \delta + \delta + (2K\varepsilon\delta/\sigma) \leq 3\delta < \infty. \tag{3.4}$$

Since $U_2(t) = \mathrm{diag}(0, e^{B_2 t})$ and $-B_2$ is a stable matrix, this is only possible when $(\tau_1, \xi_1) \in S$. Hence S is positively invariant.

To see that any solution starting on S tends to the origin as $t \to \infty$, let $(\tau, \xi) \in S$ and let ψ_j be the successive approximation defined above. Then clearly

$$|\psi_1(t, \tau, \xi)| \leq K|\xi|e^{-2\sigma(t-\tau)} \leq 2K|\xi|e^{-\sigma(t-\tau)}.$$

If $|\psi_j(t, \tau, \xi)| \leq 2K|\xi|e^{-\sigma(t-\tau)}$, then

$$|\psi_{j+1}(t, \tau, \xi)| \leq K|\xi|e^{-\sigma(t-\tau)} + \int_\tau^t Ke^{-2\sigma(t-s)}\varepsilon(2K|\xi|e^{-\sigma(s-\tau)})\,ds$$

$$+ \int_t^\infty Ke^{\sigma(t-s)}\varepsilon(2K|\xi|e^{-\sigma(s-\tau)})\,ds$$

$$\leq K|\xi|e^{-\sigma(t-\tau)} + 2K|\xi|(\varepsilon K/\sigma)e^{-\sigma(t-\tau)} + 2K|\xi|(\varepsilon K/2\sigma)e^{-\sigma(t-\tau)}$$

$$\leq 2K|\xi|e^{-\sigma(t-\tau)}$$

since $(4\varepsilon K)/\sigma < 1$. Hence in the limit as $j \to \infty$ we have $|\psi(t, \tau, \xi)| \leq 2K|\xi|e^{-\sigma(t-\tau)}$ for all $t \geq \tau$ and for all $\xi \in B(\delta/(2K))$.

Suppose that $\phi(t, \tau, \xi)$ solves (PE′) but (τ, ξ) does not belong to S. If $|\phi(t, \tau, \xi)| \leq \delta$ for all $t \geq \tau$, then (3.4) is true. Hence $(\tau, \xi) \in S$, a contradiction.

Equation (S) can be rearranged as

$$(\xi_{k+1}, \ldots, \xi_n)^\mathrm{T} = -\int_\tau^\infty U(\tau - s)g(s, \psi(s, \tau, \xi))\,ds. \tag{3.5}$$

6.3 The Stable Manifold

Utilizing estimates of the type used above, we see that the function on the right side of (3.5) is Lipschitz continuous in ξ with Lipschitz constant $L \leq \frac{1}{4}$. Hence, successive approximations can be used to solve (3.5), say

$$(\xi_{k+1}, \ldots, \xi_n)^T = h(\tau, \xi_1, \ldots, \xi_k) \tag{3.6}$$

with h continuous. If F is of class C^1 in (t, x), then the partial derivatives of the right-hand side of (3.5) with respect to $\xi_1, \ldots, \xi_n$ all exist and are zero at $\xi_1 = \cdots = \xi_n = 0$. The Jacobian with respect to $(\xi_{k+1}, \ldots, \xi_n)$ on the left side of (3.5) is one. By the implicit function theorem, the solution (3.6) is C^1 smooth, indeed h is at least as smooth as F is. Since $\partial h / \partial \xi_j = 0$ for $k < j \leq n$ at $\xi_1 = \cdots = \xi_n = 0$, then S is tangent to the hyperplane $\xi_{k+1} = \cdots = \xi_n = 0$ at $\xi = 0$, i.e., S is tangent to the stable manifold of the linear system (L) at $\xi = 0$. ∎

If we reverse time in (PE) to obtain

$$y' = -Ay - F(-t, y)$$

and then apply Theorem 3.2, we obtain the following result.

Theorem 3.3. *If the hypotheses of Theorem 3.2 are satisfied, then there is an $(n - k + 1)$-dimensional local hypersurface U based at the origin, called the **unstable manifold** of (PE), such that U is negatively invariant with respect to (PE), and for any solution ϕ of (PE) and any $\tau \in R$ such that $(\tau, \phi(\tau)) \in U$, we have $\phi(t) \to 0$ as $t \to -\infty$. Moreover, there is a $\delta > 0$ such that if $(\tau, \phi(\tau)) \in R \times B(\delta)$ but $(\tau, \phi(\tau)) \notin U$, then $\phi(t)$ must leave the ball $B(\delta)$ at some finite time $t_1 < \tau$.*

The surface U has the same degree of smoothness as F and is tangent at the origin to the unstable manifold U^* of the linear system (L).

If $F(t, x) = F(x)$ is independent of t in (PE), then it is not necessary to keep track of the initial time in Theorems 3.1 and 3.2. Indeed, it can be shown that if (S) is true for some (τ, ξ), then (S) is true for all (t, ξ) for t varying over all of R. In this case, one usually dispenses with time and one defines S and U in the x space R^n. This is what was done in Figs. 6.1 and 6.2.

Example 3.4. Consider the Volterra population model given in Example 1.2.13. Assume that in Eq. (1.2.32), $c = f = 0$ while all other constants are positive. Then these equations reduce to

$$x_1' = ax_1 - bx_1x_2, \qquad x_1(0) = \xi_1 \geq 0,$$
$$x_2' = dx_2 - ex_1x_2, \qquad x_2(0) = \xi_2 \geq 0.$$

There are two equilibrium points, namely,

$$E_1: \quad x_{1e} = x_{2e} = 0, \qquad E_2: \quad x_{1e} = d/e, \qquad x_{2e} = a/b.$$

The eigenvalues of the linear part at equilibrium E_1 are $\lambda = a$ and $\lambda = d$. Since both are positive, this equilibrium is completely unstable. At the second equilibrium point, the eigenvalues are $\lambda = \sqrt{ab} > 0$ and $\lambda = -\sqrt{ab} < 0$. Hence, ignoring time, the stable and unstable manifolds each have dimension one. These manifolds are tangent at E_2 to the lines

$$\sqrt{ad}\,x_1 + (bd/e)x_2 = 0, \qquad -\sqrt{ad}\,x_1 + (bd/e)x_2 = 0.$$

Notice that if $x_2 = a/b$ and $0 < x_1 < d/e$, then $x_1' = 0$ and $x_2' > 0$. If $x_2 > a/b$ and $0 < x_1 < d/e$, then $x_1' < 0$ and $x_2' > 0$. If $x_1(0) = 0$, then $x_1(t) = 0$ for all $t \geq 0$. Hence, the set $G_1 = \{(x_1, x_2): 0 < x_1 < d/e, x_2 > a/b\}$ is a positively invariant set. Moreover, all solutions $(x_1(t), x_2(t))$ which enter this set must satisfy the condition that $x_2(t) \to \infty$ as $t \to \infty$. Similarly, the set $G_2 = \{(x_1, x_2): x_1 > d/e, 0 < x_2 < a/b\}$ is a positive invariant set and all solutions which enter G_2 must satisfy the condition that $x_1(t) \to \infty$ as $t \to \infty$.

Since the unstable manifold U of E_2 is tangent to the line $\sqrt{ad}\,x_1 + (bd/e)x_2 = 0$, then one branch of U enters G_1 and one enters G_2 (see Fig. 6.3). The stable manifold S of E_2 cannot meet either G_1 or G_2. Hence, the phase portrait is completely determined as shown in Fig. 6.3. We see that for almost all initial conditions one species will eventually die

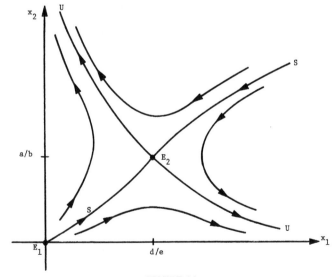

FIGURE 6.3

6.4 Stability of Periodic Solutions

out while the second will grow. Moreover, the outcome is unpredictable in the sense that near S a slight change in initial conditions can radically alter the outcome.

6.4 STABILITY OF PERIODIC SOLUTIONS

We begin by considering a T-periodic system

$$x' = f(t, x), \tag{P}$$

where $f \in C^1(R \times D)$, D is a domain in R^n, and $f(t + T, x) = f(t, x)$ for all $(t, x) \in R \times D$. Let p be a nonconstant, T-periodic solution of (P) satisfying $p(t) \in D$ for all $t \in R$. Now define $y = x - p(t)$ so that

$$y' = f_x(t, p(t))y + h(t, y), \tag{4.1}$$

where

$$h(t, y) \triangleq f(t, y + p(t)) - f(t, p(t)) - f_x(t, p(t))$$

satisfies hypothesis (3.1). From (4.1) we now obtain the corresponding linear system

$$y' = f_x(t, p(t))y. \tag{4.2}$$

By the Floquet theory (see Chapter 3) there is a periodic, nonsingular matrix $V(t)$ such that the transformation $y = V(t)z$ transforms (4.1) to a system of the form

$$z' = Az + V^{-1}(t)h(t, V(t)z).$$

This system satisfies the hypotheses of Theorem 3.2 if A is noncritical. This argument establishes the following result.

Theorem 4.1. Let $f \in C^1(R \times D)$ and let (P) have a nonconstant periodic solution p of period T. Suppose that the linear variational system (4.2) for $p(t)$ has k characteristic exponents with negative real parts and $(n - k)$ characteristic exponents with positive real parts. Then there exist two hypersurfaces S and U for (P), each containing $(t, p(t))$ for all $t \in R$, where S is positively invariant and U is negatively invariant with respect to (P) and where S has dimension $(k + 1)$ and U has dimension $(n - k + 1)$ such that for any solution ϕ of (P) in a δ neighborhood of p and any $\tau \in R$ we have

    (i) $\phi(t) - p(t) \to 0$ as $t \to \infty$ if $(\tau, \phi(\tau)) \in S$,
    (ii) $\phi(t) - p(t) \to 0$ as $t \to -\infty$ if $(\tau, \phi(\tau)) \in U$, and
    (iii) ϕ must leave the δ neighborhood of p in finite time as t increases from τ and as t decreases from τ if $(\tau, \phi(\tau))$ is not on S and not on U.

The sets S and U are the stable and the unstable manifolds associated with p. When $k = n$, then S is $(n + 1)$-dimensional, U consists only of the points $(t, p(t))$ for $t \in R$, and p is asymptotically stable. If $k < n$, then clearly p is unstable.

This simple and appealing stability analysis breaks down completely if p is a T-periodic solution of an autonomous system

$$x' = f(x), \tag{A}$$

where $f \in C^1(D)$. In this case the variational equation obtained from the transformation $y = x - p(t)$ is

$$y' = f_x(p(t))y + h(t, y), \tag{4.3}$$

where $h(t, y) \triangleq f(y + p(t)) - f(p(t)) - f_x(p(t))y$ satisfies hypothesis (3.1). In this case, the corresponding linear first approximation is

$$y' = f_x(p(t))y. \tag{4.4}$$

Note that since $p(t)$ solves (A), $p'(t)$ is a T-periodic solution of (4.4). Hence, Eq. (4.4) cannot possibly satisfy the hypotheses that no characteristic exponent has zero real part. Indeed, one characteristic multiplier is one. The hypotheses of Theorem 4.1 can never be satisfied and hence, the preceding analysis must be modified. Even if the remaining $(n - 1)$ characteristic exponents are all negative, p cannot possibly be asymptotically stable. To see this, note that for τ small, $p(t + \tau)$ is near $p(t)$ at $t = 0$, but $|p(t + \tau) - p(t)|$ does not tend to zero as $t \to \infty$. However, p will satisfy the following more general stability condition.

Definition 4.2. A T-periodic solution p of (A) is called **orbitally stable** if there is a $\delta > 0$ such that any solution ϕ of (A) with $|\phi(\tau) - p(\tau)| < \delta$ for some τ tends to the orbit

$$C(p(\tau)) = \{p(t) : 0 \le t \le T\}$$

as $t \to \infty$. If in addition for each such ϕ there is a constant $\alpha \in [0, T)$ such that $\phi(t) - p(t + \alpha) \to 0$ as $t \to \infty$, then ϕ is said to have **asymptotic phase** α.

We can now prove the following result.

Theorem 4.3. Let p be a nonconstant periodic solution of (A) with least period $T > 0$ and let $f \in C^1(D)$, where D is a domain in R^n.

6.4 Stability of Periodic Solutions

If the linear system (4.4) has $(n-1)$ characteristic exponents with negative real parts, then p is orbitally stable and nearby solutions of (A) possess an asymptotic phase.

Proof. By a change of variables of the form $x = Qw + p(0)$, where Q is assumed to be nonsingular, so that

$$w' = Q^{-1}f(Qw + p(0)),$$

Q can be so arranged that $w(0) = 0$ and $w'(0) = Q^{-1}f(p(0)) = (1, 0, \ldots, 0)^T$. Hence, without loss of generality, we assume in the original problem (A) that $p(0) = 0$ and $p'(0) = e_1 \triangleq (1, 0, \ldots, 0)^T$.

Let Φ_0 be a real fundamental matrix solution of (4.4). There is a real nonsingular matrix C such that $\Phi_0(t + T) = \Phi_0(t)C$ for all $t \in R$. Since p' is a solution of (4.4), one eigenvalue of C is equal to one [see Eq. (3.4.8)]. By hypothesis, all other eigenvalues of C have magnitude less than one, i.e., all other characteristic exponents of (4.4) have negative real parts. Thus, there is a real $n \times n$ matrix R such that

$$R^{-1}CR = \begin{bmatrix} 1 & 0 \\ 0 & D_0 \end{bmatrix},$$

where D_0 is an $(n-1) \times (n-1)$ matrix and all eigenvalues of D_0 have absolute value less than one.

Now define $\Phi_1(t) = \Phi_0(t)R$ so that Φ_1 is a fundamental matrix for (4.4) and

$$\Phi_1(t + T) = \Phi_0(t + T)R = \Phi_0(t)CR = \Phi_0(t)R(R^{-1}CR) = \Phi_1(t)\begin{bmatrix} 1 & 0 \\ 0 & D_0 \end{bmatrix}.$$

The first column $\phi_1(t)$ of $\Phi_1(t)$ necessarily must satisfy the relation

$$\phi_1(t + T) = \phi_1(t) \quad \text{for all} \quad t \in R,$$

i.e., it must be T periodic. Since $(n-1)$ characteristic exponents of (4.4) have negative real parts, there cannot be two linearly independent T periodic solutions of (4.4). Thus, there is a constant $k \neq 0$ such that $\phi_1 = kp'$. If $\Phi_1(t)$ is replaced by

$$\Phi(t) \triangleq \operatorname{diag}(k^{-1}, 1, \ldots, 1)\Phi_1(t),$$

then Φ satisfies the same conditions as Φ_1 but now $k = 1$.

There is a T periodic matrix $P(t)$ and a constant matrix B such that

$$e^{TB} = \begin{bmatrix} 1 & 0 \\ 0 & D_0 \end{bmatrix}, \quad \Phi(t) = P(t)e^{Bt}.$$

[Both $P(t)$ and B may be complex valued.] The matrix B can be taken in the block diagonal form

$$B = \begin{bmatrix} 0 & 0 \\ 0 & B_1 \end{bmatrix},$$

where $e^{B_1 T} = D_0$ and B_1 is a stable $(n-1) \times (n-1)$ matrix. Define

$$U_1(t,s) = P(t) \begin{bmatrix} 1 & 0 \\ 0 & 0 \end{bmatrix} P^{-1}(s)$$

and

$$U_2(t,s) = P(t) \begin{bmatrix} 0 & 0 \\ 0 & e^{B_1(t-s)} \end{bmatrix} P^{-1}(s)$$

so that

$$U_1(t,s) + U_2(t,s) = P(t)e^{B(t-s)}P^{-1}(s) = \Phi(t)\Phi^{-1}(s).$$

Clearly $U_1 + U_2$ is real valued. Since

$$P(t)\begin{bmatrix} 1 & 0 \\ 0 & 0 \end{bmatrix} = (\phi_1, 0, \ldots, 0),$$

this matrix is real. Similarly, the first row of

$$\begin{bmatrix} 1 & 0 \\ 0 & 0 \end{bmatrix} P^{-1}(s)$$

is the first row of $\Phi^{-1}(s)$ and the remaining rows are zero. Thus,

$$U_1(t,s) = P(t)\begin{bmatrix} 1 & 0 \\ 0 & 0 \end{bmatrix}\begin{bmatrix} 1 & 0 \\ 0 & 0 \end{bmatrix} P^{-1}(s)$$

is a real matrix. Hence,

$$U_2(t,s) = \Phi(t)\Phi^{-1}(s) - U_1(t,s)$$

is also real.

Pick constants $K > 1$ and $\sigma > 0$ such that $|U_1(t,s)| \leq K$ and $|U_2(t,s)| \leq Ke^{-2\sigma(t-s)}$ for all $t \geq s \geq 0$. As in the proof of Theorem 3.1, we utilize an integral equation. In the present case, it assumes the form

$$\psi(t) = U_2(t,\tau)\xi + \int_\tau^t U_2(t,s)h(s,\psi(s))\,ds - \int_t^\infty U_1(t,s)h(s,\psi(s))\,ds, \quad (4.5)$$

where h is the function defined in (4.3). This integral equation is again solved by successive approximations to obtain a unique, continuous solution

6.4 Stability of Periodic Solutions

$\psi(t, \tau, \xi)$ for $t \geq \tau$, $\tau \in R$, and $|\xi| \leq \delta$ and with

$$|\psi(t + \tau, \tau, \xi)| \leq 2K|\xi|e^{-\sigma t}.$$

Solutions of (4.5) will be solutions of (4.3) provided that the condition

$$U_1(t,\tau)\xi + \int_\tau^\infty U_1(t,s)h(s, \psi(s,\tau,\xi))\,ds = 0 \tag{4.6}$$

is satisfied. Since

$$U_1(t,s) = P(t)\begin{bmatrix} 1 & 0 \\ 0 & 0 \end{bmatrix}P^{-1}(s),$$

equivalently one can write

$$\begin{bmatrix} 1 & 0 \\ 0 & 0 \end{bmatrix}\left(P^{-1}(\tau)\xi + \int_\tau^\infty P^{-1}(s)h(s, \psi(s,\tau,\xi))\,ds\right) = 0.$$

Since h_x and ψ_ξ exist and are continuous with $h_x(t,0) = 0$, then by the implicit function theorem one can solve for some ξ_j in terms of τ and the other ξ_m's. Hence, the foregoing equation determines a local hypersurface. For any τ, let G_τ be the set of all points ξ such that (τ, ξ) is on this hypersurface.

The set of points (τ, ξ) which satisfy (4.6) is positively invariant with respect to (4.3). Hence G_τ is mapped to $G_{\tau'}$ under the transformation determined by (A) as t varies from τ to τ'. As τ varies over $0 \leq \tau \leq T$, the surface G traces out a neighborhood N of the orbit $C(p(0))$. Any solution which starts within N will tend to $C(p(0))$ as $t \to \infty$. Indeed, for $|\phi(\tau) - p(\tau')|$ sufficiently small, we define $\phi_1(t) = \phi(t + \tau - \tau')$. Then ϕ_1 solves (A), $|\phi_1(\tau') - p(\tau')|$ is small, and so, by continuity with respect to initial conditions, $\phi_1(t)$ will remain near $p(t)$ long enough to intersect G_τ at $\tau = 0$ at some t_1. Then as $t \to \infty$,

$$\phi_1(t + t_1) - p(t) = \psi(t) \to 0,$$

or

$$\phi(t - \tau' + \tau + t_1) - p(t) \to 0. \quad\blacksquare$$

Theorem 3.3 can be extended to obtain stable and unstable manifolds about a periodic solution in the fashion indicated in the next result, Theorem 4.4. The reader will find it instructive to make reference to Fig. 6.4.

Theorem 4.4. Let $f \in C^1(D)$ for some domain D in R^n and let p be a nonconstant T-periodic solution of (A). Suppose k characteristic exponents of (4.4) have negative real parts and $(n - k - 1)$ characteristic

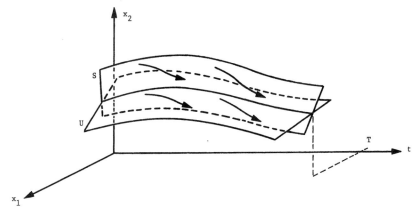

FIGURE 6.4

exponents of (4.4) have positive real parts. Then there exist T-periodic C^1-smooth manifolds S and U based at $p(t)$ such that S has dimension $k + 1$ and is positively invariant, U has dimension $(n - k)$ and is negatively invariant, and if ϕ is a solution of (A) with $\phi(0)$ sufficiently close to $C(p(0))$, then

(i) $\phi(t)$ tends to $C(p(0))$ as $t \to \infty$ if $(0, \phi(0)) \in S$,
(ii) $\phi(t)$ tends to $C(p(0))$ as $t \to -\infty$ if $(0, \phi(0)) \in U$, and
(iii) $\phi(t)$ must leave a neighborhood of $C(p(0))$ as t increases and as t decreases if $(0, \phi(0)) \notin S \cup U$.

The proof of this theorem is very similar to the proof of Theorem 4.3. The matrix R can be chosen so that

$$R^{-1}CR = \begin{bmatrix} 1 & 0 & 0 \\ 0 & D_2 & 0 \\ 0 & 0 & D_3 \end{bmatrix},$$

where D_2 is a $k \times k$ matrix with eigenvalues which satisfy $|\lambda| < 1$ and D_3 is an $(n - k - 1) \times (n - k - 1)$ matrix whose eigenvalues satisfy $|\lambda| > 1$. Define B so that

$$B = \begin{bmatrix} 0 & 0 & 0 \\ 0 & B_2 & 0 \\ 0 & 0 & B_3 \end{bmatrix}, \quad e^{BT} = R^{-1}CR.$$

Define U_1 as before and define U_2 and U_3 using $e^{B_2 t}$ and $e^{B_3 t}$. The rest of the proof involves similar modifications.

6.4 Stability of Periodic Solutions

Example 4.5. If $g \in C^1(R)$ and if $xg(x) > 0$ for all $x \neq 0$, then we have seen (cf. Example 5.11.3) that near the origin $x = x' = 0$, all solutions of

$$x'' + g(x) = 0$$

are periodic. Since one periodic solution will neither approach nor recede from nearby periodic solutions, we see that the characteristic multipliers of a given periodic solution p must both be one.

The task of computing the characteristic multipliers of a periodic linear system is complicated and difficult, in fact, little is known at this time about this problem except in certain rather special situations such as second order problems and certain Hamiltonian systems (see the problems in Chapter 3 and Example 4.5). Perturbations of certain linear, autonomous systems will be discussed in Chapter 8. It will be seen from that analysis how complicated this type of calculation can be. Nevertheless, the analysis of stability of periodic solutions of nonlinear systems by the use of Theorems 4.2 and 4.3 is of great theoretical importance. Moreover, the hypotheses of these theorems can sometimes be checked numerically. For example, if $p(t)$ is known, then numerical solution of the $(n^2 + n)$-dimensional system

$$x' = f(x), \quad x(0) = p(0),$$
$$Y' = f_x(x)Y, \quad Y(0) = E$$

over $0 \leq t \leq T$ yields $C_1 = Y(T)$ to good approximation. The eigenvalues of C_1 can usually be determined numerically with enough precision to answer stability questions.

As a final note, we point out that our conditions for asymptotic stability and for instability are sufficient but not necessary as the following example shows.

Example 4.6. Consider the system

$$x' = xf(x^2 + y^2) - y,$$
$$y' = yf(x^2 + y^2) + x,$$

where $f \in C^1[0, \infty)$, $f(1) = 0 = f'(1)$, and $f(r)(r - 1) < 0$, $r \neq 1$. Clearly $x = \cos t$, $y = \sin t$ is a solution whose linear variational equation is

$$z' = \begin{bmatrix} 0 & -1 \\ 1 & 0 \end{bmatrix} z.$$

The characteristic multipliers are both one. Using polar coordinates $x = r \cos \theta$ and $y = r \sin \theta$, this system becomes

$$\theta' = 1 \quad \text{and} \quad r' = rf(r^2).$$

Since $f(r^2) > 0$ if r is less than 1 but near one and negative if r is greater than one but near one, then clearly the periodic solution $r = 1$ is asymptotically stable.

6.5 ASYMPTOTIC EQUIVALENCE

Let us next consider a linear system

$$x' = A(t)x \tag{LH}$$

and a corresponding perturbed system

$$y' = A(t)y + F(t, y), \tag{LP}$$

where $A(t)$ and $F(t, y)$ are defined and continuous for $t \geq 0$ and $y \in B(h)$ for some $h > 0$. We wish to study the asymptotic equivalence of these two systems. This property is useful in characterizing the asymptotic behavior in certain situations where (LH) need not be asymptotically stable.

Definition 5.1. Systems (LH) and (LP) are called **asymptotically equivalent** if there is a δ, $0 < \delta < h$, such that for any solution of (LH) with $|x(t_0)| \leq \delta$ there is a solution $y(t)$ of (LP) such that

$$\lim_{t \to \infty} [x(t) - y(t)] = 0 \tag{5.1}$$

and for each solution y of (LP) with $|y(t_0)| \leq \delta$ there is a solution x of (LH) such that (5.1) is true.

Let us consider some specific cases.

Example 5.2. Consider $x' = a$ and $y' = a + f(t)$. Since $x(t) = c_1 + at$ and $y(t) = c_2 + at + \int_0^t f(s)\,ds$ for some constants c_1 and c_2, the two equations are asymptotically equivalent if and only if

$$\lim_{t \to \infty} \int_0^t f(s)\,ds = d \tag{5.2}$$

exists and is finite. This is most easily accomplished when f is absolutely integrable, i.e., when

$$\int_0^\infty |f(s)|\,ds < \infty.$$

6.5 Asymptotic Equivalence

Example 5.3. Consider the equations $x' = ax$ and $y' = ay + f(t)y$. The solutions are

$$x(t) = c_1 e^{at} \quad \text{and} \quad y(t) = c_2 \exp\left(at + \int_0^t f(s)\,ds\right).$$

Here again when $a \leq 0$ condition (5.2) is sufficient for asymptotic equivalence of the two systems.

Example 5.4. Consider the equations $x' = -x$ and $y' = -y + y^2$. In this case, both $x = 0$ and $y = 0$ are uniformly asymptotically stable equilibrium points and as such they are automatically asymptotically equivalent. Notice that here (5.1) is possible only when $|y(t_0)| < 1$. This example demonstrates why in general the constant δ in Definition 5.1 will be needed.

Example 5.5. Consider the equations $x'' + \omega^2 x = 0$ and $y'' + \omega^2 y + f(t) = 0$. It is easy to check that if f is absolutely integrable, then

$$y(t) = c_1 \cos \omega t + c_2 \sin \omega t + \omega^{-2} \int_t^\infty \sin \omega(t-s) f(s)\,ds$$

is the general solution of the second equation and the two equations are asymptotically equivalent.

Let Φ be a fundamental matrix for (LH). Then $y = \Phi(t)v$ transforms (LP) into the system

$$v' = \Phi^{-1}(t) F(t, \Phi(t)v). \tag{5.3}$$

We are now in a position to prove the following result.

Theorem 5.6. Let Φ be a fundamental matrix for (LH). If Φ and Φ^{-1} are uniformly bounded on $t \geq 0$, then (LH) and (LP) are asymptotically equivalent if and only if there is a δ such that for any $c \in B(\delta)$, there is a solution v of (5.3) such that

$$\lim_{t \to \infty} v(t) = c \tag{5.4}$$

and for any solution v with $|v(t_0)| \leq \delta$ there is a $c \in R^n$ such that (5.4) is true.

Proof. We first prove sufficiency. Let $K > 0$ be chosen so that $|\Phi(t)| \leq K$ and $|\Phi^{-1}(t)| \leq K$ for all $t \in R^+ = [0, \infty)$. In order to show asymptotic equivalence of (LH) and (LP), fix $x(t) = \Phi(t)\Phi^{-1}(\tau)\xi$ and let $c = \Phi^{-1}(\tau)\xi$. Pick v so that (5.4) is true for this c. Then $y(t) = \Phi(t)v(t)$ satisfies

$$|x(t) - y(t)| = |\Phi(t)c - \Phi(t)v(t)| \leq K|c - v(t)| \to 0$$

as $t \to \infty$. On the other hand, given $y(t) = \Phi(t)v(t)$ we can choose c such that (5.4) is true and then let $x(t) = \Phi(t)c$. Then (5.1) is true.

Conversely, let (LH) and (LP) be asymptotically equivalent. Given $c \in R^n$ with $|c|$ small, let $x(t) = \Phi(t)c$ and choose $y(t)$ such that (5.1) is true. Then $v(t) = \Phi^{-1}(t)y(t)$ satisfies

$$|v(t) - c| = |\Phi^{-1}(t)y(t) - c| = |\Phi^{-1}(t)[y(t) - x(t)]| \leq K|y(t) - x(t)| \to 0$$

as $t \to \infty$. Given v with $|v(t_0)|$ small, let $y(t) = \Phi(t)v(t)$ and choose $x(t) = \Phi(t)c$ such that (5.1) is true. Then again $|v(t) - c| \to 0$ as $t \to \infty$. ∎

We can now also prove the next result.

Corollary 5.7. Let Φ and Φ^{-1} be uniformly bounded on R^+. If there is a continuous function such that $\int_0^\infty b(t)\,dt < \infty$ and

$$|F(t, y) - F(t, \bar{y})| \leq b(t)|y - \bar{y}|$$

for all (t, y) and $(t, \bar{y})$ in $R^+ \times B(h)$ and if $F(t, 0) \equiv 0$, then (LH) and (LP) are asymptotically equivalent.

Proof. Let $|\Phi(t)| \leq K$ and $|\Phi^{-1}(t)| \leq K$ for all $t \geq 0$. Then for any solution v of (5.3) we have

$$|v'(t)| = |\Phi^{-1}(t)[F(t, \Phi(t)v) - F(t, 0)]|.$$

By the comparison theorem (Theorem 2.8.4) it follows that if $w(\tau) \geq |v(\tau)|$ for some $\tau \geq 0$ and if

$$w' = K^2 b(t)w,$$

then $w(t) \geq |v(t)|$. Hence, for any $\tau \geq 0$ and $t \geq \tau$, we have

$$|v(t)| \leq |v(\tau)| \exp\left(K^2 \int_0^\infty b(\tau)\,d\tau\right) = M.$$

Fix $\varepsilon > 0$ and pick $T > 0$ such that

$$\int_T^\infty b(s)\,ds < \varepsilon(K^2 M)^{-1}.$$

Then

$$|v(t) - v(T)| \leq \int_T^t K^2 b(s)|v(s)|\,ds \leq \int_T^\infty K^2 b(s)M\,ds < \varepsilon$$

for $t > T$. Hence, $v(t)$ has a limit $c \in R^n$ as $t \to \infty$.

Given $c \in R^n$ with $|c|$ small, consider the integral equation

$$v(t) = c - \int_t^\infty \Phi^{-1}(s)F(s, \Phi(s)v(s))\,ds.$$

Pick $T > 0$ so large that

$$2K^2 \int_T^\infty b(s)\,ds < 1.$$

6.5 Asymptotic Equivalence

With $v_0(t) \equiv c$ and an argument using successive approximations, we see that this integral equation has a solution $v \in C[T, \infty)$ with $|v(t)| \leq 2|c|$. On differentiating this integral equation, we see that v solves (5.3) on $T \leq t < \infty$. Moreover,

$$|v(t) - c| \leq \int_t^\infty K^2 b(s)|v(s)|\, ds \leq 2K^2|c| \int_t^\infty b(s)\, ds \to 0$$

as $t \to \infty$. Hence, Theorem 5.6 applies. This concludes the proof. ∎

Let us consider a specific case.

Example 5.8. Let a scalar function f satisfy the hypotheses of Corollary 5.7. Consider the equation

$$y'' + \omega^2 y = f(t, y),$$

where $\omega > 0$ is fixed. By Corollary 5.7 this equation (written as a system of first order differential equations) is asymptotically equivalent to

$$x'' + \omega^2 x = 0$$

(also written as a system of first order differential equations).

Corollary 5.7 will not apply when (LH) is, e.g., of the form

$$x' = \begin{bmatrix} 0 & -1 & 0 \\ 1 & 0 & 0 \\ 1 & 0 & -1 \end{bmatrix} x.$$

This coefficient matrix has eigenvalues $\pm i$ and -1. Thus Φ is uniformly bounded on R^+ but Φ^{-1} is not uniformly bounded. For such linear systems, the following result applies.

Theorem 5.9. *If the trivial solution of* (LH) *is uniformly stable, $A(t) \equiv A$ and if B is a continuous $n \times n$ real matrix such that*

$$\int_0^\infty |B(t)|\, dt < \infty,$$

then (LH) *and*

$$y' = [A + B(t)]y \tag{5.5}$$

are asymptotically equivalent.

Proof. We can assume that $A = \text{diag}(A_1, A_2)$ where all eigenvalues of A_1 have zero real parts and where A_2 is stable. Define $\Phi_1(t) = \text{diag}(e^{A_1 t}, 0)$ and $\Phi_2(t) = \text{diag}(0, e^{A_2 t})$. There are constants $K > 0$ and $\sigma > 0$ such that $|\Phi_1(t)| \leq K$ for all $t \leq 0$ and $|\Phi_2(t)| \leq K e^{-\sigma t}$ for all $t \geq 0$. Let

$\Phi(t) = \Phi_1(t) + \Phi_2(t) = e^{At}$. Let x be a given solution of (LH). Given is the integral equation

$$y(t) = x(t) - \int_t^\infty \Phi_1(t-s)B(s)y(s)\,ds + \int_T^t \Phi_2(t-s)B(s)y(s)\,ds.$$

Let $T > 0$ be so large that

$$4\int_T^\infty K|B(s)|\,ds < 1.$$

Then successive approximations starting with $y_0(t) \equiv x(t)$ can be used to show that the integral equation has a solution $y \in C[T, \infty)$ with $|y(t)| \leq 2(\max_{t \geq T}|x(t)|) = M$. This y satisfies the relation $|x(t) - y(t)| \to 0$ as $t \to \infty$. Moreover, y solves (5.5) since

$$\begin{aligned} y'(t) &= Ax(t) - A\int_t^\infty \Phi_1(t-s)B(s)y(s)\,ds + A\int_T^t \Phi_2(t-s)B(s)y(s)\,ds \\ &\quad + \Phi_1(0)B(t)y(t) + \Phi_2(0)B(t)y(t) \\ &= Ay(t) + B(t)y(t). \end{aligned}$$

Let $y(t)$ solve (5.5) for $t \geq \tau$. Then by the variation of constants formula, y solves

$$y(t) = \Phi(t)y(\tau) + \int_\tau^t \Phi(t)\Phi^{-1}(s)B(s)y(s)\,ds.$$

Thus, for $t \geq \tau$ we have

$$|y(t)| \leq K|y(\tau)| + \int_\tau^t K|B(s)|\,|y(s)|\,ds.$$

By the Gronwall inequality (Theorem 2.1.6), we have

$$|y(t)| \leq K|y(\tau)|\exp\left(\int_\tau^t K|B(s)|\,ds\right).$$

Thus, $y(t)$ exists and is bounded on $[\tau, \infty)$. Let $|y(t)| \leq K_0$, on $\tau \leq t < \infty$. Then the function

$$x(t) \triangleq y(t) + \int_t^\infty \Phi_1(t-s)B(s)y(s)\,ds - \int_\tau^t \Phi_2(t-s)B(s)y(s)\,ds$$

is defined for all $t \geq \tau$. Since

$$\begin{aligned} x'(t) &= [A + B(t)]y(t) - B(t)y(t) + A\int_t^\infty \Phi_1(t-s)B(s)y(s)\,ds \\ &\quad - A\int_\tau^t \Phi_2(t-s)B(s)y(s)\,ds, \end{aligned}$$

then $x(t)$ solves (LH) and $[y(t) - x(t)] \to 0$ as $t \to \infty$. ∎

PROBLEMS

1. Let $f \in C^1(D)$, where D is a domain in R^n and let x_e be a critical point of (A). Let the matrix A be defined by $A = f_x(x_e)$. Prove the following:
    (a) If A is a stable matrix, then the equilibrium x_e is exponentially stable.
    (b) If A has an eigenvalue with positive real part, then the equilibrium is unstable.
Show by example that if A is critical, then x_e can be either stable or unstable.

2. Analyze the stability properties of each equilibrium point of the following equations using Problem 1.
    (a) $x'' + \varepsilon(x^2 - 1)x' + x = 0$, $\varepsilon \neq 0$,
    (b) $x'' + x' + \sin x = 0$,
    (c) $x'' + x' + x(x^2 - 4) = 0$,
    (d) $3x''' - 7x'' + 3x' + e^x - 1 = 0$,
    (e) $x'' + cx' + \sin x = x^3$, $c \neq 0$, and
    (f) $x'' + 2x' + x = x^3$.

3. For each equilibrium point in Problems 2(a)–2(d), determine the dimension of the stable and unstable manifolds (ignore the time dimension).

4. Analyze the stability properties of the trivial solution of each of the following equations:

    (a) $x' = \begin{bmatrix} 2 & 1 \\ 7 & 3 \end{bmatrix} x + \begin{bmatrix} (e^{x_1} - 1)\sin(x_2 t) \\ e^{-t} x_1 x_2 \end{bmatrix}$, $x = (x_1, x_2)^T$.

    (b) $x' = \begin{bmatrix} (\arctan x_1) + x_2 \\ \sin(x_1 - x_2) \end{bmatrix}$, $x = (x_1, x_2)^T$.

    (c) $x' = -\begin{bmatrix} 3 & 1 & -1 \\ 1 & 4 & 0 \\ -1 & 0 & 4 \end{bmatrix} x + \begin{bmatrix} x_1 x_2 \\ x_1 x_3 \\ \sin(x_1 x_2 x_3) \end{bmatrix}$,

    (d) $x' = -a_0 y - a_1 z$, $y' = b_0(e^x - 1)$,
         $z' = -\lambda z + b_1(e^x - 1)$,

where $\lambda > 0$, $b_i \neq 0$, and $a_i/b_i > 0$ for $i = 0, 1$.

5. In Problem 4, when possible, compute a set of basis vectors for the stable manifold of each associated linearized equation.

6. Prove the following result: Let A be a stable $n \times n$ matrix, let F satisfy hypothesis (3.1), let $G \in C^1(R^+ \times B(h))$, and let $G(t, x) \to 0$ as $t \to \infty$ uniformly for $x \in B(h)$. Then for any $\varepsilon > 0$, there exist constants $\delta > 0$ and $T > 0$ such that if ϕ solves

$$x' = Ax + G(t, x) + F(t, x), \qquad x(\tau) = \xi$$

with $\tau \geq T$ and $|\xi| \leq \delta$, then $\phi(t)$ exists for all $t \geq \tau$, $|\phi(t)| < \varepsilon$ for all $t \geq \tau$, and $\phi(t) \to 0$ as $t \to \infty$.

7. Let $f \in C^2(D)$, where D is a domain in R^n and let x_e be an equilibrium point of (A) such that $f_x(x_e)$ is a noncritical matrix. Show that there is a $\delta > 0$ such that the only solution ϕ of (A) which remains in $B(x_e, \delta)$ for all $t \in R$ is $\phi(t) \equiv x_e$.

8. Let $f \in C^2(D)$, where D is a domain in R^n, let $x_e \in D$, let $f(x_e) = 0$, and let $f_x(x_e)$ be a noncritical matrix. Let $g \in C^1(R \times D)$ and let $g(t, x) \to 0$ as $t \to \infty$ uniformly for x on compact subsets of D. Show that there exists an $\alpha > 0$ such that if $\xi \in B(x_e, \alpha)$, then for any $\tau \in R^+$ the solution ϕ of

$$x' = f(x) + g(t, x), \qquad x(\tau) = \xi$$

must either leave $B(x_e, \alpha)$ in finite time or else $\phi(t)$ must tend to x_e as $t \to \infty$.

9. Let $f \in C^2(R \times D)$, where D is a domain in R^n and let p be a nonconstant T-periodic solution of (P). Let all characteristic multipliers λ of (4.2) satisfy $|\lambda| \neq 1$. Show that there is a $\delta > 0$ such that if ϕ solves (P) and if $|\phi(t)-p(t)| < \delta$ for all $t \in R$, then $\phi(t) = p(t)$ for all $t \in R$.

10. Let $f \in C^2(D)$, where D is a domain in R^n and let p be a nonconstant T-periodic solution of (A). Let $n - 1$ characteristic multipliers λ of (4.4) satisfy $|\lambda| \neq 1$. Show that there is a $\delta > 0$ such that if ϕ is a solution of (A) and if ϕ remains in a δ neighborhood of the orbit $C(p(0))$ for all $t \in R$, then $\phi(t) = \phi(t + \beta)$ for some $\beta \in R$.

11. Let F satisfy hypothesis (3.1), let $T = 2\pi$, and consider

$$x' = \begin{bmatrix} -1 + \tfrac{3}{2}\cos^2 t & 1 - \tfrac{3}{2}\sin t \cos t \\ -1 - \tfrac{3}{2}\sin t \cos t & -1 + \tfrac{3}{2}\sin^2 t \end{bmatrix} x + F(t, x). \tag{6.1}$$

Let $P_0(t)$ denote the 2×2 periodic matrix shown in Eq. (6.1).

(a) Show that $y = (\cos t, -\sin t)^T e^{t/2}$ is a solution of

$$y' = P_0(t)y. \tag{6.2}$$

(b) Compute the characteristic multipliers of (6.2).
(c) Determine the stability properties of the trivial solution of (6.1).
(d) Compute the eigenvalues of $P_0(t)$. Discuss the possibility of using the eigenvalues of (6.2), rather than the characteristic multipliers, to determine the stability properties of the trivial solution of (6.1).

Problems

12. Prove Theorem 4.4.

13. Under the hypotheses of Theorem 3.2, let $-\alpha = \sup\{\operatorname{Re}\lambda : \lambda \text{ is an eigenvalue of } A \text{ with } \operatorname{Re}\lambda < 0\} < 0$. Show that if ϕ is a solution of (PE) and $(\tau, \phi(\tau)) \in S$ for some τ, then

$$\limsup_{t\to\infty} \frac{\log|\phi(t)|}{t} \leq -\alpha.$$

14. Under the hypotheses of Theorem 3.2, suppose there are m eigenvalues $\{\lambda_1, \ldots, \lambda_m\}$ with $\operatorname{Re}\lambda_j < -\alpha < 0$ for $1 \leq j \leq m$ and all other eigenvalues λ of A satisfy $\operatorname{Re}\lambda \geq -\beta > -\alpha$. Prove that there is an m-dimensional positively invariant local hypersurface S_m based at $x = 0$ such that if $(\tau, \phi(t)) \in S_m$ for some τ and for some solution ϕ of (PE), then

$$\limsup_{t\to\infty} \frac{\log|\phi(t)|}{t} \leq -\alpha.$$

If ϕ solves (PE) but $(\tau, \phi(\tau)) \in S - S_m$, then show that

$$\limsup_{t\to\infty} \frac{\log|\phi(t)|}{t} > -\alpha.$$

If $F \in C^1(R \times B(h))$, then S_m is C^1 smooth. *Hint:* Study $y = e^{\gamma t}x$, where $\alpha > \gamma > \beta$.

15. Consider the system

$$x' = \begin{bmatrix} -\lambda_1 & 0 \\ 0 & -\lambda_2 \end{bmatrix} x + F(x),$$

where $F \in C^2(R^2)$, $F(0) = 0$, and $F_x(0) = 0$ and λ_1, λ_2 are real numbers satisfying $\lambda_1 > \lambda_2 > 0$. Show that there exists a unique solution ϕ_1 such that, except for translation, the only solution satisfying

$$\limsup_{t\to\infty} \frac{\log|\phi(t)|}{t} = -\lambda_1$$

is ϕ_1.

16. Suppose A is an $n \times n$ matrix having k eigenvalues λ which satisfy $\operatorname{Re}\lambda \leq -\alpha < 0$, $(n-k)$ eigenvalues λ which satisfy $\operatorname{Re}\lambda \geq -\beta > -\alpha$, and at least one eigenvalue with $\operatorname{Re}\lambda = -\alpha$. Let hypothesis (2.1) be strengthened to $F \in C^1(R^+ \times B(h))$ and let $F(t,x) = O(|x|^{1+\delta})$ uniformly for $t \geq 0$ for some $\delta > 0$. Let ϕ be a solution of (PE) such that

$$\limsup_{t\to\infty} \left(\frac{\log|\phi(t)|}{t}\right) \leq -\alpha.$$

Show that there is a solution ψ of (L) such that $\limsup(\log|\psi(t)|/t) \leq -\alpha$ as $t \to \infty$ and there is an $\eta > 0$ such that

$$\phi(t) - \psi(t) = O(e^{-(\alpha+\eta)t}), \qquad t \to \infty. \tag{6.3}$$

Hint: Suppose $B = \text{diag}(B_1, B_2, B_3)$, where the B_i are grouped so that their eigenvalues have real parts less than, equal to, and greater than $-\alpha$, respectively. If $\phi(t)$ is a solution satisfying the lim sup condition, then show that ϕ can be written in the form

$$\phi(t) = e^{Bt} \begin{bmatrix} c_1 \\ c_2 \\ c_3 \end{bmatrix} + \int_\tau^t \text{diag}(e^{B_1(t-s)}, 0, 0) F(s, \phi(s)) \, ds$$

$$- \int_t^\infty \text{diag}(0, e^{B_2(t-s)}, e^{B_3(t-s)}) F(s, \phi(s)) \, ds.$$

Show that $c_3 = 0$.

17. For the system

$$x_1' = 2x_2 - e^{x_1} + 1,$$
$$x_2' = -\sin x_2 - 2 \arctan x_1,$$

show that the trivial solution is asymptotically stable. Show that if $\xi = (\phi_1(\tau), \phi_2(\tau))^T$ is in the domain of attraction of $(0,0)^T$, then there exist constants $\alpha \in R$, $\beta > 0$, and $\gamma \geq 0$ such that

$$\phi_1(t) = \gamma e^{-t} \cos(2t + \alpha) + O(e^{-(1+\beta)t}),$$
$$\phi_2(t) = -\gamma e^{-t} \sin(2t + \alpha) + O(e^{-(1+\beta)t})$$

as $t \to \infty$.

18. In problem 17 show that in polar coordinates $x_1 = r\cos\theta$, $x_2 = r\sin\theta$, we have

$$\lim_{t \to \infty} [\theta(t) - 2\log r(t)] = \alpha - 2\log \gamma.$$

19. Consider the system

$$x' = \begin{bmatrix} -\lambda & 0 \\ 1 & -\lambda \end{bmatrix} x + F(x),$$

where $\lambda > 0$, $F \in C^2(R^n)$, $F(0) = 0$, and $F_x(0) = 0$. Show that for any solution ϕ in the domain of attraction of $x = 0$ there are constants c_1 and $c_2 \in R$ and $\alpha > 0$ such that

$$\phi(t) = e^{-\lambda t} \begin{bmatrix} c_1 \\ c_2 + c_1 t \end{bmatrix} + O(e^{-(\lambda+\alpha)t}).$$

Problems

20. In Problem 16 show that for any solution ψ of (L) with
$$\limsup(\log|\psi(t)|/t) \le -\alpha$$
as $t \to \infty$, there is a solution ϕ of (PE) and an $\eta > 0$ such that (6.3) is true.

21. Suppose (LH) is stable in the sense of Lagrange (see Definition 5.3.6) and for any $c \in R^n$ there is a solution v of (5.3) such that (5.4) is true. Show that for any solution x of (LH) there is a solution y of (LP) such that $x(t) - y(t) \to 0$ as $t \to \infty$. If in addition $F(t, y)$ is linear in y, then prove that (LH) and (LP) are asymptotically equivalent.

22. Let the problem $x' = Ax$ be stable in the sense of Lagrange (see Definition 5.3.6) and let $B \in C[0, \infty)$ with $|B(t)|$ integrable on R^+. Show that
$$x' = Ax \quad \text{and} \quad y' = Ay + B(t)y$$
are asymptotically equivalent or find a counter example.

23. Let A be an $n \times n$ complex matrix which is self-adjoint, i.e., $A = A^*$. Let $F \in C^1(C^n)$ with $F(0) = 0$ and $F_x(0) = 0$. Show that the systems
$$x' = iAx,$$
$$y' = iAy + F(e^{-t}y)$$
are asymptotically equivalent.

24. What can be said about the behavior as $t \to \infty$ of solutions of the Bessel equation
$$x'' + t^{-1}x' + \alpha^2 x = t^{-2}x?$$
Hint: Let $y = \sqrt{t}x$.

25. Show that the equation
$$t^2 x'' + tx' + 4x = x/\sqrt{t}$$
has solutions of the form $x = c_1 \cos(2 \log t) + c_2 \sin(2 \log t) + o(1)$ as $t \to 0^+$ for any constants c_1 and c_2. *Hint*: Use the change of variables $\tau = \log t$.

7 | PERIODIC SOLUTIONS OF TWO-DIMENSIONAL SYSTEMS

In this chapter, we study the existence of periodic solutions for autonomous two-dimensional systems of ordinary differential equations. In Section 1 we recall several concepts and results that we shall require in the remainder of the chapter. Section 2 contains an account of the Poincaré–Bendixson theory. In Section 3 this theory is applied to a second order Lienard equation to establish the existence of a limit cycle. (The concept of limit cycle will be made precise in Section 3.)

7.1 PRELIMINARIES

In this section and in the next section we concern ourselves with autonomous systems of the form

$$x' = f(x), \qquad (A)$$

where $f: R^2 \to R^2$, f is continuous on R^2, and f is sufficiently smooth to ensure the existence of unique solutions to the initial value problem $x' = f(x)$, $x(\tau) = \xi$.

We recall that a **critical point** (or equilibrium point) ξ of (A) is a point for which $f(\xi) = 0$. A point is called a **regular point** if it is not a critical point.

7.1 Preliminaries

We also recall that if $\xi \in R^2$ and if ϕ is a solution of (A) such that $\phi(0) = \xi$, then the **positive semiorbit** through ξ is defined as

$$C^+(\xi) = \{\phi(t) : t \geq 0\},$$

the **negative semiorbit** through ξ is defined as

$$C^-(\xi) = \{\phi(t) : t \leq 0\},$$

and the **orbit** through ξ is defined as

$$C(\xi) = \{\phi(t) : -\infty < t < \infty\}$$

when ϕ exists on the interval in question. When ξ is understood or is not important, we shall often shorten the foregoing notation to C^+, C^-, and C, respectively.

Given ξ, suppose the solution ϕ of (A), with $\phi(0) = \xi$, exists in the future (i.e., for all $t \geq 0$) so that $C^+ = C^+(\xi)$ exists. Recall that the **positive limit set** $\Omega(\phi)$ is defined as the set

$$\Omega(\phi) = \bigcap_{\tau > 0} \overline{C^+(\phi(\tau))},$$

and the **negative limit set** $\mathscr{A}(\phi)$ is similarly defined. Frequently, we shall find it convenient to use the notation $\Omega(\phi) \triangleq \Omega(C^+)$ for this set. We further recall Lemma 5.11.9 which states that if C^+ is a bounded set, then $\Omega(C^+)$ is a nonempty, compact set which is invariant with respect to (A). Since $C^+(\phi(\tau))$ is connected for each $\tau > 0$, so is its closure. Hence $\Omega(C^+)$ is also connected.

We collect these facts in the following result.

Theorem 1.1. If C^+ is bounded, then $\Omega(C^+)$ is a nonempty, compact, connected set which is invariant with respect to (A).

In what follows, we shall also require the **Jordan curve theorem**. Recall that a **Jordan curve** is a one-to-one, bicontinuous image of the unit circle.

Theorem 1.2. A Jordan curve in the Euclidean plane R^2 separates the plane into two disjoint sets P_i and P_e called the interior and the exterior of the curve, respectively. Both sets are open and arcwise connected, P_i is bounded, and P_e is unbounded.

We close this section by establishing and clarifying some additional nomenclature. Recall that a vector $b = (b_1, b_2)^T \in R^2$ determines a **direction**, namely, the direction from the origin $(0, 0)^T$ to b. Recall also that a **closed line segment** is determined by its two endpoints which we denote by ξ_1 and ξ_2. (The labeling of ξ_1 and ξ_2 is arbitrary. However, once a labeling has been chosen, it has to remain fixed in a given discussion.) The **direction** of the line segment L is determined by the vector $b = \xi_2 - \xi_1$ and L is the set of all points

$$\xi = \xi_1 + bt, \quad 0 \le t \le 1.$$

Let $a \in R^2$ be a nonzero vector perpendicular to b, i.e., $\langle a, b \rangle = 0$, $a \ne (0, 0)^T$. A continuous map $\phi:(\alpha, \beta) \to R^2$ is said to **cross** L at time t_0 if $\phi(t_0) \in L$ and if there is a $\delta > 0$ such that either $\langle \phi(t) - \xi_1, a \rangle$ is positive for $t_0 - \delta < t < t_0$ and negative for $t_0 < t < t_0 + \delta$ or else $\langle \phi(t) - \xi_1, a \rangle$ is negative for $t_0 - \delta < t < t_0$ and positive for $t_0 < t < t_0 + \delta$. The sign of $\langle \phi(t) - \xi_1, a \rangle$ as t varies over $t_0 - \delta < t < t_0 + \delta$ determines the direction in which ϕ is said to cross L. If ϕ_i crosses L at t_{0i} for $i = 1, 2$, then ϕ_1 and ϕ_2 **cross** L **in the same direction** if there is a $\gamma > 0$ such that $[\langle \phi_1(t + t_{01}) - \xi_1, a \rangle \langle \phi_2(t + t_{02}) - \xi_1, a \rangle] > 0$ for all t in the interval $0 < |t| < \gamma$.

7.2 POINCARÉ–BENDIXSON THEORY

We shall construct Jordan curves with the aid of transversals. A **transversal** with respect to the continuous function $f: R^2 \to R^2$ is a closed line segment L in R^2 such that every point of L is a regular point and for each point $\xi \in L$, the vector $f(\xi)$ is not parallel to the direction of the line segment L. We note that since f is continuous, given any regular point $\xi \in R^2$ and any direction $\eta \in R^2$ which is not parallel to $f(\xi)$ [i.e., $\eta \ne \alpha f(\xi)$ for any nonzero constant $\alpha \in R$], there is a transversal through ξ in the direction of η. Note also that if an orbit of (A) meets a transversal L, it must cross L. Moreover, all such crossings of L are in the same direction. A deeper property of transversals is summarized in the following result.

Lemma 2.1. If ξ_0 is an interior point of a transversal L, then for any $\varepsilon > 0$ there is a $\delta > 0$ such that any orbit passing through the ball $B(\xi_0, \delta)$ at $t = 0$ must cross L at some time $t \in (-\varepsilon, \varepsilon)$.

Proof. Suppose the transversal L has direction $\eta = (\eta_1, \eta_2)^T$. Then points $x = (x_1, x_2)^T$ of L will satisfy an equation of the form

$$g(x) \triangleq a_1 x_1 + a_2 x_2 - c = 0,$$

where c is a constant and $a = (a_1, a_2)^T$ is a vector such that $a^T \eta = 0$ and $|a| \ne 0$. Let $\phi(t, \xi)$ be the solution of (A) such that $\phi(0, \xi) = \xi$ and define G by

$$G(t, \xi) = g(\phi(t, \xi)).$$

7.2 Poincaré–Bendixson Theory

Then $G(0, \xi_0) = 0$ since $\xi_0 \in L$ and

$$\frac{\partial G}{\partial t}(0, \xi_0) = a^\mathrm{T} f(\xi_0) \neq 0$$

since L is a transversal. By the implicit function theorem, there is a C^1 function $t: B(\xi_0, \delta) \to R$ for some $\delta > 0$ such that $t(\xi_0) = 0$ and $G(t(\xi), \xi) \equiv 0$. By possibly reducing the size of δ, it can be assumed that $|t(\xi)| < \varepsilon$ when $\xi \in B(\xi_0, \delta)$. Hence $\phi(t, \xi)$ will cross L at $t(\xi)$ and $-\varepsilon < t(\xi) < \varepsilon$. ∎

In the next result, we establish some important monotonicity properties of a transversal.

Lemma 2.2. *If a compact segment $S = \{\phi(t): \alpha \leq t \leq \beta\}$ of an orbit intersects a transversal L, then $L \cap S$ consists of finitely many points whose order on L is monotone with respect to t. If in addition ϕ is periodic, then $L \cap S$ consists of exactly one point.*

Proof. The proof is by contradiction. Assume that S intersects L in infinitely many points. Then there is an infinite sequence of distinct points $\{t_m\}$ and a point $t_0 \in [\alpha, \beta]$ such that $\phi(t_m) \in L$ and such that $t_m \to t_0$. By continuity we have $\phi(t_m) \to \phi(t_0)$ and

$$\lim_{m \to \infty} \frac{\phi(t_m) - \phi(t_0)}{t_m - t_0} = f(\phi(t_0)). \tag{2.1}$$

Since $\phi(t_m) \in L$ and $\phi(t_0) \in L$, the quotients on the left side of (2.1) are all parallel to the direction of L. Hence, $f(\phi(t_0))$ is parallel to the direction of L, a contradiction. Hence $S \cap L$ contains only finitely many points.

Let $\phi(t_1)$ and $\phi(t_2)$ be two successive points of intersection of S and L as t increases with $\alpha \leq t_1 < t_2 \leq \beta$ and $\phi(t_1) \neq \phi(t_2)$. Then the Jordan curve J consisting of the arc $\{\phi(t): t_1 \leq t \leq t_2\}$ and that part of L between $P = \phi(t_1)$ and $Q = \phi(t_2)$ separates the plane into two pieces. There are two cases (see, e.g., Figs. 7.1a and 7.1b), depending on whether the solution ϕ enters the interior P_i (Fig. 7.1a) or the exterior P_e of J (Fig. 7.1b) for $t > t_2$. We consider the first case. (The second case is handled similarly.) By uniqueness of solutions, no solution can cross the arc $\{\phi(t): t_1 \leq t \leq t_2\}$. Since L is a transversal, solutions can cross L in only one direction. Hence, solutions will enter P_i along the segment of L between P and Q, but may never exit from P_i. This means the solution ϕ will remain in P_i for all $t \in (t_2, \alpha]$ and any further intersections of S and L must occur in the interior of J. This establishes the monotonicity.

Suppose now that ϕ is periodic but $\phi(t_1) \neq \phi(t_2)$. By the foregoing argument, $\phi(t)$ will remain for $t > t_2$ on the opposite side of J

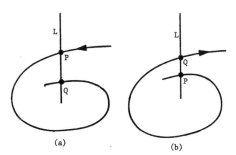

FIGURE 7.1 *Intersection of a transversal.*

from $\phi(t_1)$. But by periodicity $\phi(t) = \phi(t_1)$ for some $t > t_2$. This is a contradiction. Hence, $\phi(t_1)$ must equal $\phi(t_2)$. ∎

The next result is concerned with transversals and limit sets.

Lemma 2.3. *A transversal L cannot intersect a positive limit set $\Omega(\phi)$ of a bounded solution ϕ in more than one point.*

Proof. Let $\Omega(\phi)$ intersect L at ξ'. Let $\{t'_m\}$ be a sequence of points such that $t'_m \to \infty$, $t'_{m+1} > t'_m + 2$, and $\phi(t'_m) \to \xi'$. By Lemma 2.1 there is an $M \geq 1$ such that if $m \geq M$, then ϕ must cross L at some time t_m where $|t_m - t'_m| < 1$. By Lemma 2.2, the sequence $\{\phi(t_m)\}$ is monotone on L. Hence it tends to a point $\xi \in L \cap \Omega(\phi)$. We see from Lemma 2.1 that $|t_m - t'_m| \to 0$ as $m \to \infty$. Since $\phi'(t) = f(\phi(t))$ is bounded, it follows that

$$\lim_{m \to \infty} \phi(t_m) = \lim_{m \to \infty} (\phi(t'_m) + [\phi(t_m) - \phi(t'_m)]) = \xi' + 0.$$

Hence $\xi' = \xi$.

If η is a second point in $L \cap \Omega(\phi)$, then by the same argument there is a sequence $\{s_m\}$ such that $s_m \nearrow \infty$ and $\phi(s_m)$ tends monotonically on L to η. By possibly deleting some points s_m and t_m, we can assume that the sequences $\{t_m\}$ and $\{s_m\}$ interlace, i.e., $t_1 < s_1 < t_2 < s_2 < \cdots$, so that the sequence $\{\phi(t_1), \phi(s_1), \phi(t_2), \phi(s_2), \ldots\}$ is monotone on L. Thus ξ and η must be the same point. ∎

We can also prove the next result.

Corollary 2.4. *Let ϕ be a bounded nonconstant solution of* (A) *with $\phi(0) = \xi$.*

(a) *If $\Omega(\phi)$ and $C^+(\xi)$ intersect, then ϕ is a periodic solution.*
(b) *If $\Omega(\phi)$ contains a nonconstant periodic orbit C, then $\Omega(\phi) = C$.*

7.2 Poincaré–Bendixson Theory

Proof. Let $\eta \in \Omega(\phi) \cap C^+(\xi)$. Then η must be a regular point of (A) and thus there is a transversal L through η and there is a τ such that $\eta = \phi(\tau)$. Since $\Omega(\phi)$ is invariant with respect to (A), it follows that $C(\eta) = C(\xi) \subset \Omega(\phi)$. Since $\eta \in \Omega(\phi)$, there are points $\{t'_m\}$ such that $t'_m \nearrow \infty$ and $\phi(t'_m) \to \eta$. By Lemma 2.1, there are points t_m near t'_m with $\phi(t_m) \in L$, $|t_m - t'_m| \to 0$, and $\phi(t_m) \to \eta$ as $m \to \infty$. By Lemma 2.3, we must have $\phi(t_m) = \eta$ on the sequence $\{t_m\}$. But the solutions of the initial value problem (A) with $x(0) = \eta$ are unique. Hence, if $\phi(t_m) = \phi(t_k) = \eta$ for $t_m > t_k$, then $\phi(t + t_m) \equiv \phi(t + t_k)$ for all $t \geq 0$ or $\phi(t) \equiv \phi(t + [t_m - t_k])$. Thus ϕ is periodic with period $T = t_m - t_k$.

Assume there is a periodic orbit $C \subset \Omega(\phi)$ and assume, for purposes of contradiction, that $C \neq \Omega(\phi)$. Since $\Omega(\phi)$ is connected, there are points $\xi_m \in \Omega(\phi) - C$ and a point $\xi_0 \in C$ such that $\xi_m \to \xi_0$. Let L be a transversal through ξ_0. By Lemma 2.1, for m sufficiently large, the orbit through ξ_m must intersect L, say $\phi(\tau_m, \xi_m) = \xi'_m \in L$. By Lemma 2.3, it follows that $\{\xi'_m\}$ is a constant sequence. But from this it follows that $\xi_m = \phi(-\tau_m, \xi'_m) \in C$ which is a contradiction to our earlier assumption that $\xi_m \notin C$. This concludes the proof. ∎

Having established the foregoing preliminary results, we are in a position to prove the main result of this section.

Theorem 2.5. (*Poincaré–Bendixson*). Let ϕ be a bounded solution of (A) with $\phi(0) = \xi$. If $\Omega(\phi)$ contains no critical points, then either

(a) ϕ is a periodic solution [and $\Omega(\phi) = C^+(\xi)$], or
(b) $\Omega(\phi)$ is a periodic orbit.

If (b) is true, but not (a), then $\Omega(\phi)$ is called a **limit cycle**.

Proof. If ϕ is periodic, then clearly $\Omega(\phi)$ is the orbit determined by ϕ. So let us assume that ϕ is not periodic. Since $\Omega(\phi)$ is nonempty, invariant, and free of singular points, it contains a nonconstant and bounded semiorbit C^+. Hence, there is a point $\xi \in \Omega(\psi)$ where ψ is the solution which generates C^+. Since $\Omega(\phi)$ is closed, it follows that $\xi \in \Omega(\psi) \subset \Omega(\phi)$.

Let L be a transversal through ξ. By Lemma 2.1, we see that points of C^+ must meet L. Since Lemma 2.3 states that C^+, which is a subset of $\Omega(\phi)$, can meet L only once, it follows that $\xi \in C^+$. By Corollary 2.4, we see that C^+ is the orbit of a periodic solution. Again applying Corollary 2.4, we see that since $\Omega(\phi)$ contains a periodic orbit C, it follows that $\Omega(\phi) = C$. ∎

Example 2.6. Consider the system, in spherical coordinates, given by

$$\theta' = 1, \qquad \phi' = \pi, \qquad \rho' = \rho(\sin\theta + \pi\sin\phi).$$

We see that if $\theta(0) = \phi(0) = 0$, $\rho(0) = 1$, then

$$\theta(t) = t, \qquad \phi(t) = \pi t, \qquad \rho(t) = \exp(-\cos t - \cos \pi t).$$

This solution is bounded in R^3 but is not periodic nor does it tend to a periodic solution. The hypothesis that (A) be two-dimensional is absolutely essential.

The argument used to prove Theorem 2.5 is also sufficient to prove the following result.

Corollary 2.7. Suppose that all critical points of (A) are isolated. If $C^+(\phi)$ is bounded and if C is a nonconstant orbit in $\Omega(\phi)$, then either $C = \Omega(\phi)$ is periodic or else the limit sets $\Omega(C)$ and $\mathscr{A}(C)$ each consist of a single critical point.

Example 2.8. Consider the system, in polar coordinates, given by

$$r' = rf(r^2), \qquad \theta' = (r^2 - 1)^2,$$

where $f(1) = 0$ and $f'(1) < 0$. This example illustrates the necessity of the hypothesis that (A) can have only isolated critical points. Solutions of this system which start near the curve $r = 1$ tend to that curve. All points on $r = 1$ are critical points.

Example 2.9. Consider the system, in polar coordinates, given by

$$r' = rf(r^2), \qquad \theta' = (r^2 - 1)^2 + \sin^2 \theta + \varepsilon,$$

where $f(1) = 0$ and $f'(1) < 0$. This example illustrates the fact that either conclusion in Corollary 2.7 is possible. Solutions which start near $r = 1$ tend to $r = 1$ if $\varepsilon > 0$. When $\varepsilon = 0$, this circle consists of two trajectories whose Ω- and $\mathscr{A}$-limit sets are at $r = 1$, $\theta = 0, \pi$. If $\varepsilon > 0$, then $r = 1$ is a limit cycle.

In the next result, we consider stability properties of the periodic orbits predicted by Theorem 2.5.

Theorem 2.10. Let ϕ be a bounded solution of (A) with $\phi(0) = \xi_0$ such that $\Omega(\phi)$ contains no singular points and $\Omega(\phi) \cap C^+(\xi_0)$ is empty. If ξ_0 is in the exterior (respectively, the interior) of $\Omega(\phi)$, then $C^+(\xi_0)$ spirals around the exterior (respectively, interior) of $\Omega(\phi)$ as it approaches $\Omega(\phi)$. Moreover, for any point η exterior (respectively, interior) to $\Omega(\phi)$ but close to $\Omega(\phi)$, we have $\phi(t, \eta) \to \Omega(\phi)$ as $t \to \infty$.

Proof. By Theorem 2.5 the limit set $\Omega(\phi)$ is a periodic orbit. Let $T > 0$ be the least period, let $\xi \in \Omega(\phi)$, and let L be a transversal at ξ. Then we can argue as in the proof of Lemma 2.3 that there is a sequence $\{t_m\}$

7.2 Poincaré–Bendixson Theory

such that $t_m \to \infty$ and $\phi(t_m) \in L$ with $\phi(t_m)$ tending monotonically on L to ξ. Since $C^+(\xi)$ does not intersect $\Omega(\phi)$, the points $\phi(t_m)$ are all distinct. Let s_m be the first time greater than t_m when ϕ intersects L. Let R_m be the region bounded between $\Omega(\phi)$ on one side and the curve consisting of $\{\phi(t): t_m \leq t \leq s_m\}$ and the segment of L between $\phi(t_m)$ and $\phi(s_m)$ on the other side (see Fig. 7.2). By continuity with respect to initial conditions, R_m is contained in any ε neighborhood of $\Omega(\phi)$ when $\varepsilon > 0$ is fixed and then m is chosen sufficiently large. Hence, R_m will contain no critical points when m is sufficiently large. Thus, any solution of (A) starting in R_m must remain in R_m and must, by Theorem 2.5, approach a periodic solution as $t \to \infty$.

By continuity, for m large, a solution starting in R_m at time τ must intersect the segment of L between ξ and $\phi(s_m)$ at some time t between τ and $\tau + 2T$. Thus, a solution starting in R_m must enter R_{m+1} in finite time (for all m sufficiently large). Hence, the solution $\phi(t, \eta)$ must approach $\Omega(\phi) = \bigcap \{\overline{R}_m : m = 1, 2, \ldots\}$ as $t \to \infty$. ∎

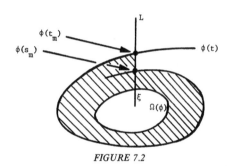

FIGURE 7.2

Example 2.11. Consider the system in R^2, written in polar coordinates, given by

$$r' = r(r-1)(r-2)^2(3-r), \qquad \theta' = 1.$$

There are three periodic orbits at $r = 1$, $r = 2$, and $r = 3$. At $r = 3$ the hypotheses of Theorem 2.10 are satisfied from both the interior and the exterior, at $r = 2$ the hypotheses are satisfied from the interior but not the exterior, while at $r = 1$ the hypotheses are satisfied on neither the interior nor the exterior.

We now introduce the concept of orbital stability.

Definition 2.12. A periodic orbit C in R^2 is called **orbitally stable from the outside** (respectively, **inside**) if there is a $\delta > 0$ such that if η

is within δ of C and on the outside (inside) of C, then the solution $\phi(t,\eta)$ of (A) spirals to C as $t \to \infty$.

C is called **orbitally unstable from the outside (inside)** if C is orbitally stable from the outside (inside) with respect to (A) with time reversed, i.e., with respect to

$$y' = -f(y). \tag{2.2}$$

We call C **orbitally stable (unstable)** if it is orbitally stable (unstable) from both inside and outside.

Now consider the system

$$x' = -y + xf(x^2 + y^2),$$
$$y' = x + yf(x^2 + y^2)$$

which can be expressed in polar coordinates by

$$r' = rf(r^2), \qquad \theta' = 1.$$

We can generate examples to demonstrate the various types of stability given above by appropriate choices of f. For instance, in Example 2.11 the periodic orbit $r = 3$ is orbitally stable, $r = 2$ is orbitally stable from the inside and unstable from the outside, and $r = 1$ is orbitally unstable.

7.3 THE LEVINSON–SMITH THEOREM

The purpose of this section is to prove a result of Levinson and Smith concerning limit cycles of Lienard equations of the form

$$x'' + f(x)x' + g(x) = 0 \tag{3.1}$$

when f and g satisfy the following assumptions:

$$\begin{aligned} &f:R \to R \quad \text{is even and continuous, and} \\ &g:R \to R \quad \text{is odd, is in } C^1(R), \text{ and } xg(x) > 0 \text{ for all } x \neq 0; \end{aligned} \tag{3.2}$$

there is a constant $a > 0$ such that $F(x) \triangleq \int_0^x f(s)\,ds < 0$ on $0 < x < a$, $F(x) > 0$ on $x > a$, and $f(x) > 0$ on $x > a$; \hfill (3.3)

$G(x) \triangleq \int_0^x g(s)\,ds \to \infty$ as $|x| \to \infty$ and $F(x) \to \infty$ as $x \to \infty$. \hfill (3.4)

We now prove the following result.

7.3 The Levinson–Smith Theorem

Theorem 3.1. *If Eq. (3.1) satisfies hypotheses (3.2)–(3.4), then there is a nonconstant, orbitally stable periodic solution $p(t)$ of Eq. (3.1). This periodic solution is unique up to translations $p(t + \tau)$, $\tau \in R$.*

Proof. Under the change of variables $y = x' + F(x)$, Eq. (3.1) is equivalent to

$$x' = y - F(x), \qquad y' = -g(x). \tag{3.5}$$

The coefficients of (3.5) are smooth enough to ensure local existence and uniqueness of the initial value problem determined by (3.5). Hence, existence and uniqueness conditions are also satisfied by a corresponding initial value problem determined by (3.1).

Now define a Lyapunov function for (3.5) by

$$v(x, y) = y^2/2 + G(x).$$

The derivative of v with respect to t along solutions of Eq. (3.5) is given by

$$dv/dt = v'_{(3.5)}(x, y) = -g(x)F(x). \tag{3.6}$$

Also, the derivative of v with respect to x along solutions of (3.5) is given by

$$dv/dx = -g(x)F(x)/(y - F(x)), \qquad 0 < x < a, \tag{3.7}$$

and the derivative with respect to y along solutions of (3.5) is determined as

$$dv/dy = F(x), \qquad x \geq a. \tag{3.8}$$

From (3.5) we see that $x(t)$ is increasing in t when $y > F(x)$ and decreasing for $y < F(x)$ while $y(t)$ is decreasing for $x > 0$ and increasing for $x < 0$. Thus for any initial point $A = (0, \alpha)$ on the positive y axis, the orbit of (3.5) issuing from A is of the general shape shown in Fig. 7.3.

Note that by symmetry (i.e., by oddness) of F and g, if $(x(t), y(t))$ is a solution of (3.5), then so is $(-x(t), -y(t))$. Hence, if the distance OA in Fig. 7.3 is larger than the distance OD, then the positive semiorbit through any point A' between O and A must be bounded. Moreover, the orbit through A will be periodic if and only if the distance OD and OA are equal.

Referring to Fig. 7.3, we note that for any fixed x on $0 \leq x \leq a$, the y coordinate on AB is greater than the y coordinate on $A'B'$. Thus from (3.7) we can conclude that $v(B) - v(A) < v(B') - v(A')$. From (3.8) and (3.3) we see that $v(E) - v(B) < 0$. From (3.2), (3.3), and (3.8) we see that $v(G) - v(E) < v(C') - v(B')$. Similar arguments show that $v(C) - v(G) < 0$ and $v(D) - v(C) < v(D') - v(C')$. Thus we see that $v(D) - v(A) < v(D') - v(A')$. Hence, if $A = (0, \alpha)$ and $t(\alpha)$ is the first positive t for which the x coordinate of the orbit through A is zero, then $\alpha^2 - y(t(\alpha))^2 = 2(v(A) - v(D))$ is an increasing function of α. (The same result is true by essentially the same argument for $\alpha > 0$, α small.)

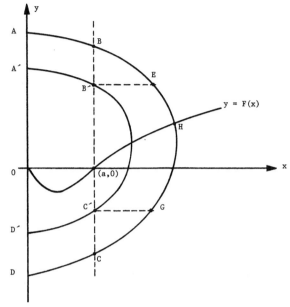

FIGURE 7.3

For α small, let $(x(t), y(t))$ be the orbit through $A = (0, \alpha)$. When $x(t) \neq 0$, then by (3.6) it follows that $dv/dt > 0$. Thus $\alpha^2 - y(t(\alpha))^2 < 0$ near $\alpha = 0$. We wish to show that $\alpha^2 - y(t(\alpha))^2 > 0$ for α sufficiently large.

For α large we note that

$$v(D) - v(A) = -\int_A^B \frac{g(x)F(x)}{y - F(x)}\,dx - \int_C^D \frac{g(x)F(x)}{y - F(x)}\,dx - \int_C^B F(x)\,dy, \quad (3.9)$$

where the integrals are line integrals. Let $x(\alpha)$ be the x coordinate of the first point H where the semiorbit $C^+(A)$ intersects the curve $y = F(x)$. Then $x(\alpha)$ is an increasing function of α. If $x(\alpha)$ is bounded, say $0 < x(\alpha) < B$ on $0 < \alpha < \infty$, then by continuity with respect to initial conditions, $y(t(\alpha))$ is also bounded. Hence $\alpha^2 - y(t(\alpha))^2 > 0$ for α sufficiently large. Therefore, we may assume that $x(\alpha) \to \infty$ and $y(t(\alpha)) \to \infty$ as $\alpha \to \infty$.

Since for $0 < x < a$ and y large we have

$$dy/dx = -g(x)/(y - F(x)), \qquad y(0) = \alpha,$$

bounded, then $y(x, \alpha) \to \infty$ as $\alpha \to \infty$ uniformly for $0 < x < a$. Similarly, the y coordinate of the curve from D to C tends uniformly to $-\infty$ as $\alpha \to \infty$.

7.3 The Levinson–Smith Theorem

This implies that the integrals

$$\int_A^B \frac{g(x)F(x)}{y - F(x)}\,dx \quad \text{and} \quad \int_C^D \frac{g(x)F(x)}{y - F(x)}\,dx$$

from (3.9) tend to zero as $\alpha \to \infty$.

Fix α so large that $C^+(A)$ intersects the x axis at some point to the right of $(a, 0)$. By Green's theorem (in the plane), we have

$$\int_C^B F(x)\,dy = \iint_R f(x)\,dx\,dy, \tag{3.10}$$

where R is the region bounded by the curve $C^+(A)$, between B and C, and the line $x = a$ (see Fig. 7.4). The integral on the right in (3.10) is clearly positive and is monotone increasing with α. Thus it is now clear that in (3.9), $v(D) - v(A) \to -\infty$ as $\alpha \to \infty$. Hence, there is a unique point α^* where $v(D) = v(A)$. The corresponding orbit $C(A^*)$, $A^* = (0, \alpha^*)$, is periodic. Since $v(D) - v(A)$ changes sign from positive to negative exactly once, it is clear from Theorem 2.10 that the periodic orbit is orbitally stable. ∎

We conclude this section with the following example.

Example 3.2. Perhaps the most widely known example which can be used to illustrate the applicability of Theorem 3.1 is the van der Pol equation [see Eq. (1.2.18)] given by

$$x'' + \varepsilon(x^2 - 1)x' + x = 0,$$

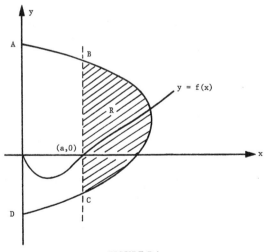

FIGURE 7.4

where ε is any positive constant. In this case

$$f(x) = \varepsilon(x^2 - 1), \qquad g(x) = x$$

and

$$F(x) = \varepsilon(\tfrac{1}{3}x^3 - x), \qquad G(x) = \tfrac{1}{2}x^2.$$

Thus (3.2)–(3.4) are easy to verify.

PROBLEMS

1. Find the periodic orbits and determine their orbital stability for the system

$$r' = rf(r^2), \qquad \theta' = 1$$

when

(a) $f(s) = (s - \pi^2)(s - 4\pi^2)$,
(b) $f(s) = \sin(s)$, and
(c) $f(s) = |\sin(s)|$.

2. Prove that any nonconstant periodic solution of a two-dimensional system (A) must contain a critical point in its interior.

3. For system (A) show that two adjacent periodic orbits cannot both be orbitally stable on the sides facing each other if there are no critical points in the annular region between these two periodic orbits.

4. Show that the equation

$$x'' + (3x^5 - 8x^3 - 12x)x' + x = 0$$

has a unique nontrivial periodic solution.

5. Show that the problem

$$x'' + (1 - 400\cos 4x)x' + x = 0$$

has at least one nontrivial periodic solution. *Hint*: Generalize Theorem 3.1.

6. Assume that $F \in C^1(R)$, that F is odd, that $F(0) = 0$, that $F(x) \to \infty$ as $x \to \infty$ and that F satisfies Eq. (3.3). Show that

$$y'' + F(y') + y = 0$$

has a unique, nonconstant, orbitally stable periodic solution. *Hint*: Let $x = y'$.

Problems

7. Let ϕ be the solution of
$$x'' + g(x) = 0, \quad x(0) = -A < 0, \quad x'(0) = 0,$$
where g satisfies hypotheses (3.2) and (3.4). Show that when $\phi'(t) > 0$, then $\phi(t)$ solves the equation
$$dx/dt = \sqrt{2[G(A) - G(x)]}.$$

8. For the problem
$$x'' + ax + bx^3 = 0 \quad \text{with} \quad a > 0 \quad \text{and} \quad b > 0,$$
show that the solution satisfying $x(0) = A$, $x'(0) = 0$ is periodic with period
$$T = 4\sqrt{2} \int_0^{\pi/2} \frac{d\theta}{\sqrt{2a + bA^2(1 + \sin^2\theta)}}.$$

Hint: Use Problem 7.

9. Let $f: R \times R \to R$, assume that $f(t + T, x) = f(t, x)$ for all $t \in R$, $x \in R$, and for some $T > 0$, and assume that $f \in C^1(R \times R)$. Show that if $x' = f(t, x)$ has a solution ϕ bounded on R^+, then it has a T-periodic solution. *Hint*: If ϕ is not periodic, then $\{\phi(nT): n = 0, 1, 2, \ldots\}$ is a monotone sequence.

10. Assume that $f \in C^1(R^n)$, let D be a subset of R^n with finite area, and let $\phi(t, \xi)$ be the solution of
$$x' = f(x), \quad x(0) = \xi$$
for $\xi \in D \subset R^n$. Define $F(\xi) = \phi(\tau, \xi)$ for all $\xi \in D$. Show that the area of the set $F(D) = \{y = F(\xi): \xi \in D\}$ is given by
$$\iint_D \exp\left(\int_0^\tau \operatorname{tr} \frac{\partial f}{\partial x}(\phi(s, \xi))\, ds\right) d\xi.$$

Hint: From advanced calculus, we know that under the change of variables $x = F(\xi)$
$$\iint_{F(D)} d\xi = \iint_D |\det F_\xi(\xi)|\, d\xi,$$
where F_ξ is the Jacobian matrix.

11. Consider the equation
$$x'' + x' + x(x^2 - 1) = 0$$
or equivalently, the system of equations
$$\begin{aligned} x' &= y, \\ y' &= -y - x(x^2 - 1). \end{aligned} \quad (4.1)$$

(a) Show that this system is uniformly ultimately bounded.

(b) Find all critical points and compute the dimension of their stable and unstable manifolds.

(c) Show that there are no compact sets D with positive area which are invariant with respect to Eq. (4.1). (See Problem 9.) In particular, there are no periodic solutions.

(d) Show that any solution ϕ on the stable manifold of the unstable critical point of (4.1) must spiral outward with $|\phi(t)| + |\phi'(t)| \to \infty$ as $t \to -\infty$. [Use part (c) and the Poincaré–Bendixson theorem.]

(e) In the (x, y) plane, sketch the domain of attraction of each stable critical point.

8 | PERIODIC SOLUTIONS OF SYSTEMS

In this chapter we introduce the interesting and complicated topics of existence and stability of periodic solutions of autonomous and of periodic systems of ordinary differential equations of general order n. In Section 1 we establish some required notation. In Section 2 we study in detail existence and nonexistence of periodic solutions of periodically forced linear periodic systems of equations. These results are interesting and important in their own right and they are also required in the study of certain nonlinear problems. In Section 3 we investigate a periodic system of the form

$$x' = f(t, x), f(t + T, x) = f(t, x) \tag{P}$$

and perturbations of this system in the case where (P) has a known periodic solution. In Section 4 we study the autonomous system

$$x' = f(x) \tag{A}$$

and perturbations of (A) in the case where (A) has a known periodic solution. In Section 5 we consider perturbation problems of the form

$$x' = Ax + \varepsilon g(t, x, \varepsilon)$$

where $|\varepsilon|$ is small and $y' = Ay$ has nontrivial T-periodic solutions. In Section 6 we study the stability, in certain simple situations, of the periodic solutions whose existence was established in Section 5. Section 7 contains a brief introduction to the important topic of averaging and Section 8 contains a brief introduction to Hopf bifurcation. In Section 9 we prove a nonexistence result for certain autonomous systems. We note here that even though a large theory for existence of periodic solutions has been developed, in certain applications, the very difficult question of nonexistence of periodic solutions is more interesting and useful. In Section 9 we give one result which will serve to introduce the idea of nonexistence.

8.1 PRELIMINARIES

We consider the periodic system (P) where $f \in C(R \times D)$ and D is a domain in R^n. Throughout, we assume for (P) that f is sufficiently smooth to ensure the uniqueness of solutions of initial value problems. A simple but very useful fact that motivates most of the work presented in this chapter is that a solution ϕ of (P) is T periodic if and only if $\phi(0) = \phi(T)$. Indeed, if we define $\phi_0(t) \triangleq \phi(t + T)$, then ϕ_0 will solve (P), i.e.,

$$\phi_0'(t) = \phi'(t + T) = f(t + T, \phi(t + T)) = f(t, \phi_0(t)),$$

and $\phi_0(0) = \phi(0)$. By uniqueness of solutions of the initial value problem, we have $\phi_0(t) = \phi(t)$ for $T \geq t \geq 0$, i.e., $\phi(t + T) = \phi(t)$ for $0 \leq t \leq T$. By a simple induction argument it can be shown that ϕ exists for all $t \in R$ and that $\phi(t + T) = \phi(t)$ for all $t \in R$. Hence, in order to find a T-periodic solution of (P), it is sufficient to find fixed points of a period map. Specifically, if $\phi(t, \tau, \xi)$ solves (P) with $\phi(\tau, \tau, \xi) = \xi$, then

$$F(\xi) \triangleq \phi(T + \tau, \tau, \xi)$$

for $\xi \in D$ (and any fixed τ), is called a **period map**. We need to find $\xi_0 \in D$ such that $F(\xi_0) = \xi_0$.

We define the set $\mathscr{P}_T$ by

$$\mathscr{P}_T = \{g \in C(R) : g \text{ is } T \text{ periodic}\}.$$

The range of the function g can be R^n, C^n, or the real or complex $n \times n$ matrices. The particular range required in a given situation will always be clear from context.

8.2 NONHOMOGENEOUS LINEAR SYSTEMS

Consider a real nonhomogeneous linear periodic system

$$x' = A(t)x + f(t), \qquad \text{(LN)}$$

8.2 Nonhomogeneous Linear Systems

where $A(t)$ is an $n \times n$ matrix and both A and f are in $\mathscr{P}_T$. Let $\Phi(t)$ be a fundamental matrix solution of the corresponding homogeneous system

$$x' = A(t)x \qquad \text{(LH)}$$

with $\Phi(0) = E$ so that $y = [\Phi^{-1}(t)]^T$ is a fundamental matrix solution of the adjoint system

$$y' = -A(t)^T y \qquad (2.1)$$

(refer to Section 3.2 for further details).

Lemma 2.1. Systems (LH) and (2.1) have the same number of linearly independent solutions in $\mathscr{P}_T$.

Proof. A solution $p(t) = \Phi(t)\xi$ of (LH) is in $\mathscr{P}_T$ if and only if $\Phi(T)\xi = \xi$, or equivalently,

$$(\Phi(T) - E)\xi = 0. \qquad (2.2)$$

Solutions of (2.1) have the form $q(t) = \Phi^{-1}(t)^T \eta$, where η is an n-dimensional column vector. Hence, $q \in \mathscr{P}_T$ if and only if

$$\eta^T(\Phi^{-1}(T) - E) = 0. \qquad (2.3)$$

The number of linearly independent solutions of (2.3) is the same as the number of linearly independent solutions of

$$\eta^T(\Phi^{-1}(T) - E)\Phi(T) = \eta^T(E - \Phi(T)) = 0. \qquad (2.4)$$

The number of linearly independent solutions of (2.2) and of (2.4) is the same. ∎

We are now in a position to prove the following result.

Theorem 2.2. If A and f are in $\mathscr{P}_T$, then (LN) has a solution $p \in \mathscr{P}_T$ if and only if

$$\int_0^T y(t)^T f(t)\,dt = 0 \qquad (2.5)$$

for all solutions y of (2.1) which are in $\mathscr{P}_T$. If (2.5) is true and if (2.1) has k linearly independent solutions in $\mathscr{P}_T$, then (LN) has a k-parameter family of solutions in $\mathscr{P}_T$.

Proof. Solutions of (LN) have the form

$$p(t) = \Phi(t)\xi + \int_0^t \Phi(t)\Phi^{-1}(s)f(s)\,ds. \qquad (2.6)$$

Thus, $p(T) = \xi$ if and only if

$$(\Phi(T) - E)\xi = -\int_0^T \Phi(T)\Phi^{-1}(s)f(s)\,ds$$

or

$$(\Phi^{-1}(T) - E)\xi = \int_0^T \Phi^{-1}(s)f(s)\,ds.$$

This equation has the form

$$(\Phi^{-1}(T) - E)\xi = a$$

where $(\Phi^{-1}(T) - E)$ may be singular. Hence, there is a solution ξ if and only if $\eta^T a = 0$ for all nontrivial row vectors η^T which solve (2.4). This condition coincides with (2.5). If p is one periodic solution of (LN) and x is any solution of the homogeneous problem (LH) which is in $\mathscr{P}_T$, then $x + p$ solves (LN) and is in $\mathscr{P}_T$. Hence, by Lemma 2.1 there is a k-parameter family of solutions of (LN) in $\mathscr{P}_T$. ∎

Let us consider a specific case.

Example 2.3. Consider the problem

$$x'' + x = \sin \omega t. \tag{2.7}$$

Two linearly independent solutions of the adjoint system for (2.7) are $\sin t$ and $\cos t$. They are not periodic of period $T = 2\pi/\omega$ if $\omega \neq 1, \frac{1}{2}, \frac{1}{3}, \ldots$. If ω is not one of these exceptional values, then (2.7) has the unique T-periodic solution

$$p(t) = (1 - \omega^2)^{-1} \sin \omega t.$$

If $\omega = 1$, then (2.5) is not satisfied and moreover, (2.7) has no 2π-periodic solution. Indeed, in this case the general solution of (2.7) is easily seen to be

$$x = c_1 \sin t + c_2 \cos t - (t \cos t)/2.$$

If $\omega = 1/m$ for some integer $m > 1$, then (2.5) reduces to

$$\int_0^{2\pi m} (\sin t, \cos t) \begin{bmatrix} 0 \\ \sin(t/m) \end{bmatrix} dt = 0,$$

and

$$\int_0^{2\pi m} (\cos t, -\sin t) \begin{bmatrix} 0 \\ \sin(t/m) \end{bmatrix} dt = 0.$$

Since (2.5) is satisfied in this case, we know that (2.7) has a two-parameter family of $2m\pi$-periodic solutions. Indeed, it is easy to compute

$$p(t) = c_1 \sin t + c_2 \cos t + (m^2/(m^2 - 1)) \sin(t/m).$$

8.2 Nonhomogeneous Linear Systems

Corollary 2.4. If $A \in \mathscr{P}_T$ and $f \in \mathscr{P}_T$ and if (2.1) has no solutions in $\mathscr{P}_T$, then (LN) has a unique solution p in $\mathscr{P}_T$ and

$$p(t) = \int_t^{t+T} [\Phi(s)(\Phi^{-1}(T) - E)\Phi^{-1}(t)]^{-1} f(s) \, ds. \tag{2.8}$$

Proof. According to the proof of Theorem 2.2, the periodic solution is the unique solution of (LN) which satisfies the initial condition $p(0) = \xi$, where

$$\xi = (\Phi^{-1}(T) - E)^{-1} \int_0^T \Phi^{-1}(s) f(s) \, ds.$$

Substituting this expression into (2.6) and simplifying, we obtain

$$p(t) = \Phi(t)(\Phi^{-1}(T) - E)^{-1} \int_0^T \Phi(s)^{-1} f(s) \, ds + \Phi(t) \int_0^t \Phi^{-1}(s) f(s) \, ds$$

$$= \Phi(t)(\Phi^{-1}(T) - E)^{-1} \left(\int_0^T \Phi^{-1}(s) f(s) \, ds \right.$$

$$\left. + \int_0^t (\Phi^{-1}(T) - E)\Phi^{-1}(s) f(s) \, ds \right)$$

$$= \Phi(t)(\Phi^{-1}(T) - E)^{-1} \left(\int_0^T \Phi^{-1}(s) f(s) \, ds \right.$$

$$\left. + \int_0^t [\Phi^{-1}(T+s) - \Phi^{-1}(s)] f(s) \, ds \right)$$

$$= \Phi(t)(\Phi^{-1}(T) - E)^{-1} \left(\int_t^T \Phi^{-1}(s) f(s) \, ds + \int_T^{T+t} \Phi^{-1}(s) f(s - T) \, ds \right)$$

$$= \Phi(t)(\Phi^{-1}(T) - E)^{-1} \int_t^{T+t} \Phi^{-1}(s) f(s) \, ds$$

$$= \int_t^{t+T} [\Phi(s)(\Phi^{-1}(T) - E)\Phi^{-1}(t)]^{-1} f(s) \, ds.$$

In the foregoing calculations we have used the fact that $f(s + T) \equiv f(s)$ and that $\Phi^{-1}(T)\Phi^{-1}(s) = \Phi^{-1}(T + s)$. ∎

If $A(t)$ is independent of t, then the representation (2.8) can be replaced in the following fashion.

Corollary 2.5. Consider the real n-dimensional system

$$x' = Ax + f(t), \tag{2.9}$$

where $f \in \mathscr{P}_T$ and where the time-independent $n \times n$ matrix A has no eigenvalues with zero real part. Then (2.9) has a unique solution p in $\mathscr{P}_T$. Moreover, there is a matrix $G(t)$ (independent of f in $\mathscr{P}_T$) which is piecewise continuous in t and satisfies

$$\int_{-\infty}^{\infty} |G(t)|\, dt < \infty,$$

such that the periodic solution p can be written as

$$p(t) = \int_{-\infty}^{\infty} G(t-s)f(s)\, ds. \tag{2.10}$$

Proof. Let $x = By$ where B is an $n \times n$ real matrix chosen so that

$$B^{-1}AB = \begin{bmatrix} A_1 & 0 \\ 0 & A_2 \end{bmatrix}$$

is block diagonal and both $-A_1$ and A_2 are Hurwitzian. By Corollary 2.4, the system

$$y' = B^{-1}ABy + B^{-1}f(t) \tag{2.11}$$

has a unique solution in $\mathscr{P}_T$. Moreover, if

$$G_0(t) \triangleq \begin{cases} \operatorname{diag}(-e^{A_1 t}, 0) & \text{for } t < 0, \\ \operatorname{diag}(0, e^{A_2 t}) & \text{for } t \geq 0, \end{cases}$$

then

$$q(t) = \int_{-\infty}^{\infty} G_0(t-s) B^{-1} f(s)\, ds$$

is defined for all $t \in R$ and

$$|q(t)| \leq \int_{-\infty}^{\infty} |G_0(t-s)|\, |B^{-1}|\, |f(s)|\, ds$$

$$\leq K \left(\int_{-\infty}^{0} |e^{A_1 t}|\, dt + \int_{0}^{\infty} |e^{A_2 t}|\, dt \right)$$

for some constant $K > 0$. Clearly

$$q(t+T) = \int_{-\infty}^{\infty} G_0(t+T-s) B^{-1} f(s)\, ds$$

$$= \int_{-\infty}^{\infty} G_0(t-s) B^{-1} f(s-T)\, ds = q(t)$$

so that $q \in \mathscr{P}_T$. We can now express $q(t)$ in the form

$$q(t) = \int_{-\infty}^{t} \begin{bmatrix} 0 & 0 \\ 0 & e^{A_2(t-s)} \end{bmatrix} B^{-1} f(s)\, ds - \int_{t}^{\infty} \begin{bmatrix} e^{A_1(t-s)} & 0 \\ 0 & 0 \end{bmatrix} B^{-1} f(s)\, ds$$

8.2 Nonhomogeneous Linear Systems

so that

$$q'(t) = \begin{bmatrix} 0 & 0 \\ 0 & E_2 \end{bmatrix} B^{-1} f(t) + \int_{-\infty}^{t} \begin{bmatrix} A_1 & 0 \\ 0 & A_2 \end{bmatrix} \begin{bmatrix} 0 & 0 \\ 0 & e^{A_2(t-s)} \end{bmatrix} B^{-1} f(s)\,ds$$

$$+ \begin{bmatrix} E_1 & 0 \\ 0 & 0 \end{bmatrix} B^{-1} f(t) - \int_{t}^{\infty} \begin{bmatrix} A_1 & 0 \\ 0 & A_2 \end{bmatrix} \begin{bmatrix} e^{A_1(t-s)} & 0 \\ 0 & 0 \end{bmatrix} B^{-1} f(s)\,ds,$$

where E_1 and E_2 are identity matrices of appropriate orders. Thus, q is a solution of (2.11). Now if we define $G(t) = BG_0(t)B^{-1}$ and $p(t) = Bq(t)$, then $p \in \mathscr{P}_T$ and (2.10) is true. ∎

We conclude this section with a specific example.

Example 2.6. Consider the linear problem

$$x' = \begin{bmatrix} 1 & 3 \\ 0 & -2 \end{bmatrix} x + \begin{bmatrix} \sin \omega t \\ 1 \end{bmatrix},$$

where $\omega = 2\pi/T > 0$. Since

$$A = \begin{bmatrix} 1 & 3 \\ 0 & -2 \end{bmatrix} = B \begin{bmatrix} 1 & 0 \\ 0 & -2 \end{bmatrix} B^{-1}, \quad B = \begin{bmatrix} 1 & -1 \\ 0 & 1 \end{bmatrix},$$

the eigenvalues of A are 1 and -2. Since A has no eigenvalue of the form $\lambda = 2m\pi i/T = im\omega$ for any integer m, the homogeneous system $x' = Ax$ has no nontrivial solution in $\mathscr{P}_T$. Hence, Corollary 2.4 can be applied. Since

$$[\Phi(s)(\Phi^{-1}(T) - E)\Phi^{-1}(t)]^{-1}$$

$$= \left\{ B \begin{bmatrix} e^s & 0 \\ 0 & e^{-2s} \end{bmatrix} \begin{bmatrix} e^{-T} - 1 & 0 \\ 0 & e^{2T} - 1 \end{bmatrix} \begin{bmatrix} e^{-t} & 0 \\ 0 & e^{2t} \end{bmatrix} B^{-1} \right\}^{-1}$$

$$= B \begin{bmatrix} e^{t-s}(e^{-T} - 1)^{-1} & 0 \\ 0 & e^{-2(t-s)}(e^{2T} - 1)^{-1} \end{bmatrix} B^{-1}$$

$$= \begin{bmatrix} e^{t-s}(e^{-T} - 1)^{-1} & e^{t-s}(e^{-T} - 1)^{-1} - e^{-2(t-s)}(e^{2T} - 1)^{-1} \\ 0 & e^{-2(t-s)}(e^{2T} - 1)^{-1} \end{bmatrix},$$

then (2.8) reduces in this case to

$$p(t) = \begin{bmatrix} p_1(t) \\ p_1(t) \end{bmatrix}$$

$$= \begin{bmatrix} \int_{t}^{t+T} [e^{t-s}(e^{-T} - 1)^{-1}(\sin \omega s + 1) - e^{-2(t-s)}(e^{2T} - 1)^{-1}]\,ds \\ \int_{t}^{t+T} e^{-2(t-s)}(e^{2T} - 1)^{-1}\,ds \end{bmatrix}.$$

Since $\omega T = 2\pi$, this expression reduces to

$$p_1(t) = \frac{-3}{2} - \frac{1}{1+\omega^2}(\sin \omega t + \omega \cos \omega t) \tag{2.12a}$$

and

$$p_2(t) = \tfrac{1}{2}. \tag{2.12b}$$

We note that this solution could be obtained more readily by other methods (e.g., by making use of Laplace transforms). The usefulness of (2.8) is in its theoretical applicability rather than in its use to produce actual solutions in specific cases.

Since the eigenvalues of A have nonzero real parts, the more stringent hypotheses of Corollary 2.5 are also satisfied in this case. An elementary computation yields in this case

$$G(t) = \begin{cases} -B\begin{bmatrix} e^t & 0 \\ 0 & 0 \end{bmatrix}B^{-1} = -\begin{bmatrix} e^t & e^t \\ 0 & 0 \end{bmatrix} & \text{if } t < 0, \\[6pt] B\begin{bmatrix} 0 & 0 \\ 0 & e^{-2t} \end{bmatrix}B^{-1} = \begin{bmatrix} 0 & -e^{-2t} \\ 0 & e^{-2t} \end{bmatrix} & \text{if } t \geq 0, \end{cases}$$

so that

$$p(t) = \int_{-\infty}^{t} \begin{bmatrix} 0 & -e^{-2(t-s)} \\ 0 & e^{-2(t-s)} \end{bmatrix} \begin{bmatrix} \sin \omega s \\ 1 \end{bmatrix} ds - \int_{t}^{\infty} \begin{bmatrix} e^{t-s} & e^{t-s} \\ 0 & 0 \end{bmatrix} \begin{bmatrix} \sin \omega s \\ 1 \end{bmatrix} ds$$

$$= \int_{-\infty}^{t} e^{-2(t-s)} \begin{bmatrix} -1 \\ 1 \end{bmatrix} ds - \int_{t}^{\infty} e^{t-s} \begin{bmatrix} \sin \omega s + 1 \\ 0 \end{bmatrix} ds$$

is the unique solution in $\mathscr{P}_T$. This is the same solution as computed earlier in (2.12).

8.3 PERTURBATIONS OF NONLINEAR PERIODIC SYSTEMS

The behavior of the system

$$x' = g(t, x) + \varepsilon h(t, x, \varepsilon), \tag{3.1}$$

for $|\varepsilon|$ small, can often be predicted, in part, from an analysis of this system when $\varepsilon = 0$. We are interested in the case where g and h are T periodic in t and where the reduced system with $\varepsilon = 0$ has a nontrivial solution $p \in \mathscr{P}_T$.

8.3 Perturbations of Nonlinear Periodic Systems

This situation may be viewed as a perturbation of the linear problem which we studied in Section 2 when g and h are sufficiently smooth. Indeed, if $y = x - p$, then y satisfies

$$y' = g(t, y + p(t)) - g(t, p(t)) + \varepsilon[h(t, y + p(t), \varepsilon) - h(t, p(t), \varepsilon)]$$
$$= g_x(t, p(t))y + O(|y|^2) + \varepsilon[h(t, y + p(t), \varepsilon) - h(t, p(t), \varepsilon)]$$

or

$$y' = g_x(t, p(t))y + H(t, y, \varepsilon), \quad (3.2)$$

where $g_x(t, p(t))$ is a periodic matrix, $H \in C^1$ near $\varepsilon = 0$, $y = 0$, and

$$H(t, 0, 0) = 0, \qquad H_y(t, 0, 0) = 0. \quad (3.3)$$

If the linear system (LH) with $A(t) = g_x(t, p(t))$ has no nontrivial solutions in $\mathscr{P}_T$, then the perturbed system (2.2) is not hard to analyze (see the problems at the end of this chapter). However, when (LH) has nontrivial solutions in $\mathscr{P}_T$, the analysis of the behavior of solutions of (3.2) for $|\varepsilon|$ small is extremely complex. For this reason, we shall adopt a somewhat different approach based on the implicit function theorem (cf. Theorem 6.1.1).

Theorem 3.1. Consider the real, n-dimensional system

$$x' = f(t, x, \varepsilon), \quad (3.4)$$

where $f: R \times R^n \times [-\varepsilon_0, \varepsilon_0] \to R^n$, for some $\varepsilon_0 > 0$, where f, f_ε, and f_x are continuous and where f is T periodic in t. At $\varepsilon = 0$, assume that (3.4) has a solution $p \in \mathscr{P}_T$ whose first variational equation

$$y' = f_x(t, p(t), 0)y \quad (3.5)$$

has no nontrivial solution in $\mathscr{P}_T$. Then for $|\varepsilon|$ sufficiently small, say $|\varepsilon| < \varepsilon_1$, system (3.4) has a solution $\psi(t, \varepsilon) \in \mathscr{P}_T$ with ψ continuous in $(t, \varepsilon) \in R \times [-\varepsilon_1, \varepsilon_1]$ and $\psi(t, 0) = p(t)$.

In a neighborhood $N = \{(t, x): 0 \le t \le T, |p(t) - x| < \varepsilon_2\}$ for some $\varepsilon_2 > 0$ there is only one T-periodic solution of (3.4), namely $\psi(t, \varepsilon)$. If the real parts of the characteristic exponents of (3.5) all have negative real parts, then the solution $\psi(t, \varepsilon)$ is asymptotically stable. If at least one characteristic root has positive real part, then $\psi(t, \varepsilon)$ is unstable.

Proof. We shall consider initial values $x(0)$ of the form $x(0) = p(0) + \eta$ where $\eta \in R^n$ and where $|\eta|$ is small. Let $\phi(t, \varepsilon, \eta)$ be the solution of (3.4) which satisfies $\phi(0, \varepsilon, \eta) = p(0) + \eta$. For a solution ϕ to be in $\mathscr{P}_T$ it is necessary and sufficient that

$$\phi(T, \varepsilon, \eta) - p(0) - \eta = 0. \quad (3.6)$$

We shall solve (3.6) by using the implicit function theorem. At $\varepsilon = 0$ there is a solution of (3.6), i.e., $\eta = 0$. The Jacobian matrix of (3.6) with respect to η will be $\phi_\eta(T, \varepsilon, \eta) - E$. We require that this Jacobian be nonsingular at $\varepsilon = 0$, $\eta = 0$. Since ϕ satisfies

$$\phi'(t, \varepsilon, \eta) = f(t, \phi(t, \varepsilon, \eta), \varepsilon),$$

we see by Theorem 2.7.1 that ϕ_η solves

$$\phi'_\eta(t, 0, 0) = f_x(t, \phi(t, 0, 0), 0)\phi_\eta(t, 0, 0), \qquad \phi_\eta(0, 0, 0) = E.$$

Now (3.5) has no nontrivial solutions in $\mathscr{P}_T$ if and only if $\phi_\eta(T, 0, 0) - E$ is nonsingular. Hence, this matrix is nonsingular.

By the implicit function theorem, there are constants $\varepsilon_1 > 0$ such that for $|\varepsilon| < \varepsilon_1$ (3.6) has a one-parameter family of solutions $\eta(\varepsilon)$ with $\eta(0) = 0$. Moreover, $\eta \in C^1[-\varepsilon_1, \varepsilon_1]$. In a neighborhood $|\eta| < \alpha_1$, $|\varepsilon| < \varepsilon_1$, these are the only solutions of (3.6). Define $\psi(t, \varepsilon) = \phi(t, \varepsilon, \eta(\varepsilon))$ for $(t, \varepsilon) \in R \times [-\varepsilon_1, \varepsilon_1]$. Clearly, ψ is the family of periodic solutions which has been sought.

To prove the stability assertions, we check the characteristic exponents of the variational equation

$$y' = f_x(t, \psi(t, \varepsilon), \varepsilon)y \tag{3.7}$$

for $|\varepsilon| < \varepsilon_1$ and invoke the Floquet theory (see Corollary 6.2.5). Solutions of (3.7) are continuous functions of the parameter ε and (3.7) reduces to (3.5) when $\varepsilon = 0$. Thus, if $\Phi(t, \varepsilon)$ is the fundamental matrix for (3.7) such that $\Phi(0, \varepsilon) = E$, then the characteristic roots of $\Phi(T, \varepsilon)$ are all less than one in magnitude for $|\varepsilon|$ small if those of $\Phi(T, 0)$ are. Similarly, if $\Phi(T, 0)$ has at least one characteristic root λ with $|\lambda| > 1$, then $\Phi(T, \varepsilon)$ has the same property for $|\varepsilon|$ sufficiently small. This proves the stability assertions. ∎

When $f(t, x, \varepsilon)$ is holomorphic in (x, ε) for each fixed $t \in R$, the proof of the above theorem actually shows that

$$\psi(t, \varepsilon) = \phi(t, \varepsilon, \eta(\varepsilon))$$

is holomorphic in ε near $\varepsilon = 0$. In this case, we can expand ψ in a power series of the form

$$\psi(t, \varepsilon) = \sum_{j=0}^{\infty} \psi_j(t)\varepsilon^j,$$

where each $\psi_j \in \mathscr{P}_T$. Since

$$\psi'(t, \varepsilon) = \sum_{j=0}^{\infty} \psi'_j(t)\varepsilon^j = f(t, \psi(t, \varepsilon), \varepsilon),$$

8.3 Perturbations of Nonlinear Periodic Systems

by equating like powers of ε, we see that
$$\psi_0'(t) = f(t, \psi_0(t), 0),$$
$$\psi_1'(t) = f_x(t, \psi_0(t), 0)\psi_1(t) + f_\varepsilon(t, \psi_0(t), 0),$$
$$\psi_2'(t) = f_x(t, \psi_0(t), 0)\psi_2(t) + \tfrac{1}{2}\{f_{\varepsilon\varepsilon}(t, \psi_0(t), 0) + 2f_{x\varepsilon}(t, \psi_0(t), 0)\psi_1(t) + h(t)\},$$
$$\vdots$$

where $h_i(t) = \psi_1^T(t) f_{ixx}(t, \psi_0(t), 0)\psi_1(t)$. The first equation above has solution $\psi_0(t) = p(t)$. This solution can be used in the f_ε term of the second equation above. Since (3.5) has no nontrivial solution in $\mathscr{P}_T$, then we see from Corollary 2.4 that the second equation has a unique solution $\psi_1 \in \mathscr{P}_T$. Continuing in this manner, these equations can, in theory, be successively solved.

Example 3.2. For fixed $a > 0$ consider the second order problem

$$x'' + a(x^2 - 1)x' + x = \varepsilon \sin t, \tag{3.8}$$

or the equivalent system

$$x' = y - a(x^3/3 - x),$$
$$y' = -x + \varepsilon \sin t.$$

At $\varepsilon = 0$ the periodic solution p of Theorem 3.1 is the singular point $x = y = 0$. The variational system (3.5) reduces here to

$$x' = y + ax,$$
$$y' = -x.$$

The characteristic equation of this system is $\lambda^2 - a\lambda + 1 = 0$ and both roots of the characteristic equation have positive real parts. Hence, all hypotheses of Theorem 3.1 are satisfied. We conclude that in a neighborhood of $x = y = 0$ there is a unique T-periodic family of solutions $x(t, \varepsilon)$. These solutions are unstable.

Next, we expand $x(t, \varepsilon)$ as

$$x(t, \varepsilon) = \sum_{j=0}^{\infty} \psi_j(t)\varepsilon^j.$$

We see that $x_0(t) \equiv 0$ and $x_1(t)$ is the unique solution of the equation

$$z'' - az' + z = g(t), \tag{3.9}$$

in $\mathscr{P}_T$, with $g(t) = \sin t$. Hence, $x_1(t) = a^{-1} \cos t$. Additional computations show that $x_2(t)$ solves the same equation with $g(t) \equiv 0$. Hence, $x_2(t) \equiv 0$.

Further computations show that $x_3(t)$ is a solution of (3.9) with

$$g(t) = -ax_1^2(t)x_1'(t) = \frac{1}{a^2}\cos^2 t \sin t = \frac{1}{4a^2}(\sin t + \sin 3t).$$

Thus, we obtain

$$x_3(t) = \frac{1}{4a^3}\cos t - \frac{2}{a^2(9a^2+64)}\sin 3t + \frac{3}{4a(9a^2+64)}\cos 3t.$$

Next, we can verify that $x_4(t) \equiv 0$. Thus, there is a family of 2π-periodic solutions of the form

$$x(t,\varepsilon) = \left[\frac{\varepsilon}{a} + \frac{1}{4}\left(\frac{\varepsilon}{a}\right)^3\right]\cos t + \frac{\varepsilon^3}{a(9a^2+64)}\left(\frac{3}{4}\cos 3t - \frac{2}{a}\sin 3t\right) + O(\varepsilon^5).$$

It is important not only to understand what situations the theory which we are developing will cover, but also what situations it will not cover. For example, when $\varepsilon = 0$, Eq. (3.8) reduces to the van der Pol equation. By the results in Section 7.3, we know that when $\varepsilon = 0$ this equation has a nontrivial stable limit cycle. However, our theory gives no clue to the behavior of the solutions of (3.8) near the limit cycle when ε is small but not zero.

The Duffing equation

$$x'' + x + \varepsilon x^3 = 0 \tag{3.10}$$

provides another interesting example which is not covered by the theory developed here. We know that all solutions of (3.10) are periodic and that the period varies with amplitude. With initial conditions $x(0) = A$, $x'(0) = 0$, let us try to solve for a solution of the form

$$x(t,\varepsilon) = \sum_{j=0}^{\infty} \varepsilon^j x_j(t).$$

Substituting this expression into (3.10) and equating the coefficients of like powers of ε, we find that

$$x_0'' + x_0 = 0, \quad x_0(0) = A, \quad x_0'(0) = 0.$$

Thus, $x_0(t) = A\cos t$. Next, we compute

$$x_1'' + x_1 = -\tfrac{3}{4}A^3\cos t - \tfrac{1}{4}A^3\cos 3t, \quad x_1(0) = x_1'(0) = 0,$$

and thus,

$$x_1(t) = -\tfrac{3}{8}A^3 t\sin t - \tfrac{1}{32}A^3(\cos t - \cos 3t).$$

8.4 Perturbations of Nonlinear Autonomous Systems

Note that the coefficient $x_1(t)$ contains a secular term, i.e., a term containing multiplication by t. If the procedure is continued [to obtain $x_2(t)$, and so forth], then secular terms containing higher powers of t will occur. These secular terms occur because the solution $x(t,\varepsilon)$, though periodic, does not have period 2π. Thus, the chosen representation is simply not appropriate if accuracy is desired from a few terms of the series. This idea is nicely illustrated by the series

$$\sin(1+\varepsilon)t = \sin t + \varepsilon t \cos t - \frac{(\varepsilon t)^2}{2!}\sin t - \frac{(\varepsilon t)^3}{3!}\cos t + \cdots.$$

8.4 PERTURBATIONS OF NONLINEAR AUTONOMOUS SYSTEMS

Now consider an autonomous system

$$x' = f(x,\varepsilon) \tag{4.1}$$

which has a nontrivial periodic solution $p(t)$. On linearizing (4.1) about $p(t)$, we obtain the variational equation

$$y' = f_x(p(t),0)y. \tag{4.2}$$

This equation always has at least one Floquet multiplier equal to one since $y = p'(t)$ is always a nontrivial periodic solution. Thus we see that the hypotheses of Theorem 3.1 can never be satisfied. For this autonomous case, a somewhat different approach is needed and a slightly different result is proved.

Theorem 4.1. Let $f:R^n \times [-\varepsilon_0,\varepsilon_0] \to R^n$ with f, $\partial f/\partial \varepsilon$ and $\partial f/\partial x$ continuous on $R^n \times [-\varepsilon_0,\varepsilon_0]$. At $\varepsilon = 0$, suppose that (4.1) has a nontrivial periodic solution $p(t)$ of period T_0 and suppose that $(n-1)$ Floquet multipliers of (4.2) are different from one. Then for $|\varepsilon|$ sufficiently small, there is a continuous function $T(\varepsilon)$ and a continuous family $\psi(t,\varepsilon)$ of $T(\varepsilon)$-periodic solutions of (4.1) such that $\psi(t,0) = p(t)$ and $T(0) = T_0$.

Proof. Under the change of variables $x = Bz + p(0)$, Eq. (4.1) at $\varepsilon = 0$ takes the form

$$z' = B^{-1}f(Bz + p(0), 0).$$

We now choose the $n \times n$ nonsingular constant matrix B such that if $P(t) \triangleq B^{-1}(p(t) - p(0))$, then

$$P'(0) = B^{-1}p'(0) = (1,0,\ldots,0)^T \triangleq e_1.$$

Hence, without loss of generality we can assume for $p(t)$ and for (4.1) that $p(0) = 0$ and $p'(0) = e_1$.

By continuity with respect to parameters, if follows that for $|\varepsilon|$ and $|\eta|$ sufficiently small, the solution $\phi(t, \eta, \varepsilon)$ of (4.1) which satisfies $\phi(0, \eta, \varepsilon) = \eta = (\eta_1, \ldots, \eta_n)^T$ with $\eta_1 = 0$ must return to and cross the plane determined by $x_1 = 0$ within time $2T_0$. In order to prove the theorem, we propose to find solutions $\tau(\varepsilon)$ and $\eta(\varepsilon)$ of the equation

$$\phi(T_0 + \tau, \eta, \varepsilon) - \eta = 0. \tag{4.3}$$

The point $\tau = 0, \eta = 0, \varepsilon = 0$ is a solution of (4.3). The Jacobian of (4.3) with respect to the variables $(\tau, \eta_2, \ldots, \eta_n)$ is the determinant of the matrix

$$\begin{bmatrix} 1 & \partial\phi_1/\partial\eta_2 & \cdots & \partial\phi_1/\partial\eta_n \\ 0 & \partial\phi_2/\partial\eta_2 - 1 & \cdots & \partial\phi_2/\partial\eta_n \\ \vdots & \vdots & & \vdots \\ 0 & \partial\phi_n/\partial\eta_2 & \cdots & \partial\phi_n/\partial\eta_n - 1 \end{bmatrix} \tag{4.4}$$

evaluated at $\tau = \varepsilon = 0, \eta = 0$.

Note that $\partial\phi(t, 0, 0)/\partial\eta_1$ is a solution of (4.2) which satisfies the initial condition $y(0) = e_1$ (see Theorem 2.7.1). Since $y = p'(t)$ satisfies the same conditions, then by uniqueness $\partial\phi(t, 0, 0)/\partial\eta_1 = p'(t)$. By periodicity, $\partial\phi(T_0, 0, 0)/\partial\eta_1 = e_1$. But $(n - 1)$ Floquet multipliers of (4.2) are not one. Hence, the cofactor of the matrix

$$\frac{\partial\phi}{\partial\eta}(T_0, 0, 0) - E \tag{4.5}$$

obtained by deleting the first row and first column is not zero. Since this cofactor is the same for both (4.4) and (4.5), then clearly the matrix (4.5) is nonsingular.

This nonzero Jacobian implies, via the implicit function theorem, that (4.3) has a unique continuous pair of solutions $\tau(\varepsilon)$ and $\eta_j(\varepsilon)$ for $j = 2, \ldots, n$ in a neighborhood of $\varepsilon = 0$ such that $\tau(0) = 0$ and $\eta_j(0) = 0$. We conclude the proof by defining $\eta_1(\varepsilon) \equiv 0$, $T(\varepsilon) = T_0 + \tau(\varepsilon)$ and $\psi(t, \varepsilon) = \phi(t, \eta(\varepsilon), \varepsilon)$. ∎

We now study the stability question for $\psi(t, \varepsilon)$.

Theorem 4.2. *If in Theorem 4.1 there are $(n - 1)$ Floquet multipliers with magnitude less than one, then the periodic solution $\psi(t, \varepsilon)$ of (4.1) is orbitally stable.*

Proof. Let $\Phi(t, \varepsilon)$ be the matrix which solves the equation

$$y' = f_x(\psi(t, \varepsilon), \varepsilon)y, \qquad y(0) = E. \tag{4.6}$$

Then $\Phi(T(\varepsilon),\varepsilon)$ is a continuous matrix valued function of ε and $\Phi(T(0),0)$ has $(n-1)$ eigenvalues λ with $|\lambda|<1$. By continuity $\Phi(T(\varepsilon),\varepsilon)$ will have, for $|\varepsilon|$ sufficiently small, $(n-1)$ eigenvalues with magnitude less than one. ∎

The functions $\tau(\varepsilon)$ and $\eta(\varepsilon)$ obtained in the proof of Theorem 4.1 will be as smooth as is $f(x,\varepsilon)$. In particular, if f is holomorphic in (x,ε), then $\tau(\varepsilon)$ and $\eta(\varepsilon)$ will be holomorphic in ε near $\varepsilon=0$. In this case we can expand $T(\varepsilon)$ as

$$T(\varepsilon) = T_0 + T_1\varepsilon + T_2\varepsilon^2 + \cdots. \tag{4.7}$$

If we define $t = T(\varepsilon)s$, then (4.1) reduces to

$$\frac{dx}{ds} = T(\varepsilon)f(x,\varepsilon) \tag{4.8}$$

with a family $q(s,\varepsilon) = \psi(T(\varepsilon)s,\varepsilon)$ of periodic solutions of period one. Moreover,

$$q(s,\varepsilon) = \sum_{m=0}^{\infty} \varepsilon^m q_m(s) \tag{4.9}$$

and the periodic functions $q_m(s)$ can be computed by substituting (4.7) and (4.9) into (4.8) and equating the coefficients of like powers of ε.

8.5 PERTURBATIONS OF CRITICAL LINEAR SYSTEMS

Consider a periodic system of the form

$$x' = Ax + \varepsilon g(t,x,\varepsilon), \tag{5.1}$$

where A is a real $n \times n$ matrix, $g:R \times R^n \times [-\varepsilon_0,\varepsilon_0] \to R^n$ is continuous and g is 2π periodic in t. In this section we shall be interested in the case where A has an eigenvalue iN for some nonnegative integer N. A real linear change of variables $x = By$ will leave the form of (5.1) unaltered. Hence, without loss of generality we shall assume that A is in real Jordan canonical form (cf. Problem 3.25). The following example is typical and is general enough to illustrate the method involved. We consider the case where

$$A = \begin{bmatrix} S & 0 & 0 & \cdots & 0 & 0 & 0 \\ E & S & 0 & \cdots & 0 & 0 & 0 \\ 0 & E & S & \cdots & 0 & 0 & 0 \\ \vdots & \vdots & \vdots & \ddots & \vdots & \vdots & \vdots \\ 0 & 0 & 0 & \cdots & E & S & 0 \\ 0 & 0 & 0 & \cdots & 0 & 0 & C \end{bmatrix}, \tag{5.2}$$

where

$$S = \begin{bmatrix} 0 & -N \\ N & 0 \end{bmatrix}, \quad E = \begin{bmatrix} 1 & 0 \\ 0 & 1 \end{bmatrix}$$

and C is an $(n - 2k) \times (n - 2k)$ matrix with no eigenvalue of the form iM for $M = 0, 1, 2, \ldots$. [The form of (5.2) could be generalized considerably; however, that would serve no purpose in our discussion.]

Notice that

$$e^{St} = \begin{bmatrix} \cos Nt & -\sin Nt \\ \sin Nt & \cos Nt \end{bmatrix}, \tag{5.3}$$

$$e^{At} = \begin{bmatrix} e^{St} & 0 & 0 & \cdots & 0 & 0 & 0 \\ te^{St} & e^{St} & 0 & \cdots & 0 & 0 & 0 \\ t^2 e^{St}/2! & te^{St} & e^{St} & \cdots & 0 & 0 & 0 \\ \vdots & \vdots & \vdots & \ddots & \vdots & \vdots & \vdots \\ t^{k-1} e^{St}/(k-1)! & \cdot & \cdot & \cdots & te^{St} & e^{St} & 0 \\ 0 & \cdot & \cdot & \cdots & 0 & 0 & e^{Ct} \end{bmatrix}. \tag{5.4}$$

Moreover, $e^{2\pi S} = E$ and $e^{2\pi C} - E$ is not singular.

Solutions of (5.1) can be written in the form

$$\phi(t, b, \varepsilon) = e^{tA} b + \varepsilon \int_0^t e^{A(t-s)} g(s, \phi(s, b, \varepsilon), \varepsilon) \, ds.$$

By uniqueness, a solution ϕ is in $\mathscr{P}_{2\pi}$ if and only if b is a solution of

$$(e^{2\pi A} - E)b + \varepsilon \int_0^{2\pi} e^{(2\pi - s)A} g(s, \phi(s, b, \varepsilon), \varepsilon) \, ds = 0. \tag{5.5}$$

Now suppose that (5.5) has a solution $b(\varepsilon)$ which we put in the form

$$b(\varepsilon) = (b_1(\varepsilon)^T, b_2(\varepsilon)^T, \ldots, b_k(\varepsilon)^T, b_{k+1}(\varepsilon)^T)^T,$$

where b_j is a 2 vector for $1 \leq j \leq k$ and b_{k+1} is an $(n - 2k)$ vector. Similarly, we write $g = (g_1^T, g_2^T, \ldots, g_k^T, g_{k+1}^T)^T$. From (5.4) it is clear that for any possible solution $b(\varepsilon)$ we must have

$$b_1(0) = b_2(0) = \cdots = b_{k-1}(0) = 0, \quad b_{k+1}(0) = 0. \tag{5.6}$$

The first pair of equations in (5.5) can be replaced by

$$G_k(b, \varepsilon) \triangleq \int_0^{2\pi} e^{(2\pi - s)S} g_1(s, \phi(s, b, \varepsilon), \varepsilon) \, ds = 0. \tag{5.7}$$

8.5 Perturbations of Critical Linear Systems

The remaining equations can be grouped as follows:

$$G_1(b, \varepsilon) \triangleq 2\pi b_1 + O(\varepsilon),$$

$$G_2(b, \varepsilon) \triangleq \frac{(2\pi)^2}{2} b_1 + 2\pi b_2 + O(\varepsilon),$$

$$\vdots$$

$$G_{k-1}(b, \varepsilon) \triangleq \frac{(2\pi)^{k-1}}{(k-1)!} b_1 + \cdots + 2\pi b_{k-2} + O(\varepsilon),$$

and

$$G_{k+1}(b, \varepsilon) = (e^{2\pi C} - E)b_{k+1} + O(\varepsilon).$$

The terms $O(\varepsilon)$ involve integrals of the g_j's. The above equations define an n-vector valued function $G(b, \varepsilon)$.

We are now in a position to prove the first result of this section.

Theorem 5.1. Let A have the form (5.2), let g and g_x be in $C(R \times R^n \times [-\varepsilon_0, \varepsilon_0])$ with g 2π periodic in t. Suppose there is a 2 vector α such that if $b(0) = (0, 0, \ldots, \alpha^T, 0)^T$, then $b(0)$ solves (5.7) at $\varepsilon = 0$ and such that $\det(\partial G_k/\partial \alpha)(b(0), 0) \neq 0$. Then for $|\varepsilon|$ sufficiently small, (5.1) has a continuous isolated 2π-periodic family $\psi(t, \varepsilon)$ of solutions such that

$$\psi(t, 0) = e^{At}b(0) = (0, 0, \ldots, 0, (e^{St}\alpha)^T, 0)^T.$$

Proof. The assumptions in the theorem are sufficient in order to solve $G(b, \varepsilon)$ using the implicit function theorem. Clearly $b(0)$, as defined above, solves $G(b(0), 0) = 0$. Notice that

$$\frac{\partial G}{\partial b}(b(0), 0) = \begin{bmatrix} 2\pi E & 0 & 0 & \cdots & 0 & 0 \\ \frac{(2\pi)^2}{2} E & 2\pi E & 0 & \cdots & 0 & 0 \\ \vdots & \vdots & \vdots & & \vdots & \vdots \\ \frac{\partial G_k}{\partial b_1}(b(0), 0) & \cdot & \cdots & \frac{\partial G_k}{\partial b_k}(b(0), 0) & 0 \\ 0 & \cdot & \cdot & \cdots & 0 & e^{2\pi C} - E \end{bmatrix}.$$

Thus

$$\det \frac{\partial G}{\partial b}(b(0), 0) = (2\pi)^{2k-2} \det\left(\frac{\partial G_k}{\partial \alpha}(b(0), 0)\right) \det(e^{2\pi C} - E) \neq 0.$$

By the implicit function theorem there is a unique solution $b(\varepsilon)$ in a neighborhood of $\varepsilon = 0$ and $b(\varepsilon)$ is continuous in ε. Finally, define $\psi(t, \varepsilon) = \phi(t, b(\varepsilon), \varepsilon)$. ∎

Let us now apply the preceding theorem to a specific case.

Example 5.2. For α, β, and A real, $\beta \neq 0$ and $\alpha \neq 0$, let $h(t, y) \triangleq -\alpha y - \beta y^3 + A \cos t$ and consider the equation

$$y'' + y = \varepsilon h(t, y), \qquad y(0) = b_1, \quad y'(0) = b_2. \tag{5.8}$$

This equation can be expressed as the system of equations

$$y' = z,$$
$$z' = -y + \varepsilon h(t, y).$$

In order to apply Theorem 5.1 it is necessary to find values $b_1(0)$ and $b_2(0)$. Since the transformation $b_1 = \gamma \cos \delta$ and $b_2 = \gamma \sin \delta$ is nonsingular for $\gamma > 0$, we can just as well find γ and δ. Replacing in (5.8) t by $t + \delta$ and letting $Y(t) = y(t + \delta)$, we obtain

$$Y'' + Y = \varepsilon h(t + \delta, Y), \qquad Y(0) = \gamma, \quad Y'(0) = 0. \tag{5.9}$$

For this problem, Eq. (5.7) at $\varepsilon = 0$ reduces to

$$-\int_0^{2\pi} \sin(2\pi - s) h(s + \delta, \gamma \cos s) \, ds = 0,$$

$$\int_0^{2\pi} \cos(2\pi - s) h(s + \delta, \gamma \cos s) \, ds = 0,$$

or

$$\alpha \gamma + (3\beta/4)\gamma^3 - A \cos \delta = 0, \qquad \sin \delta = 0.$$

Clearly, this means that $\delta = 0$ while $\gamma = \gamma_0$ must be a positive solution of

$$\alpha \gamma + (3\beta \gamma^3 / 4) - A = 0.$$

The Jacobian condition in Theorem 5.1 reduces in this example to

$$\det \begin{bmatrix} \alpha + (9/4)\beta \gamma_0^2 & \gamma_0 A \sin 0 \\ 0 & \cos 0 \end{bmatrix} = \alpha + (9/4)\beta \gamma_0^2 \neq 0.$$

If a positive γ_0 exists such that this Jacobian is not zero, then there is a continuous, 2π-periodic family of solutions $y(t, \varepsilon)$ such that

$$y(t, 0) = \gamma_0 \cos t, \qquad y'(t, 0) = -\gamma_0 \sin t.$$

As in the earlier sections, the solution $\psi(t, \varepsilon)$ is as smooth as the terms of $g(t, y, \varepsilon)$ are. If g is holomorphic in (y, ε), then $\psi(t, \varepsilon)$ will be

8.5 Perturbations of Critical Linear Systems

holomorphic in ε near $\varepsilon = 0$. In this case, ψ has a convergent power series expansion

$$\psi(t, \varepsilon) = \sum_{m=0}^{\infty} \psi_m(t)\varepsilon^m.$$

Substituting this series into Eq. (5.1) and equating coefficients of like powers in ε, we can successively determine the periodic coefficients $\psi_m(t)$. For example, in the case of Eq. (5.8), a long but elementary computation yields

$$y(t + \delta(\varepsilon), \varepsilon) = \gamma_0 \cos t + \varepsilon[\gamma_1 \cos t + (\beta\gamma_0^3/32)\cos 3t] + O(\varepsilon^2), \quad (5.10)$$

where $\gamma_1 = -3\beta^2\gamma_0^5/[128(\alpha + (9/4)\beta\gamma_0^2)]$ and where $\delta(\varepsilon) = O(\varepsilon^2)$.

Theorem 5.1 does not apply when g is independent of t since then the Jacobian in Theorem 5.1 can never be nonsingular. [For example, one can check that in (5.8) with h independent of t that the right-hand side of (5.9) is independent of δ so that the Jacobian must be zero.] Hence, our analysis does not cover such interesting examples as

$$y'' + y + \varepsilon y^3 = 0 \quad \text{or} \quad y'' + \varepsilon(y^2 - 1)y' + y = 0.$$

We shall modify the previous results so that such situations can be handled.
Consider the autonomous system

$$x' = Ax + \varepsilon g(x, \varepsilon), \quad (5.11)$$

where A is a real $n \times n$ matrix of the form (5.2) and where $g: R^n \times [-\varepsilon_0, \varepsilon_0] \to R^n$. We seek a periodic solution $\psi(t, \varepsilon)$ with period $T(\varepsilon) = 2\pi + \tau(\varepsilon)$, where $\tau(\varepsilon) = \varepsilon\eta(\varepsilon) = O(\varepsilon)$. In this case (5.5) is replaced by

$$(e^{2\pi A} - E)b + e^{2\pi A}(e^{\tau A} - E)b + \varepsilon \int_0^{2\pi+\tau} e^{(2\pi+\tau-s)A} g(\phi(s, b, \varepsilon), \varepsilon) ds = 0.$$
$$(5.12)$$

The initial conditions $b(\varepsilon)$ still need to satisfy (5.6). The problem is to find $b_k(\varepsilon) = (\alpha(\varepsilon), \beta(\varepsilon))^T$. Since solutions of (5.11) are invariant under translation and since at $\varepsilon = 0$ we have

$$e^{At}b(0) = (0, 0, \ldots, 0, [e^{St}b_k(0)]^T, 0)^T,$$

there is no loss of generality in assuming that $\beta(\varepsilon) = 0$. Thus, the first two components of (5.12) are

$$G_k(c, \varepsilon) = \begin{bmatrix} \cos(N\eta\varepsilon) - 1 \\ \sin(N\eta\varepsilon) \end{bmatrix} \frac{\alpha}{\varepsilon} + \int_0^{2\pi+\tau} e^{(2\pi+\tau-s)S} g_1(\phi(s, b, \varepsilon), \varepsilon) ds = \begin{bmatrix} 0 \\ 0 \end{bmatrix}$$

if $\varepsilon \neq 0$, and

$$G_k(c,0) = \begin{bmatrix} 0 \\ N\eta(0)\alpha \end{bmatrix} + \int_0^{2\pi} e^{(2\pi-s)S} g_1(\phi(s,b,0),0)\,ds = \begin{bmatrix} 0 \\ 0 \end{bmatrix}$$

at $\varepsilon = 0$, where $c = (b_1, b_2, \ldots, b_{k-1}, \alpha, \eta, b_{k+1}) \in R^n$.

Theorem 5.3. Assume that A satisfies (5.2) and that g, g_ε, and g_x are in $C(R^n \times [-\varepsilon_0, \varepsilon_0])$. Suppose there exist η_0 and α_0 such that $c_0 = (0, \ldots, 0, \alpha_0, \eta_0, 0)$ satisfies $G_k(c_0, 0) = 0$ and that $\det(\partial G_K/\partial c)(c_0, 0) \neq 0$. Then there exist a continuous function $c(\varepsilon)$ and solutions $\psi(t, \varepsilon)$ such that $c(0) = c_0$, $\psi(t, \varepsilon) \in \mathscr{P}_{T(\varepsilon)}$ where $T(\varepsilon) = 2\pi + \varepsilon\eta(\varepsilon)$, and

$$\psi(t, 0) = e^{At}(0, \ldots, 0, (\alpha_0, 0), 0)^T.$$

We remark that $\psi(t, \varepsilon)$ and $T(\varepsilon)$ will be holomorphic in ε when g is holomorphic in (x, ε). The proof of Theorem 5.3 and of this remark are left to the reader.

8.6 STABILITY OF SYSTEMS WITH LINEAR PART CRITICAL

The present section consists of two parts. In Section A, we consider time varying systems and in the second part we consider autonomous systems.

A. Time Varying Case

Let $g: R \times R^n \times [-\varepsilon_0, \varepsilon_0] \to R^n$ be 2π periodic in t and assume that g is of class C^2 in (x, ε). Suppose that $x' = Ax$ has a 2π-periodic solution $p(t)$ and suppose that

$$x' = Ax + \varepsilon g(t, x, \varepsilon) \tag{6.1}$$

has a continuous family of solutions $\psi(t, \varepsilon) \in \mathscr{P}_{2\pi}$ with $\psi(t, 0) = p(t)$. To simplify matters, we specify the form of A to be

$$A = \begin{bmatrix} S & 0 \\ 0 & C \end{bmatrix}, \quad S = \begin{bmatrix} 0 & -N \\ N & 0 \end{bmatrix}, \tag{6.2}$$

where N is a positive integer and C is an $(n-2) \times (n-2)$ constant matrix with no eigenvalues of the form iM for any integer M.

8.6 Stability of Systems with Linear Part Critical

The stability of the solution $\psi(t, \varepsilon)$ can be investigated using the linearization of (6.1) about ψ, i.e.,

$$y' = Ay + \varepsilon g_x(t, \psi(t, \varepsilon), \varepsilon)y,$$

and Corollary 6.2.5. Let $Y(t, \varepsilon)$ be that fundamental matrix for this linear system which satisfies $Y(0, \varepsilon) = E$. Our problem is to determine whether or not all eigenvalues of $Y(2\pi, \varepsilon)$ have magnitudes less than one for ε near zero.

By the variation of constants formula, we can write

$$Y(t, \varepsilon) = e^{At} + \varepsilon \int_0^t e^{A(t-s)} g_x(s, \psi(s, \varepsilon), \varepsilon) Y(s, \varepsilon) \, ds. \tag{6.3}$$

At $t = 2\pi$ we have $Y(2\pi, \varepsilon) = e^{2\pi R(\varepsilon)}$ for some $R(\varepsilon)$ so that

$$e^{2\pi R(\varepsilon)} = e^{2\pi A} \left\{ E + \varepsilon \int_0^{2\pi} e^{-sA} g_x(s, \psi(s, \varepsilon), \varepsilon) Y(s, \varepsilon) \, ds \right\}. \tag{6.4}$$

Using (6.3) and (6.4), we obtain

$$e^{2\pi R(\varepsilon)} = e^{2\pi A} \left\{ E + \varepsilon \int_0^{2\pi} e^{-sA} g_x(s, \psi(s, \varepsilon), \varepsilon) e^{As} \, ds \right.$$

$$\left. + \varepsilon^2 \int_0^{2\pi} e^{-sA} g_x(s, \psi(s, \varepsilon), \varepsilon) e^{sA} \left(\int_0^s g_x(u, \psi(u, \varepsilon), \varepsilon) Y(u, \varepsilon) \, du \right) ds \right\}.$$

By the mean value theorem, there exists ε^* between 0 and ε such that

$$\psi(t, \varepsilon) = p(t) + \varepsilon \psi_\varepsilon(t, \varepsilon^*)$$

so that

$$\frac{\partial g_i}{\partial x_j}(t, \psi(t, \varepsilon), \varepsilon) = \frac{\partial g_i}{\partial x_j}(t, p(t), 0) + O(\varepsilon).$$

This means that

$$e^{2\pi R(\varepsilon)} = e^{2\pi A}\{E + \varepsilon D + \varepsilon^2 G(\varepsilon)\},$$

where $G(\varepsilon)$ is a continuous matrix valued function and

$$D = \int_0^{2\pi} e^{-sA} g_x(s, p(s), 0) e^{sA} \, ds.$$

By (6.2) it is easy to compute

$$e^{2\pi R(\varepsilon)} = \left[\begin{array}{cc|c} 1 + \varepsilon d_{11} + O(\varepsilon^2) & \varepsilon d_{12} + O(\varepsilon^2) & O(\varepsilon) \\ \varepsilon d_{21} + O(\varepsilon^2) & 1 + \varepsilon d_{22} + O(\varepsilon^2) & \\ \hline O(\varepsilon) & & e^{2\pi C} + O(\varepsilon) \end{array} \right],$$

where

$$D_2 \triangleq \begin{bmatrix} d_{11} & d_{12} \\ d_{21} & d_{22} \end{bmatrix} = \int_0^{2\pi} e^{(2\pi-s)S} \left\{ \begin{bmatrix} \partial g_1/\partial x_1 & \partial g_1/\partial x_2 \\ \partial g_2/\partial x_1 & \partial g_2/\partial x_2 \end{bmatrix}_{(s,p(s),0)} \right\} e^{sS} ds.$$

We wish to compare

$$f(\lambda, \varepsilon) \triangleq \det[\lambda E - e^{2\pi R(\varepsilon)}]$$

and

$$h(\lambda, \varepsilon) \triangleq \det\left[\lambda E - \begin{bmatrix} E_2 + \varepsilon D_2 & 0 \\ 0 & e^{2\pi C} \end{bmatrix}\right].$$

To do this, we need the following result from complex variables (which we give here without proof).

Theorem 6.1 (*Rouché*). If $F(z)$ and $H(z)$ are holomorphic in a simply connected region D containing a closed contour Γ and if $|H(z)| < |F(z)|$ on Γ, then F and $F + H$ have the same number of zeros inside of Γ.

Invoking this result, we now prove the following result.

Theorem 6.2. Suppose, as above, ψ is a family of 2π-periodic solutions of (6.1), $g \in C^2$ in (x, ε), and A satisfies (6.2). Suppose that

(i) both eigenvalues of D_2 have negative real parts, and
(ii) all eigenvalues of C have negative real parts.

Then for ε positive and sufficiently small, the periodic solution $\psi(t, \varepsilon)$ is uniformly asymptotically stable.

Proof. The function $f(\lambda, \varepsilon) = \det(\lambda E - e^{2\pi R(\varepsilon)})$ can be evaluated by first expanding by cofactors down the first row and then expanding each cofactor down its first remaining row. This process yields

$$\begin{aligned} f(\lambda, \varepsilon) &= \det[(\lambda - 1)E_2 - \varepsilon D_2] \det[\lambda E_{n-2} - e^{2\pi C} + O(\varepsilon)] \\ &\quad + O((\lambda - 1)\varepsilon^2) + O(\varepsilon^3) \\ &= h(\lambda, \varepsilon) + O(\varepsilon(\lambda - 1)^2 + (\lambda - 1)\varepsilon^2 + \varepsilon^3). \end{aligned}$$

As $\varepsilon \to 0^+$ there are $(n - 2)$ zeros $\lambda_j(\varepsilon)$, $3 \leq j \leq n$, of $f(\lambda, \varepsilon)$ which approach eigenvalues $\lambda_j(0)$ of $e^{2\pi C}$. These numbers $\lambda_j(0)$ are inside of the unit disk $|\lambda| < 1$. The remaining two zeros $\lambda_1(\varepsilon)$ and $\lambda_2(\varepsilon)$ approach one. We wish to show that for ε small and positive, $\lambda_1(\varepsilon)$ and $\lambda_2(\varepsilon)$ are inside of the unit disk.

The zeros of $\det[(\lambda - 1)E_2 - \varepsilon D_2]$ have the form

$$\lambda_j^*(\varepsilon) = 1 + \varepsilon \delta_j, \qquad j = 1, 2,$$

8.6 Stability of Systems with Linear Part Critical

where δ_j is an eigenvalue of D_2. From (i) it follows that $[\lambda_j^*(\varepsilon) - 1]/\varepsilon = \delta_j$ has negative real part. Thus $|\lambda_j^*(\varepsilon)| < 1$ for ε positive and sufficiently small. We now consider two cases:

Case 1. $\delta_1 \neq \delta_2$, $\operatorname{Im} \delta_1 > 0$. Let Γ be the circle in the complex λ plane with center at $\lambda_1^*(\varepsilon)$ and radius εr. Let $\delta_1 = -a_1 + ib_1$, where $a_1 > 0$ and $b_1 > 0$. Fix $r > 0$ so small that $0 < r < \min\{a_1, b_1\}$. Notice that

$$|\lambda_1^*(\varepsilon)|^2 = |1 + \varepsilon\delta_1|^2 = 1 - 2\varepsilon a_1 + O(\varepsilon^2),$$

and thus

$$\begin{aligned}(|\lambda_1^*(\varepsilon)| + r\varepsilon)^2 &= |\lambda_1^*(\varepsilon)|^2 + 2\varepsilon r|\lambda_1^*(\varepsilon)| + r^2\varepsilon^2 \\ &= 1 - 2\varepsilon a_1 + 2\varepsilon r(1 + O(\varepsilon)) + O(\varepsilon^2) \\ &= 1 + 2\varepsilon(r - a_1) + O(\varepsilon^2) < 1\end{aligned}$$

for ε positive and sufficiently small. Hence, Γ contains exactly one zero $\lambda_1^*(\varepsilon)$ and Γ is entirely inside of the unit disk.

If $\lambda \in \Gamma$, then $\lambda = \lambda_1^*(\varepsilon) + r\varepsilon e^{i\theta} = 1 + \varepsilon\delta_1 + r\varepsilon e^{i\theta}$ and so

$$\begin{aligned}|h(\lambda, \varepsilon)| &= |\det[(\lambda - 1)E_2 - \varepsilon D_2]| |\det(\lambda E_{n-2} - e^{2\pi C})| \\ &\geq \varepsilon^2 |\det[(re^{i\theta} + \delta_1)E_2 - D_2]| |\det[\lambda E_{n-2} - e^{2\pi C}]| \\ &\geq \varepsilon^2 K\end{aligned}$$

for some constant K. Also, on Γ we have

$$|f(\lambda, \varepsilon) - h(\lambda, \varepsilon)| \leq O((1-\lambda)^2 \varepsilon + (1-\lambda)\varepsilon^2 + \varepsilon^3) = O(\varepsilon^3) < K\varepsilon^2$$

where ε is sufficiently small. By Rouché's theorem, h and $f = h + (f - h)$ have the same number of zeros inside of Γ, namely one zero. The same argument applies to $\lambda_2(\varepsilon)$.

Case 2. $\delta_1 = \delta_2 = \delta < 0$. In this situation, the argument is the same except that here we choose $r < |\delta|/2$. By Rouché's theorem, f and h each have two zeros in a circle Γ inside the unit circle.

Case 3. $\delta_1 < \delta_2 < 0$. In this situation, the argument is the same as in Case 1, except that we choose $r < (\delta_1 - \delta_2)/2$. ∎

The reader is invited to prove the following result.

Theorem 6.3. If in Theorem 6.2 either $\varepsilon < 0$ or, for $\varepsilon > 0$, D_2 has a root with positive real part or C has a root with positive real part, then for $|\varepsilon|$ sufficiently small $\psi(t, \varepsilon)$ is unstable.

B. Autonomous Case

We now consider the autonomous system

$$x' = Ax + \varepsilon g(x, \varepsilon), \tag{6.5}$$

where A satisfies (6.2), $g: R^n \times [-\varepsilon_0, \varepsilon_0] \to R^n$ and $g \in C^2$. We assume the existence of a smooth family $\psi(t, \varepsilon)$ of solutions in $\mathscr{P}_{T(\varepsilon)}$, where $T(\varepsilon) = 2\pi + \tau(\varepsilon) \in C^2[-\varepsilon_0, \varepsilon_0]$, $\tau(0) = 0$ and $\psi(t, 0) = p(t)$. We can check the stability of $\psi(t, \varepsilon)$ by studying the linear system

$$y' = Ay + \varepsilon g_x(\psi(t, \varepsilon), \varepsilon)y. \tag{6.6}$$

If $Y(t, \varepsilon)$ is the fundamental matrix solution of (6.6) such that $Y(0, \varepsilon) = E$, then Floquet multipliers can be determined from

$$Y(2\pi + \tau(\varepsilon), \varepsilon) = e^{(2\pi + \tau(\varepsilon))R(\varepsilon)}.$$

One Floquet multiplier will always be one. The problem is to determine where the others lie. As before, we can compute

$$e^{(2\pi + \tau)R} = e^{(2\pi + \tau)A}\{E + \varepsilon D + O(\varepsilon^2)\},$$

where

$$D = \int_0^{2\pi} e^{-sA} g_x(p(s), 0) e^{sA}\, ds$$

and

$$e^{(2\pi+\tau)A} = \left[\begin{array}{cc|c} \cos N\tau & -\sin N\tau & 0 \\ \sin N\tau & \cos N\tau & \\ \hline 0 & & e^{(2\pi+\tau)C} \end{array}\right].$$

Thus

$$e^{(2\pi + \tau(\varepsilon))R(\varepsilon)} = \begin{bmatrix} E + \varepsilon D_2^* + O(\varepsilon^2) & O(\varepsilon) \\ O(\varepsilon) & e^{2\pi C} + O(\varepsilon) \end{bmatrix},$$

where

$$D_2^* = \begin{bmatrix} d_{11} & d_{12} - N\tau'(0) \\ d_{21} + N\tau'(0) & d_{22} \end{bmatrix}, \quad \tau' = \frac{d\tau}{d\varepsilon}.$$

We are now in a position to state the following result.

Theorem 6.4. If A, g, ψ, and τ are defined as above, if all eigenvalues of C have negative real parts, and if one eigenvalue of D_2^* has negative real part, then for ε positive and sufficiently small, $\psi(t, \varepsilon)$ is orbitally

8.6 Stability of Systems with Linear Part Critical

stable. If instead ε is negative or for $\varepsilon > 0$, D_2^* or C has an eigenvalue with positive real part, then $\psi(t, \varepsilon)$ is unstable.

The proof of this theorem is similar to the proof of Theorem 6.2 and is left to the reader as an exercise.

The technique of proof of Theorem 6.2 can be generalized considerably. We shall give one example of such a generalization since we shall need this generalization later. Consider the system

$$\begin{aligned} x' &= \varepsilon f(x) + \varepsilon g_1(t, x, y, \varepsilon), \\ y' &= By + \varepsilon g_2(t, x, y, \varepsilon), \end{aligned} \tag{6.7}$$

where $g_i: R \times R^{n_1+n_2} \times [-\varepsilon_0, \varepsilon_0] \to R^{n_i}$, g_i is 2π periodic in t, $g_i \in C^2$, and $g_{1x}(t, x_0, 0, 0) = 0$. Assume $z(t, \varepsilon) = (x(t, \varepsilon)^T, y(t, \varepsilon)^T)^T$ is a continuous family of solutions in $\mathscr{P}_{2\pi}$ such that there is an x_0 with $f(x_0) = 0$ and

$$x(t, 0) \equiv x_0, \qquad y(t, 0) \equiv 0.$$

Let $f \in C^1(R^{n_1})$ and define $C = f_x(x_0)$. Under this condition we now prove the following result.

Theorem 6.5. Suppose no eigenvalue of B or of C has zero real part. For ε positive and sufficiently small, $z(t, \varepsilon)$ is uniformly asymptotically stable if all eigenvalues of B and C have negative real parts and is unstable if at least one eigenvalue has a positive real part.

Proof. On linearizing (6.7) about $z(t, \varepsilon)$, we obtain a coefficient matrix of the form

$$\begin{bmatrix} \varepsilon f_x(x(t, \varepsilon)) + \varepsilon G_{11}(t, z(t, \varepsilon), \varepsilon) & \varepsilon G_{12}(t, z(t, \varepsilon), \varepsilon) \\ \varepsilon G_{21}(t, z(t, \varepsilon), \varepsilon) & B + \varepsilon G_{22}(t, z(t, \varepsilon), \varepsilon) \end{bmatrix}$$

for some functions G_{ij} where $G_{11}(t, 0, 0) = 0$. If we follow the proof of Theorem 6.2 we find that

$$A = \begin{bmatrix} 0 & 0 \\ 0 & B \end{bmatrix}, \qquad e^{At} = \begin{bmatrix} E & 0 \\ 0 & e^{Bt} \end{bmatrix}$$

and

$$e^{2\pi R(\varepsilon)} = e^{2\pi A}(E + \varepsilon D + O(\varepsilon^2)) = \begin{bmatrix} E + \varepsilon D_{11} + O(\varepsilon^2) & O(\varepsilon) \\ O(\varepsilon) & e^{2\pi B} + O(\varepsilon) \end{bmatrix}.$$

Here

$$D_{11} = \int_0^{2\pi} f_x(x_0) \, ds = 2\pi C.$$

As before, m eigenvalues $\lambda_j(\varepsilon)$ tend, as $\varepsilon \to 0^+$, to eigenvalues $\lambda_j(0)$ of $e^{2\pi B}$. These eigenvalues satisfy $|\lambda_j(\varepsilon)| \neq 1$ for ε nearly zero. The other $\lambda_j(\varepsilon)$ can be shown, by Rouché's theorem, to be close to $1 + 2\pi\varepsilon\delta_j$ where δ_j is the corresponding eigenvalue of C. ∎

8.7 AVERAGING

We now study periodic systems of equations which can be decomposed into the form

$$\begin{aligned} x' &= \varepsilon F(t, x, y, \varepsilon), \\ y' &= By + \varepsilon G(t, x, y, \varepsilon), \end{aligned} \quad (7.1)$$

where $x \in R^n$, $y \in R^m$, B is a constant $m \times m$ matrix and F and G are smooth functions defined on a neighborhood of $x = 0$, $y = 0$, $\varepsilon = 0$ and are 2π periodic in t. For $|y|$ and $|\varepsilon|$ small, we conjecture that y has little effect on the first equation in (7.1). Indeed, it seems likely that the constant term in the Fourier series for F provides a good approximation for $F(t, x, y, \varepsilon)$. Therefore, as an approximation we replace (7.1) by

$$\begin{aligned} x' &= \varepsilon F_0(x), \\ y' &= By \end{aligned} \quad (7.2)$$

where

$$F_0(x) = \frac{1}{2\pi} \int_0^{2\pi} F(u, x, 0, 0)\, du.$$

If (7.2) has a critical point $(x_0, 0)$ whose stability can be determined by linearization, then we expect (7.1) to have a 2π-periodic solution which is near $(x_0, 0)$ and which has the same stability properties as $(x_0, 0)$. The following result shows that this approximate analysis is indeed valid.

Theorem 7.1. Let F and G be continuous in $(t, x, y, \varepsilon) \in R \times B(x_0, h) \times B(h) \times [-\varepsilon_0, \varepsilon_0]$, 2π periodic in t, and of class C^2 in (x, y). Suppose that $F_y(t, x_0, 0, 0) = 0$. Let $(x_0, 0)$ be a critical point of (7.2) such that all eigenvalues of the linearized system

$$x' = \varepsilon \frac{\partial F_0}{\partial x}(x_0)x, \qquad y' = By \qquad (7.3)$$

have nonzero real parts for $\varepsilon \neq 0$. Then for ε positive and sufficiently small, system (7.1) has a unique 2π-periodic solution $z(t, \varepsilon) = (x(t, \varepsilon), y(t, \varepsilon))$ in a

8.7 Averaging

neighborhood of $(x_0, 0)$ which is continuous in (t, ε) and which satisfies $z(t, \varepsilon) \to (x_0, 0)$ as $\varepsilon \to 0^+$. Moreover, the stability properties of $z(t, \varepsilon)$ are the same as those of $(x_0, 0)$.

Proof. Since $F(t, x, 0, 0)$ is 2π periodic in t, we can subtract its mean value and integrate the resulting difference to obtain a 2π-periodic function of t. Thus, if we define

$$u(t, x) = \int_0^t \left\{ F(v, x, 0, 0) - \frac{1}{2\pi} \int_0^{2\pi} F(s, x, 0, 0) \, ds \right\} dv,$$

then u is 2π periodic in t, and C^1 in (t, x), and C^2 in x.

For ε small, we can invert the change of variables $x = w + \varepsilon u(t, w)$, $y = y$, to obtain $w = x - \varepsilon u(t, x) + O(\varepsilon^2)$. Under this change of variables, we can obtain a new equation to replace (7.1) as follows:

$$x' = w' + \varepsilon(u_t + u_w w') = \varepsilon F(t, w + \varepsilon u, y, \varepsilon)$$

or

$$(1 + \varepsilon u_w) w' = \varepsilon \{ F(t, w + \varepsilon u, y, \varepsilon) - u_t \}$$
$$= \varepsilon F_0(w) + \varepsilon \{ F(t, w + \varepsilon u, y, \varepsilon) - F(t, w, 0, 0) \}$$
$$+ \varepsilon (F(t, w, 0, 0) - F_0(w) - u_t).$$

By the choice of $u(t, w)$, the last term is zero. Hence, (7.1) is replaced by

$$w' = \varepsilon F_0(w) + \varepsilon F_1(t, w, y, \varepsilon),$$
$$y' = By + \varepsilon G_1(t, w, y, \varepsilon), \qquad (7.4)$$

where $G_1(t, w, y, \varepsilon) = G(t, w + \varepsilon u(t, w), y, \varepsilon)$ and

$$F_1(t, w, y, \varepsilon) = [1 + \varepsilon u_w(t, w)]^{-1} [F(t, w + \varepsilon u, y, \varepsilon) - F(t, w, 0, 0)$$
$$- \varepsilon F_0(w) u_w(t, w)].$$

Thus, $F_1(t, w, 0, 0) = 0$ and $F_{1w}(t, w, 0, 0) = 0$.

We now generate a sequence with elements $(w_m(t, \varepsilon), y_m(t, \varepsilon)) \in \mathscr{P}_{2\pi}$ as follows. Let $w_0(t, \varepsilon) = x_0$ and let $y_0(t, \varepsilon) = 0$. Given w_m and y_m, let w_{m+1} and y_{m+1} be the unique 2π-periodic solutions of

$$w' = \varepsilon F_0'(x_0)(w - x_0) = \varepsilon F_2(t, w_m, y_m, \varepsilon),$$
$$y' - By = \varepsilon G_1(t, w_m, y_m, \varepsilon), \qquad (7.5)$$

where $F_2(t, v, y, \varepsilon) = F_0(v) - F_0'(x_0)(v - x_0) + F_1(t, v, y, \varepsilon)$ and $F_0' = \partial F_0/\partial w$. Since $F_0(x_0) = 0$, it follows that $F_2(t, x_0, 0, 0) = 0$ and $F_{2z}(t, x_0, 0, 0) = 0$. By Corollary 2.5 we can write these periodic solutions in the form

$$w_{m+1}(t, \varepsilon) = x_0 + \int_{-\infty}^{\infty} \varepsilon J_1(t - s) F_2(s, w_m(s, \varepsilon), y_m(s, \varepsilon), \varepsilon) \, ds$$

and
$$y_{m+1}(t,\varepsilon) = \int_{-\infty}^{\infty} \varepsilon J_2(t-s) G_1(s, w_m(s,\varepsilon), y_m(s,\varepsilon), \varepsilon)\,ds,$$
where
$$\int_{-\infty}^{\infty} \varepsilon |J_1(s)|\,ds \le K \quad \text{and} \quad \int_{-\infty}^{\infty} |J_2(s)|\,ds \le K.$$

In a sufficiently small ball about $(x_0, 0)$, say $|w - x_0| \le \delta$ and $|y| \le \delta$, and for ε sufficiently small (say $0 < \varepsilon \le \varepsilon_1$), we can arrange things so that $|F_2(t, w, y, \varepsilon)| \le \delta K^{-1}$ and $|\varepsilon G_1(t, w, y, \varepsilon)| \le \delta K^{-1}$ and both $F_2(t, w, y, \varepsilon)$ and $\varepsilon G_1(t, w, y, \varepsilon)$ are Lipschitz continuous with constant L as small as we please, say $L \le (2K)^{-1}$. By the method of successive approximations it is easy to see that

$$|w_m(t, \varepsilon) - x_0| \le \delta, \qquad |y_m(t, \varepsilon)| \le \delta.$$

for $m = 0, 1, 2, \ldots$ and $t \in R$ while

$$|w_{m+1}(t,\varepsilon) - w_m(t,\varepsilon)| + |y_{m+1}(t,\varepsilon) - y_m(t,\varepsilon)|$$
$$\le 0.5(|w_m(t,\varepsilon) - w_{m-1}(t,\varepsilon)| + |y_m(t,\varepsilon) - y_{m-1}(t,\varepsilon)|)$$

for $m = 1, 2, 3, \ldots$ and $t \in R$. This sequence converges uniformly for $(t, x) \in R \times (0, \varepsilon_0]$ to solutions $w(t, \varepsilon)$ and $y(t, \varepsilon)$ of (7.5). These functions also solve (7.4) so that $x = w + \varepsilon u(t, w)$ and y solve (7.1). The stability properties of $(x(t, \varepsilon), y(t, \varepsilon))$ follow from Theorem 6.5. ∎

Theorem 7.1 cannot usually be directly applied to systems of interest. Typically, the problem in question must first be transformed in such a way that the result will apply to the transformed system. We shall illustrate two common methods of arranging such a transformation. In the first method, use is made of *rotating coordinates* and in the second method, *polar coordinates* are utilized. We give a simple example to demonstrate each of these methods.

Example 7.2. Consider the problem

$$y'' + y = \varepsilon f(y, y'). \tag{7.6}$$

For $\varepsilon = 0$, this problem reduces to the (linear) harmonic oscillator. For ε small but not zero, we define

$$\begin{bmatrix} u \\ v \end{bmatrix} = \begin{bmatrix} \cos t & -\sin t \\ \sin t & \cos t \end{bmatrix} \begin{bmatrix} y \\ y' \end{bmatrix}$$

so that

$$u' = -\varepsilon f(u\cos t + v\sin t, -u\sin t + v\cos t)\sin t,$$
$$v' = \varepsilon f(u\cos t + v\sin t, -u\sin t + v\cos t)\cos t.$$

This system is in the proper form to apply Theorem 7.1.

8.8 Hopf Bifurcation

Alternately, we can use the transformation
$$y = r\cos\theta, \qquad y' = r\sin\theta$$
to transform (7.6) into the system of equations
$$r' = \varepsilon f(r\cos\theta, r\sin\theta)\sin\theta,$$
$$\theta' = -1 + \varepsilon f(r\cos\theta, r\sin\theta)\cos\theta/r.$$
If we replace the independent variable t by the variable θ so that
$$\frac{dr}{d\theta} = -\varepsilon\frac{rf(r\cos\theta, r\sin\theta)\sin\theta}{r - \varepsilon f(r\cos\theta, r\sin\theta)}, \tag{7.7}$$
then Theorem 7.1 can be applied to (7.7). ∎

8.8 HOPF BIFURCATION

The results of the preceding section show how an existing periodic solution varies as a parameter varies. In contrast to this, bifurcation occurs when periodic solutions are suddenly created (or destroyed) as a parameter varies. For example, consider the following system in polar coordinates:
$$r' = r[(r-1)^2 - \varepsilon], \qquad \theta' = 1.$$
For $\varepsilon < 0$ there are no nontrivial periodic solutions, while at $\varepsilon = 0$ a 2π-periodic solution exists with $r = 1$ which immediately bifurcates into two solutions with $r = 1 - \sqrt{\varepsilon}$ and $r = 1 + \sqrt{\varepsilon}$. In general, exactly what might happen when bifurcation occurs depends very much on the form of the equation in question. The reader may want to analyze, for example, the following problem, given in polar coordinates. Notice that the form of this system is similar to that of the last example written above, namely,
$$r' = r(\varepsilon - r^2), \qquad \theta' = 1.$$
In this system, the origin is globally asymptotically stable for $\varepsilon < 0$. When ε is increased beyond $\varepsilon = 0$, this system has a periodic solution with amplitude $\varepsilon^{1/2}$. In such a case, the equilibrium $r = 0$ has been described as having suddenly "blown a smoke ring." This particular type of bifurcation is an example of what we call **Hopf bifurcation**. The purpose of this section is to study Hopf bifurcation for systems of two equations.

Consider the two-dimensional system
$$x' = A(\varepsilon)x + G(x, \varepsilon), \tag{8.1}$$

where A is a C^1 smooth real 2×2 matrix, $G: R^2 \times (-\varepsilon_0, \varepsilon_0) \to R^2$ is of class C^3, $G(0, \varepsilon) = 0$, and $G_x(0, \varepsilon) = 0$. Let $A(\varepsilon)$ have eigenvalues $\lambda(\varepsilon) = \alpha(\varepsilon) \pm i\beta(\varepsilon)$, where

$$\alpha(0) = 0, \qquad \alpha'(0) > 0, \qquad \beta(0) > 0.$$

We are now in a position to prove the following result.

Theorem 8.1. Under the foregoing assumptions, there is a continuous real valued function $\varepsilon(a)$ with $\varepsilon(0) = 0$ and a one-parameter family $x(t, a)$ of nontrivial periodic solutions of (8.1) with periods $T(a) = 2\pi/\beta(0) + O(a)$ such that $x(t, a) \to 0$ as $a \to 0^+$.

Proof. By a real and linear change of variables, we can put $A(\varepsilon)$ into the form

$$A(\varepsilon) = \begin{bmatrix} \alpha(\varepsilon) & -\beta(\varepsilon) \\ \beta(\varepsilon) & \alpha(\varepsilon) \end{bmatrix}.$$

Hence, we can utilize polar coordinates and assume that the system transforms into an equivalent system of the form

$$\frac{dr}{d\theta} = \gamma(\varepsilon)r + H(r, \theta, \varepsilon), \qquad \gamma(\varepsilon) \triangleq \alpha(\varepsilon)/\beta(\varepsilon). \tag{8.2}$$

Here $H(r, \theta, \varepsilon) = r\beta^{-1}(r\beta + \cos\theta G_2 - \sin\theta G_1)^{-1}[(\beta\cos\theta - \alpha\sin\theta)G_1 + (\beta\sin\theta + \alpha\cos\theta)G_2]$.

Let $r(\theta, a, \varepsilon)$ be the solution of (8.2) such that $r(0, a, \varepsilon) = a$. By the variation of constants formula, solutions of (8.2) have the form

$$r(\theta, a, \varepsilon) = ae^{\theta\gamma(\varepsilon)} + \int_0^\theta e^{\gamma(\varepsilon)(\theta - s)} H(r(s, a, \varepsilon), s, \varepsilon) \, ds. \tag{8.3}$$

For the existence of a periodic solution we require that $r(2\pi, a, \varepsilon) = a$. Now define

$$F(a, \varepsilon) = e^{2\pi\gamma(\varepsilon)} - 1 + a^{-1} \int_0^{2\pi} e^{\gamma(\varepsilon)(2\pi - s)} H(r(s, a, \varepsilon), s, \varepsilon) \, ds$$

for $a \neq 0$ and $F(0, \varepsilon) = e^{2\pi\gamma(\varepsilon)} - 1$. For the existence of a periodic solution we need $F(a, \varepsilon) = 0$. By (8.3) we see that $r(\theta, a, \varepsilon) = O(a)$ as $a \to 0$ uniformly on $(\theta, \varepsilon) \in [0, 2\pi] \times (-\varepsilon_0, \varepsilon_0)$. Thus $F(a, \varepsilon)$ is continuous in a neighborhood of $a = \varepsilon = 0$. Since $H(r, a, \varepsilon) = O(r^2)$ uniformly in $(a, \varepsilon) \in [0, 2\pi] \times (-\varepsilon_0, \varepsilon_0)$, a similar argument shows that $F \in C^1$ near $a = \varepsilon = 0$. But $F(0, 0) = 0$ and

$$\frac{\partial F}{\partial \varepsilon}(0, 0) = e^{2\pi\gamma(\varepsilon)} 2\pi \gamma'(\varepsilon)|_{\varepsilon=0} = (e^{2\pi \cdot 0})(2\pi)\left(\frac{\beta(0)\alpha'(0) - 0 \cdot \beta'(0)}{\beta^2(0)}\right)$$

$$= 2\pi\alpha'(0)/\beta(0) > 0.$$

8.9 A Nonexistence Result

By the implicit function theorem, there is a solution $\varepsilon(a)$ of $F(a, \varepsilon) = 0$ defined near $a = 0$ and satisfying $\varepsilon(0) = 0$. The family of periodic solutions is $r(\theta, a, \varepsilon(a))$. ∎

We close this section with a specific case.

Example 8.2. Consider the problem
$$x'' - 2\varepsilon x' + \varepsilon x + ax + bx^3 = 0$$
where $a > 0, b \neq 0$. Theorem 8.1 applies with $\alpha(\varepsilon) = \varepsilon$ and $\beta(\varepsilon) = (a + \varepsilon - \varepsilon^2)^{1/2}$.

8.9 A NONEXISTENCE RESULT*

In this section, we consider autonomous systems
$$x' = F(x), \tag{A}$$
where $F: R^n \to R^n$ and F is Lipschitz continuous on R^n with Lipschitz constant L. For $x = (x_1, \ldots, x_n)^T$ and $y = (y_1, \ldots, y_n)^T$, we let
$$\langle x, y \rangle = \sum_{i=1}^{n} x_i y_i$$
denote the usual inner product on R^n. We shall always use the Euclidean norm on R^n, i.e., $|x|^2 = \langle x, x \rangle$ for all $x \in R^n$.

Now assume that (A) has a nonconstant solution $\phi \in \mathscr{P}_T$. Then there is a simple relationship between T and L, given in the following result.

Theorem 9.1. If F and ϕ are as described above, then $T \geq 2\pi/L$.

Before proving this theorem, we need to establish the following auxiliary result.

Lemma 9.2. If $y(t) \triangleq F(\phi(t))/|F(\phi(t))|$, then y' exists almost everywhere and
$$\int_0^T |y'(t)| \, dt \geq 2\pi.$$

Proof. Since ϕ is bounded and F is Lipschitz continuous, it follows that $F(\phi(t))$ is Lipschitz continuous and hence so is y. Thus y is absolutely continuous and so y' exists almost everywhere. To prove the

above bound, choose t_1 and $t_1 + \tau$ in $[0, T]$ so that $\tau > 0$ and
$$|\phi(t_1 + \tau) - \phi(t_1)| = \sup\{|\phi(s) - \phi(u)| : s, u \in [0, T]\}.$$
Since $\phi_1(t) \triangleq \phi(t - t_1)$ is also a solution of (A) in $\mathscr{P}_T$, we can assume without loss of generality that $t_1 = 0$. Define
$$v = \phi(0) - \phi(\tau) \quad \text{and} \quad u(t) = |\phi(t) - \phi(\tau)|^2/2.$$
Then u has its maximum at $t = 0$ so that
$$u'(0) = \langle \phi(t) - \phi(\tau), \phi'(t) \rangle|_{t=0} = \langle v, \phi'(0) \rangle = \langle v, F(\phi(0)) \rangle = 0.$$
Similarly we see that $\langle v, F(\phi(\tau)) \rangle = 0$.

We now show that
$$\int_0^\tau |y'(t)|\, dt \geq \pi. \tag{9.1}$$
First note that if $y(0) = -y(\tau)$, then the shortest curve between $y(0)$ and $y(\tau)$ which remains on the unit sphere S has length π. So the length of $y(t)$ between $y(0)$ and $y(\tau)$ is at least π. If $y(0) \neq -y(\tau)$, define
$$\gamma = y(0) + y(\tau), \qquad m = y(0) - y(\tau).$$
Then $\langle \gamma, m \rangle = |y(0)|^2 - |y(\tau)|^2 = 1 - 1 = 0$. Define
$$a(t) \triangleq \langle y(t), \gamma \rangle / \langle \gamma, \gamma \rangle$$
and note that
$$a(\tau) = \langle (\gamma - m)/2, \gamma \rangle / \langle \gamma, \gamma \rangle = \tfrac{1}{2}.$$
If we define $h(t) = y(t) - a(t)\gamma$, then $\langle \gamma, h(t) \rangle = 0$. Thus $y(t) = a(t)\gamma + h(t)$ and $-a(t)\gamma + h(t)$ have the same norm, namely, one. Now
$$\phi'(t) = |F(\phi(t))| y(t) = |F(\phi(t))| (a(t)\gamma + h(t))$$
and $\langle \gamma, h \rangle = 0$. Therefore
$$\left(\int_0^\tau |F(\phi(t))| a(t)\, dt \right) \langle \gamma, \gamma \rangle + \int_0^\tau |F(\phi(t))| \cdot 0\, dt$$
$$= \langle \phi(\tau) - \phi(0), \gamma \rangle = \langle -v, y(0) + y(\tau) \rangle$$
$$= -\langle v, F(\phi(0)) \rangle |F(\phi(0))|^{-1} - \langle v, F(\phi(\tau)) \rangle |F(\phi(\tau))|^{-1} = 0.$$
Hence
$$\int_0^\tau |F(\phi(t))| a(t)\, dt = 0,$$
and a must be zero at some point $\tau_0 \in (0, \tau)$.

8.9 A Nonexistence Result

Define $y_1(t)$ on $[0, \tau_0]$ and $y_1(t) = h(t) - a(t)\gamma$ on $[\tau_0, \tau]$. Then $y_1 \in S$ for $0 \le t \le \tau$, y_1 is absolutely continuous, and

$$y_1(\tau) + y_1(0) = [y(\tau) - 2a(\tau)\gamma] + y(0) = y(\tau) + y(0) - 2(\tfrac{1}{2})\gamma = 0.$$

Thus $y_1(\tau) = -y_1(0)$. Hence, the length of the arc from $y_1(0)$ to $y_1(\tau)$ is at least π. Since $\langle h, \gamma \rangle = 0$, the arc lengths for y and y_1 are the same over $[0, \tau]$. This proves (9.1).

Finally, the length of y over $[\tau, T]$ is at least π by the same argument [by starting at $\phi(\tau)$ instead of $\phi(0)$]. This concludes the proof. ∎

Proof of Theorem 9.1. Now let $y(t)$ be as defined in Lemma 8.2. Then y is Lipschitz continuous on R, and hence, absolutely continuous and differentiable almost everywhere. Since $|y(t)| = 1$, it follows that

$$0 = \frac{d}{dt}\langle y(t), y(t)\rangle = 2\langle y(t), y'(t)\rangle$$

or $y \perp y'$ almost everywhere. This fact and the fact that $y|F(\phi)| = F(\phi)$ yield

$$F(\phi)' = y'|F(\phi)| + y|F(\phi)|',$$

or

$$|F(\phi)'|^2 = |y'|^2|F(\phi)|^2 + |y|^2||F(\phi)|'|^2.$$

Thus we have

$$|y'| \le |F(\phi)'|/|F(\phi)| \le L|\phi'|/|F(\phi)| = L,$$

and

$$\int_0^T |y'(t)|\, dt \le \int_0^T L\, dt = TL.$$

Finally, by Lemma 9.2 we have $2\pi \le TL$. ∎

We conclude this section with an example.

Example 9.3. Consider the second order equation

$$x'' + g(x)x' + x = 0$$

with $g(x)$ an odd function such that $g(x) < 0$ for $0 < x < a$ and $g(x) > 0$ for $x > a$. Assume that $|g(x)| \le \alpha$. Then the equivalent system

$$x' = y - G(x), \quad y' = -x, \quad G(x) \triangleq \int_0^x g(u)\, du$$

is Lipschitz continuous with constant $L = (\max\{2, 1 + 2\alpha^2\})^{1/2}$. Hence, there can be no periodic solution of (9.2) with period $T \ge 2\pi/L$.

PROBLEMS

1. In (P) suppose that $f: R \times R^n \to R^n$ with $f \in C^1$. Show that if (P) has a solution ϕ which is bounded on R and is uniformly asymptotically stable in the large, then $\phi \in \mathscr{P}_T$.

2. In (P) let $f \in C^1(R \times R^n)$. Suppose the eigenvalues $\lambda_j(t, x)$ of $(f_x(t, x) + f_x(t, x)^T)$ satisfy $\lambda_j(t, x) \leq -\mu < 0$ for all (t, x). Show that if (P) has at least one solution which is bounded on R^+, then (P) has a unique solution $\phi \in \mathscr{P}_T$.

3. Suppose A is a real, stable $n \times n$ matrix and $F \in C^2(R \times R^n)$ with $F(t, 0) = 0$, $F_x(t, 0) = 0$, and $F(t + T, x) \equiv F(t, x)$. Let $p \in \mathscr{P}_T$. Show that there is an $\varepsilon_0 > 0$ such that when $0 \leq |\varepsilon| < \varepsilon_0$, then

$$x' = Ax + F(t, x) + \varepsilon p(t)$$

has at least one solution $\phi(t, \varepsilon) \in \mathscr{P}_T$. Moreover $\phi(t, \varepsilon) \to 0$ as $|\varepsilon| \to 0$ uniformly for $t \in [0, T]$.

4. Consider the system

$$y' = B(y)y + p(t), \qquad (10.1)$$

where B is a C^1 matrix and $p \in \mathscr{P}_T$. Show that if

$$\max_{1 \leq i \leq n} \left(b_{ii}(y) + \sum_{i \neq j,\, j=1}^{n} |b_{ij}(y)| \right) = -c(y) \leq -\alpha < 0$$

for all $y \in R^n$, then there is a solution $\phi \in \mathscr{P}_T$ of (10.1).

5. Suppose A is a real, constant, $n \times n$ matrix which has an eigenvalue $i\omega$ with $\omega > 0$. Fix $T > 0$. Show that for any $\varepsilon_0 > 0$ there is an ε in the interval $0 < \varepsilon \leq \varepsilon_0$ such that

$$\varepsilon x' = Ax$$

has a nontrivial solution in $\mathscr{P}_T$.

6. Let A be a constant $n \times n$ matrix and let $\varepsilon_0 > 0$. Show that for $0 < \varepsilon \leq \varepsilon_0$ the system $x' = \varepsilon A x$ has no nontrivial solution in $\mathscr{P}_T$ if an only if $\det A \neq 0$.

7. In (3.1), let $g \in C^1(R \times R^n)$ and let $h \in C^1(R \times R^n \times (-\varepsilon_0, \varepsilon_0))$. Assume that p is a periodic solution of (3.1) at $\varepsilon = 0$ such that

$$y' = g_x(t, p(t))y$$

has no nontrivial solutions in $\mathscr{P}_T$ and assume h satisfies (3.3). Let $\Phi(t)$ solve $y' = g_x(t, p(t))y$, $y(0) = E$ and let H be as in (3.2). Show that for $|\varepsilon|$ small, Eq. (3.1) has a solution in $\mathscr{P}_T$ if an only if the integral solution

$$y(t) = \int_t^{t+T} [\Phi(s)(\Phi^{-1}(T) - E)\Phi(t)]^{-1} H(s, y(s), \varepsilon)\, ds \qquad (10.2)$$

Problems

has a solution in $\mathscr{P}_T$. Use successive approximations to show that for $|\varepsilon|$ sufficiently small, Eq. (10.2) has a unique solution in $\mathscr{P}_T$.

8. Suppose all solutions of (LH) are bounded on R^+. If $A(t)$ and $f(t)$ are in $\mathscr{P}_T$ and if (2.5) is not true for some solution y of (2.1) with $y \in \mathscr{P}_T$, then show that all solutions of (LN) are unbounded on R^+.

9. Express $x'' - x = \sin t$ as a system of first order ordinary differential equations. Show that Corollary 2.5 can be applied, compute $G(t)$ for this system, and compute the unique solution of this equation which is in $\mathscr{P}_{2\pi}$.

10. Find the unique 2π-periodic solution of

$$x' = \begin{bmatrix} 0 & -\pi & 0 \\ \pi & 0 & 0 \\ 0 & 0 & 2 \end{bmatrix} x + \begin{bmatrix} \cos t \\ 0 \\ \sin t \end{bmatrix}.$$

11. Find all periodic solutions of

$$x' = \begin{bmatrix} 0 & -1 \\ 1 & 0 \end{bmatrix} x + \begin{bmatrix} \cos \omega t \\ 1 \end{bmatrix}.$$

where ω is any positive number.

12. Show that for $|\varepsilon|$ small, the equation

$$x'' + x' + x + 2x^3 = \varepsilon \cos t$$

has a unique solution $\phi(t, \varepsilon) \in \mathscr{P}_{2\pi}$. Expand ϕ as

$$\phi(t, \varepsilon) = \phi_0(t) + \varepsilon \phi_1(t) + O(\varepsilon^2)$$

and compute ϕ_0 and ϕ_1. Determine the stability properties of $\phi(t, \varepsilon)$.

13. For α, β, and k positive show that the equation

$$y'' + y + \varepsilon(\alpha y + \beta y^3 + ky') = 0$$

exhibits Hopf bifurcation from the critical point $y = y' = 0$.

14. For α, β, and k positive and $A \neq 0$, consider the equation

$$y'' + y = \varepsilon(-\alpha y - \beta^3 - ky' + A \cos t).$$

Show that for ε small and positive there is a family of periodic solutions $\phi(t, \varepsilon)$ and a function $\delta(\varepsilon)$ such that

$$\phi(t + \delta(\varepsilon), \varepsilon) = \gamma_0 \cos t + \varepsilon \phi_1(t) + O(\varepsilon^2)$$

and

$$\delta(\varepsilon) = \delta_0 + \varepsilon \delta_1 + O(\varepsilon^2).$$

Find equations which δ_0 and γ_0 satisfy. Study the stability properties of $\phi(t, \varepsilon)$.

15. For $y'' + y + \varepsilon y^3 = 0$, or equivalently, for

$$y_1' = -y_2,$$
$$y_2' = y_1 + \varepsilon y_1^3,$$

show that the hypotheses of Theorem 5.3 cannot be satisfied. This equation has a continuous two-parameter family of solutions $\phi(t, \alpha, \varepsilon) \varepsilon \mathscr{P}_{T(\varepsilon)}$ with $\phi(0, \alpha, 0) = \alpha > 0$, $\phi'(0, \alpha, 0) = 0$, and $T = 2\pi/\omega$, where

$$\omega(\alpha, \varepsilon) = 1 + \frac{3}{8} \alpha^2 \varepsilon - \frac{21}{256} \alpha^4 \varepsilon^2 + O(\varepsilon^2)$$

and

$$\phi(t, \alpha, \varepsilon) = \left[\alpha - \frac{\varepsilon \alpha^3}{32} + \frac{23}{1024} \varepsilon^2 \alpha^5 + O(\varepsilon^3) \right] \cos \omega t$$
$$+ \left(\frac{1}{32} \varepsilon \alpha^3 - \frac{3}{128} \varepsilon^2 \alpha^5 \right) \cos 3\omega t + c\varepsilon^2 \cos 5\omega t + O(\varepsilon^3).$$

Compute the value of the constant c.

16. For $y'' + \varepsilon(y^2 - 1)y' + y = 0$, let $\phi(t, \varepsilon)$ be that limit cycle satisfying $\phi(0, \varepsilon) > 0$, $\phi'(0, \varepsilon) = 0$.

    (a) Show that $\phi(t, \varepsilon) \to \alpha \cos t$ as $\varepsilon \to 0$ for some constant $\alpha > 0$.

    (b) Compute the value of α.

17. In Theorem 6.4 consider the special case

$$y'' + y = \varepsilon f(y, y').$$

Let $\phi(t, \varepsilon)$ be the periodic family of solutions such that $\phi(t, 0) = \gamma_0 \cos t$ for some $\gamma_0 > 0$. Show that $\phi(t, \varepsilon)$ is orbitally stable if ε is small and positive and

$$\int_0^{2\pi} \frac{\partial f}{\partial y'} (\gamma_0 \cos t, -\gamma_0 \sin t) \, dt < 0.$$

18. Prove Theorem 5.3.

19. Prove Theorem 6.4.

20. Prove an analog of Theorem 5.1 for the coupled system

$$x'' + x = \varepsilon f(t, x, y, x', y'),$$
$$y'' + 4y = \varepsilon g(t, x, y, x', y').$$

Find sufficient conditions on f and g in order that for ε small and positive there is a 2π-periodic family of solutions.

21. Consider the system
$$\begin{bmatrix} x' \\ y' \end{bmatrix} = \begin{bmatrix} A & 0 \\ 0 & B \end{bmatrix} \begin{bmatrix} x \\ y \end{bmatrix} + \varepsilon \begin{bmatrix} \phi_{11}(t) & \varepsilon\phi_{12}(t) \\ \phi_{21}(t) & \phi_{22}(t) \end{bmatrix} \begin{bmatrix} x \\ y \end{bmatrix},$$
where A and B are square matrices of dimension $k \times k$ and $l \times l$, respectively, where the ϕ_{ij} are $2\pi/\omega$ periodic matrices of appropriate dimensions, e^{At} is a $T = 2\pi/\omega$ periodic matrix, all eigenvalues of B have negative real parts, and $\phi_{ij} \in \mathscr{P}_T$. If all eigenvalues of
$$F \triangleq \frac{\omega}{2\pi} \int_0^{2\pi/\omega} e^{-At} \phi_{11}(t) e^{At} \, dt$$
have negative real parts, show that the trivial solution $x = 0$, $y = 0$ is exponentially stable when $\varepsilon > 0$.

22. In Example 8.2, transform the equation into one suitable for averaging and compute the average equation. *Hint*: Let $x = \varepsilon x_1$ and $x' = \varepsilon\sqrt{a}\, x_2'$.

23. Find a constant $K > 0$ such that any limit cycle of
$$x'' + \varepsilon \frac{(5x^4 - 1)x'}{x^2(x^4 - 1)^2 + 1} + x = 0,$$
for $0 < \varepsilon < 1$, must have period $T(\varepsilon) \geq K$.

BIBLIOGRAPHY

GENERAL REFERENCES

The book by Simmons [39] contains an excellent exposition of differential equations at an advanced undergraduate level. More advanced texts on differential equations include Brauer and Nohel [5], Coddington and Levinson [9], Hale [19], Hartman [20], Hille [22], and Sansone and Conti [37].

Differential equations arise in many disciplines in engineering and in the sciences. The area with the most extensive applications of differential equations is perhaps the theory of control systems. Beginning undergraduate texts in this area include D'Azzo and Houpis [2] and Dorf [14]. More advanced books on linear control systems include Brockett [6], Chen [8], Desoer [13], Kailath [24], and Zadeh and Desoer [48].

Chapter 1. For further treatments of electrical, mechanical, and electromechanical systems see Refs. [2] and [14]. For biological systems, refer to Volterra [43] and Poole [36].

Chapter 2. In addition to the general references (especially Refs. [5], [9], [19], [20]), see Lakshmikantham and Leela [26] or Walter [44] for a more detailed account of the comparison theory, and Sell [38] for a detailed and general treatment of the invariance theorem.

Chapter 3. For background material on matrices and vector spaces, see Bellman [3], Gantmacher [15], or Michel and Herget [32]. For general references on systems of linear ordinary differential equations, refer, e.g., to Refs. [5], [9], [19], and [20]. For applications to linear control systems, see e.g., Refs. [6], [8], [13], [24], and [48].

Chapter 4. For further information on boundary value problems, refer to Ref. [22], Ince [23], and Yosida [47].

Chapter 5. In addition to general references on ordinary differential equations (e.g., Refs. [5], [9], [19], [20]), see Antosiewicz [1], Cesari [7], Coppel [10], Hahn [17], LaSalle and Lefschetz [27], Lefschetz [29], Michel and Miller [33], Narendra and Taylor [34], Vidyasagar [42], Yoshizawa [46], and Zubov [49], for extensive treatments and additional topics dealing with the Lyapunov stability theory.

Chapter 6. For additional information on the topics of this chapter, see especially Ref. [22] as well as the general references cited for Chapter 5.

Chapter 7. See Lefschetz [28] and the general references [9] and [20] for further material on periodic solutions in two-dimensional systems.

Chapter 8. In addition to the general references (e.g., [9] and [19]), refer to Bogoliubov and Mitropolskii [4], Cronin [11], Hale [18], Krylov and Bogoliubov [25], Marsden and McCracken [30], Mawhin and Rouche [31], Nohel [35], Stoker [40], Urabe [41], and Yorke [45] for additional material on oscillations in systems of general order. References with engineering applications on this topic include Cunningham [12], Gibson [16], and Hayashi [21].

REFERENCES

1. H. A. Antosiewicz, A survey of Lyapunov's second method, in *Contributions to the Theory of Nonlinear Oscillations* (Annals of Mathematics Studies, No. 41). Princeton Univ. Press, Princeton, New Jersey, 1958.
2. J. J. D'Azzo and C. H. Houpis, *Linear Control System Analysis and Design*. McGraw-Hill, New York, 1975.
3. R. Bellman, *Introduction to Matrix Analysis*, 2nd ed., McGraw-Hill, New York, 1970.
4. N. N. Bogoliubov and Y. A. Mitropolskii, *Asymptotic Methods in the Theory of Nonlinear Oscillations*. Gordon and Breach, New York, 1961.

5. F. Brauer and J. A. Nohel, *Qualitative Theory of Ordinary Differential Equations*. Benjamin, New York, 1969.
6. R. W. Brockett, *Finite Dimensional Linear Systems*. Wiley, New York, 1970.
7. L. Cesari, *Asymptotic Behavior and Stability Problems in Ordinary Differential Equations*, 2nd ed. Springer-Verlag, Berlin, 1963.
8. C. T. Chen, *Introduction to Linear System Theory*. Holt, New York, 1970.
9. E. A. Coddington and N. Levinson, *Theory of Ordinary Differential Equations*. McGraw-Hill, New York, 1955.
10. W. A. Coppel, *Stability and Asymptotic Behavior of Differential Equations* (Heath Mathematical Monographs), Heath, Boston, 1965.
11. J. Cronin, *Fixed Points and Topological Degree in Nonlinear Analysis*. Amer. Math. Soc., Providence, Rhode Island, 1964.
12. W. J. Cunningham, *Introduction to Nonlinear Analysis*. McGraw-Hill, New York, 1958.
13. C. A. Desoer, *A Second Course on Linear Systems*. Van Nostrand Reinhold, Princeton, New Jersey, 1970.
14. R. C. Dorf, *Modern Control Systems*. Addison-Wesley, Reading, Massachusetts, 1980.
15. F. R. Gantmacher, *Theory of Matrices*. Chelsea Publ., Bronx, New York, 1959.
16. J. E. Gibson, *Nonlinear Automatic Control*. McGraw-Hill, New York, 1963.
17. W. Hahn, *Stability of Motion*. Springer-Verlag, Berlin, 1967.
18. J. K. Hale, *Oscillations in Nonlinear Systems*. McGraw-Hill, New York, 1963.
19. J. K. Hale, *Ordinary Differential Equations*. Wiley (Interscience), New York, 1969.
20. P. Hartman, *Ordinary Differential Equations*. Wiley, New York, 1964.
21. C. Hayashi, *Nonlinear Oscillations in Physical Systems*. McGraw-Hill, New York, 1964.
22. E. Hille, *Lectures on Ordinary Differential Equations*. Addison-Wesley, Reading, Massachusetts, 1969.
23. E. L. Ince, *Ordinary Differential Equations*. Dover, New York, 1944.
24. T. Kailath, *Linear Systems*. Prentice-Hall, Englewood Cliffs, New Jersey, 1980.
25. N. Krylov and N. N. Bogoliubov, *Introduction to Nonlinear Mechanics* (Annals of Mathematics Studies, No. 11). Princeton Univ. Press, Princeton, New Jersey, 1947.
26. V. Lakshmikantham and S. Leela, *Differential and Integral Inequalities*, Vol. I. Academic Press, New York, 1969.—
27. J. P. LaSalle and S. Lefschetz, *Stability by Liapunov's Direct Method with Applications*. Academic Press, New York, 1961.
28. S. Lefschetz, *Differential Equations: Geometric Theory*, 2nd ed. Wiley (Interscience), New York, 1962.
29. S. Lefschetz, *Stability of Nonlinear Control Systems*. Academic Press, New York, 1965.
30. J. E. Marsden and M. McCracken, *The Hopf Bifurcation and Its Applications*. Springer-Verlag, Berlin, 1976.
31. J. Mawhin and N. Rouche, *Ordinary Differential Equations: Stability and Periodic Solutions*. Pitman, Boston, 1980.
32. A. N. Michel and C. J. Herget, *Mathematical Foundations in Engineering and Science: Algebra and Analysis*. Prentice-Hall, Englewood Cliffs, New Jersey, 1981.
33. A. N. Michel and R. K. Miller, *Qualitative Analysis of Large Scale Dynamical Systems*. Academic Press, New York, 1977.
34. K. S. Narendra and H. J. Taylor, *Frequency Domain Criteria for Absolute Stability*. Academic Press, New York, 1973.
35. J. A. Nohel, Stability of perturbed periodic motions, *J. Reine und Angewandte Mathematik* **203** (1960), 64–79.
36. R. W. Poole, *An Introduction to Qualitative Ecology* (Series in Population Biology) McGraw-Hill, New York, 1974.

References

37. G. Sansone and R. Conti, *Nonlinear Differential Equations*. Macmillan, New York, 1964.
38. G. R. Sell, Nonautonomous differential equations and topological dynamics, Parts I and II, *Trans. Amer. Math. Soc.* **127** (1967), 241–262, 263–283.
39. G. F. Simmons, *Differential Equations*. McGraw-Hill, New York, 1972.
40. J. J. Stoker, *Nonlinear Vibrations in Mechanical and Electrical Systems*. Wiley (Interscience), New York, 1950.
41. M. Urabe, *Nonlinear Autonomous Oscillations*. Academic Press, New York, 1967.
42. M. Vidyasagar, *Nonlinear Systems Analysis*. Prentice-Hall, Englewood Cliffs, New Jersey, 1978.
43. V. Volterra, *Lecons sûr la théorie mathématique de la lutte pour la vie*. Gauthiers-Villars, Paris, 1931.
44. W. Walter, *Differential and Integral Inequalities*. Springer-Verlag, Berlin, 1970.
45. J. A. Yorke, Periods of periodic solutions and the Lipschitz constant, *Proc. Amer. Math. Soc.* **22** (1969), 509–512.
46. T. Yoshizawa, *Stability Theory by Liapunov's Second Method*. Math. Soc. Japan, Tokyo, 1966.
47. K. Yosida, *Lectures on Differential and Integral Equations*. Wiley (Interscience), New York, 1960.
48. L. A. Zadeh and C. A. Desoer, *Linear System Theory—The State Space Approach*. McGraw-Hill, New York, 1963.
49. V. I. Zubov, *Methods of A. M. Lyapunov and Their Applications*. Noordhoff, Amsterdam, 1964.

INDEX

A

Abel formula, 90
Absolute stability, 245
 problem, 245
 of regulator systems, 243
Acceleration, 8
 angular, 11
Adjoint
 of a linear system, 99, 307
 matrix, 81
 of a matrix equation, 99
 of an nth order equation, 124
 operator, 124
Aizerman conjecture, 245
Ampere, 15
Analytic hypersurface, 267
Angular
 acceleration, 11
 displacement, 11
 velocity, 11
Ascoli Arzela lemma, 41, 48
Asymptotic behavior of eigenvalues, 147
Asymptotic equivalence, 280
Asymptotic phase, 274
Asymptotic stability, 173, 208, *see also*
 Exponential stability
 in the large, 176, 227
 linear system, 180
 uniform, 173
 uniformly in the large, 176
Attractive, 173
Autonomous differential equation, 4, 103

periodic solutions, 292, 317, 335
stability, 178
Averaging, 330

B

$B(h)$, 65, 168
$B(x,h)$, 65, 168
Banach fixed point theorem, 79
Basis, 81
Bessel equation, 289
Bessel inequality, 157
Bilinear concomitant, 125
Boundary, 43
Boundary conditions, 138, 160
 general, 160
 periodic, 138
 separated, 139
Boundary value problem, 140
 inhomogeneous, 152, 160
Bounded solution, 175, 212
Boundedness, 41, 172, 212

C

C^m hypersurface, 267
Capacitor, 15
 microphone, 33
Center, 187
Chain of generalized eigenvectors, 83
Characteristic
 equation, 82, 121

exponent, 115
polynomial, 82, 121
root, 121
Chetaev instability theorem, 216
Class K, 197
Class KR, 197
Closed line segment, 291
Closure, 43
Compact, 42
Companion
 form, 118
 matrix, 118
Comparison
 principle, 239, 255
 theorem, 73, 239, 255
 theory, 70
Complete metric space, 78
Complete orthogonal set, 154, 157
Completely unstable, 214
Complex valued equation, 7, 74
Conjugate matrix, 81
Conservative dynamical system, 25
Continuation, 49
Continued solution, 49
Continuity, 40
 Lipschitz, 53, 66
 piecewise, 41
Contraction map, 79
Controllable, 245
Convergence, 41, 66
Converse theorem, 234
Convolution, 105
Coulomb, 15
Critical
 eigenvalue, 183
 matrix, 183
 point, 169, 290
 polynomial, 184
Current source, 14

D

Damping
 term, 9
 torque, 12
Dashpot, 9, 12
Decrescent function, 197, 198
Definite function, 200
Derivative along solutions, 195
Diagonalization of a set of sequences, 42
Diagonalized matrix, 82

Differentiable function, 40
Diffusion equation, 138
Dini derivative, 72
Direct method of Lyapunov, 205
Direction
 of a line segment, 292
 of a vector, 291
Dissipation function, 29
Distance, 65
Domain, 1, 3, 45
Domain of attraction, 173, 230
Dry friction, 20, 68
Duffing equation, 23, 316, 322

E

Eigenfunction, 141, 160
Eigenvalue, 82, 141, 160
Eigenvector, 82, 83
 generalized, 82
 multiple, 141
 simple, 141
Elastance element, 8
Electric charge, 15
Electric circuits, 14
Electromechanical system, 31
Energy
 dissipated in a resistor, 15
 dissipated by viscous damping, 10, 12
 kinetic, 9, 12
 potential, 9, 12
 stored in a capacitor, 15
 stored in an inductor, 15
 stored in a mass, 9, 12
 stored in a spring, 9, 12
Epidemic model, 24
ϵ approximate solution, 46
Equicontinuous, 41
Equilibrium point, 169, 290
Euclidean norm, 65
Euler
 method, 47
 polygons, 47
Existence of solutions
 of boundary value problems, 139, 145, 161
 of initial value problems, 45, 74, 79
 periodic solutions, 290, 305
Exponential stability, 173, 176, 211, 240
 in the large, 176, 211, 242
Extended solution, 49

F

Farad, 15
Finite escape time, 224
First approximation, 264
First order ordinary differential equations, 1, 3
Fixed point of a map, 79
Floquet
 exponent, 115
 multiplier, 115
 theorem, 113, 133
Force, 8
Forced
 Duffing equation, 23
 system of equations, 103
Fourier
 coefficient, 156
 series, 157
Fredholm alternative, 307
Fundamental matrix, 90
Fundamental set of solutions, 90, 119

G

Generalized
 eigenvector, 82
 eigenvector of rank k, 82
 Fourier coefficient, 156
 Fourier series, 157
Graph, 4, 51
Green's formula, 125
Green's function, 160
Gronwall inequality, 43, 75

H

Hamiltonian, 25
 equations, 26, 35, 251
Hard spring, 21
Harmonic oscillator, 22
Henry, 15
Holomorphic function, 74
Hopf bifurcation, 333
Hurwitz
 determinant, 185
 matrix, 183
 polynomial, 184
Hypersurface, 267

I

Identity matrix, 35, 68

Implicit function theorem, 259
Indefinite function, 196
Inductor, 14
Inertial element, 8
Inhomogeneous boundary value problem, 152, 161
Initial value problem, 2
 complex valued, 75
 first order equation, 2
 first order system, 3
 nth order equation, 6
Inner product, 141, 160
Instability in the sense of Lyapunov, 176, *see also* Unstable
Integral equation, 2, 164
Invariance
 theorem, 62, 225
 theory, 221
Invariant set, 62, 222
Isolated equilibrium point, 170

J

Jacobian, 259
 matrix, 68, 171, 259
Jordan
 block, 84
 canonical form, 82, 107
 real canonical form, 134
Jordan curve, 291
 theorem, 291

K

Kalman–Yacubovich lemma, 245
Kamke function, 197
Kinetic energy, 9, 12
Kirchhoff
 current law, 14
 voltage law, 14

L

Lagrange
 equation, 28
 identity, 125, 142
 stability, 175
Lagrangian, 29
Laplace transform, 103
Least period, 5
Level curve, 201

Index

Levinson–Smith theorem, 298
Lienard equation, 19, 227, 262, 298
Lim inf, 43
Lim sup, 43
Limit cycle, 295
Linear displacement, 8
Linear independence, 81
Linear part of a system, 260
Linear system, 5, 80, 179
   constant coefficients, 100
   homogeneous, 5, 88
   nth order, 5, 117
   nonhomogeneous, 5, 88
   periodic coefficients, 5, 112, 306
   stability, 179, 218
Linearization about a solution, 260
Liouville transformation, 135, 147
Lipschitz
   condition, 53, 66
   constant 53, 66
   continuous, 53
Local hypersurface, 267
Logarithm of a matrix, 112
Loop current method, 15
Lower semicontinuous, 44
Lure result, 246
Lyapunov function, 194, 218
   vector valued, 241, 255
Lyapunov's
   first instability theorem, 214
   first method, 264
   indirect method, 264
   second instability theorem, 215
   second method, 205

M

MKS system, 9, 12, 15
Malkin theorem, 253
Mass, 8
   on a hard spring, 21
   on a linear spring, 22
   on a nonlinear spring, 21
   on a soft spring, 21
   on a square law spring, 22
Matrix, 81
   adjoint, 81
   conjugate, 81
   critical, 183
   differential equation, 90
   exponential, 100

Hurwitzian, 183
logarithm, 112
norm, 65, 168
self adjoint, 81
similar, 82
stable, 183
symmetric, 81
transpose, 81
unstable, 183
Maximal element of a set, 45
Maximal solution, 71, 77
Maxwell mesh current method, 15
Mechanical
   rotational system, 11
   translational system, 8
Minimal solution, 71
Moment of inertia, 11
Motion, 4, 222
Multiple eigenvalue, 141

N

nth order differential equation, 5
Natural basis, 81
Negative
   definite, 196, 197
   limit set, 291
   semiorbit, 4, 291
   trajectory, 4
Newton's second law, 8
Nodal analysis method, 15, 17
Noncontinuable solution, 49, 53
Norm
   of a matrix, 65, 168
   of a vector, 64

O

O notation, 147, 258
Ohm, 15
Ohm's law, 14
Ω limit set, 224
Orbit, 4, 291
Orbitally stable, 274, 298
   from inside, 297
   from outside, 297
Orbitally unstable, 298
   from inside, 298
   from outside, 298
Orthogonal, 142, 154
Orthonormal, 154

Oscillation theory, 125, 143

P

Partially ordered set, 45
Particular solution, 99
Pendulum, 22, 170
Period, 5, 112, 335
Period map, 306
Periodic solution
  of a Lienard equation, 298
  of a linear system, 306
  of a nonlinear system, 312
  of a two-dimensional system, 292
Periodic system, 5
  linear, 112, 306
  periodic solution, 312, 319, 330, 333
  stability, 178, 264, 273, 312, 324
Perturbation
  of a critical linear system, 319
  of a linear system, 258
  of a nonlinear autonomous system, 317
  of a nonlinear periodic system, 312
Planar network, 16
Poincaré–Bendixson theorem, 295
Popov criterion, 247
Popov plot, 249
Positive
  definite function, 196, 197
  limit set, 62, 76, 224, 291
  semidefinite, 196, 197
  semiorbit, 4, 291
  semitrajectory, 4, 222
Positively invariant set, 222
Potential energy, 9, 12, 15
Predator–prey model, 24, 271

Q

Quadratic form, 199
Quadratic Lyapunov function, 218
Quasimonotone, 77

R

Radially unbounded, 196, 197, 252
Rank of a matrix, 81
Rayleigh
  dissipation function, 29
  equation, 21
  quotient, 165

Reaction damping force, 9
Reactive
  force, 8, 9
  torque, 11, 12
Regular point, 290
Regulator system, 243
Resistor, 14
Rest position, 169
Right eigenvector, 82
Rotating coordinates, 332
Rouché theorem, 326
Routh array, 185
Routh–Hurwitz criterion, 184

S

Saddle, 187
Second method of Lyapunov, 205
Second order linear systems, 186
Sector condition, 245
Self adjoint
  boundary value problem, 139
  matrix, 81
  operator, 160
Semicontinuity, 44, 63
Semidefinite, 200
Separated boundary conditions, 139, 143
Servomotor, 31
Sign function, 20
Similar matrix, 82
Simple eigenvalue, 141
Singular point, 169, 290
Soft spring, 21
Solution of a differential equation, 1, 53
  nth order, 5
  particular solution, 99
  scalar equation, 2, 53
  system, 3, 53, 75
Solution of a differential inequality, 72
Spherical neighborhood, 65, 168
Spring, 9, 12, 21
  hard, 21
  linear, 22
  soft, 21
  square law, 22
Stability, 167, 172, *see also* Asymptotic stability
  of an equilibrium point, 172, 260
  from the first approximation, 264, 324
  of a linear system, 179, 218
  by linearization, 264, 324

Index

of a periodic solution, 178, 273, 286, 324, 328
  in the sense of Lyapunov, 176
Stable, 172, 179, 181, 205, 239, *see also* Unstable
  focus, 187
  manifold, 265, 267, 273
  matrix, 183
  node, 187
  polynominal, 184
State of a system, 97
State transition matrix, 95
Stationary point, 169
Stiffness element, 9
Sturm's theorem, 128, 135, 145
Successive approximations, 56
Sylvester
  inequalities, 200
  theorem, 200
Symmetric matrix, 81
System of first order differential equations, 3, 63
  autonomous, 4
  complex, 7
  homogeneous, 5, 6
  inhomogeneous, 5, 6
  linear, 5, 6
  periodic, 5

T

Tangent hypersurface, 267
Torque, 11
Total stability, 253
Trajectory, 4, 222
Transfer function, 243
Transpose matrix, 81
Transversal, 292
Trivial solution, 88, 172

U

Ultimately radially unbounded, 252
Uniform
  Cauchy sequence, 41
  convergence, 41
Uniformly asymptotically stable, 173, 181, 183, 208, 239
  in the large, 176, 209
Uniformly bounded
  functions, 41
  solutions, 176, 212, 240
Uniformly stable, 173, 180, 183, 206, 239
Uniformly ultimately bounded, 176, 212, 240
Uniqueness of solutions
  boundary value problems, 139
  initial value problems, 53
  periodic solutions, 309
Unstable, 175, 213, 215, 216, 217
  focus, 187
  manifold, 271, 273
  matrix, 183
  node, 187
  periodic solution, 327
  polynomial, 184
Upper bound of a chain, 45
Upper right Dini derivative, 72
Upper right-hand derivative of a Lyapunov function, 196
Upper semicontinuous, 44

V

Vandermonde determinant, 133
Van der Pol equation, 20, 301, 315
Variation of constants formula, 98
Vector Lyapunov function, 241, 255
Vector valued comparison equation, 241
Velocity, 8
Verhulst–Pearl equation, 25
Viscous friction, 9
  coefficient, 12
Voltage source, 14
Volterra population equation, 24, 271
Volts, 15

W

Wave equation, 137
Weierstrass comparison test, 41
Wronskian, 118

Y

Yacubovich–Kalman lemma, 245

Z

Zorn's lemma, 44, 51
Zubov's theorem, 232